Dictionary OF BIOLOGY

Covering Terms, Definitions
Principles, Laws, etc.

Dictionary OF BIOLOGY

Covering Terms, Definitions Principles, Laws, etc.

Covering all Terms of NCERTs

arihant

Arihant Prakashan (Series), Meerut

arihant

Arihant Prakashan (Series), Meerut

卐 **Administrative & Production Offices**

Regd. Office

'Ramchhaya' 4577/15, Agarwal Road, Darya Ganj, New Delhi -110002

Tele: 011- 47630600, 43518550

卐 **Head Office**

Kalindi, TP Nagar, Meerut (UP) - 250002 Tel: 0121-7156203, 7156204

卐 **Sales & Support Offices**

Agra, Ahmedabad, Bengaluru, Bareilly, Chennai, Delhi, Guwahati, Hyderabad, Jaipur, Jhansi, Kolkata, Lucknow, Nagpur & Pune.

卐 **ISBN** : 978-93-88128-95-7

卐 **PRICE**: ₹ 250.00

PO No : TXT-59-T069578-2-26

Published by Arihant Publications (India) Ltd.

For further information about the books published by Arihant, log on to www.arihantbooks.com or e-mail at info@arihantbooks.com

Follow us on

Preface

Arihant Dictionary of Biology aims to provide concise explanation of biological terms. It tries to cover all branches of Biology that are useful for students of schools & colleges along with all those who are preparing for competitive examinations.

There are many popular dictionaries in the market but they are for the reference purpose and common school & college students do not find these dictionaries easy to handle/comprehend because they are not meant for the specific purpose. Keeping all these things in mind I have tried to fill the space by preparing this dictionary. Information provided in this dictionary is apt and to the point without going into lengthy details that are not useful for the students. You are going to find each and every such word which you are looking for. Most of the entries are more than just a definition. Explanations are clear, straight forward and up-to-date. There is a network of cross references, some entries are simply referred to another entry, it may indicate that terms are synonyms.

At last, not the least, it is only your success that is our sole aim and for which we have consummated this ambitious project. The success of this dictionary rests with / lies in your success.

With Best Wishes

Contents

A

A-band The dark staining region of the myofilament that contains the thick (myosin) and thin (actin) filaments.

abaxial The surface facing away from the axis. In a leaf, surface away from the stem (lower surface).

ABC model A model to describe the biological basis of the growth of the flower's individual organ during flower development.

abdomen The part of the body between the thorax and the pelvis. It encloses stomach, intestine, liver, spleen and pancreas. In mammals, it is the portion of the body which is located below the rib cage and in arthropods below the thorax.

abducens nerve Sixth cranial nerve of vertebrates. Motor in nature, supplying lateral rectus muscles of the eye.

abductor (levator) **muscle** The muscles that moves a limb away from the body. It works antagonistically with abductors.

abiogenesis The supposed origin of living organisms from lifeless matter. It does not involve the action of living parents, i.e. it is spontaneous generation; also called biopoiesis.

abiotic factors The non-living factors or environmental factors of habitat such as temperature, cloud cover, rain, snow, hurricanes, etc. which affect, the life of living organisms.

ABO system The most important blood group system based on presence or absence of antigens A and B on red blood cells and antibodies against these in blood serum.

Blood group	Antigens present on the red blood cell	Antibodies present in the serum	genotype
A	A antigen	Anti–B	AA or AO
B	B antigen	Anti–A	BB or BO
AB	A antigen and B antigen	None	AB
O	None	Anti–A and Anti–B	OO

abomasum The fourth chamber of the stomach in ruminant animals, such as cows, sheep and deer, in which digestion takes place.

aboral Refers to side of the animal that is away from the mouth.

abortion The termination of pregnancy at an early stage.

abrasion Wearing away through some unusual or abnormal mechanical process (in pathology). A superficial injury to skin or other body tissue by rubbing or scrapping (clinical sign).

abscess Localised collection of pus within a tissue in the body.

Abscisic Acid (ABA) A plant hormone, that functions mainly as a growth inhibitor. It promotes abscission of flowers and fruits, induces dormancy of buds and seeds and causes closure of leaf pores (also called abscisin or dormin).

a

abscission The process of shedding or separating parts like leaves, fruits, flowers and bark being separated from the body of a plant at the end of a growing season. It is result of formation of abscission zone at the base, which ruptures under stress.

absorption The uptake of liquid or dissolved substances into a cell.
e.g. the digested food is absorbed in small intestine through the villi into the blood and lymphatic system.
In plants, water and mineral salts are absorbed mainly by the root hairs.

absorption spectrum All the pigments in the chloroplast absorb light or photons at definite wavelengths of visible light. The curve depicting the quantity of light absorbed at each wavelength is called absorption spectrum.

abyssal The deep ocean area roughly below 2000 m.

abzyme An antibody with catalytic activity.

Acanthodii The class of fossil fish abundant in early freshwater deposits.

acapnia A marked decrease of carbon dioxide in blood.

acarina (acari) A taxon of arachnids that contains mites and ticks.

acarology The study of mites and ticks is called acarology.

acceptor A receptor that binds a hormone without any apparent biological response (in biochemistry).

accessory bud The secondary buds formed besides a principal bud (axillary or terminal).

accessory (subsidiary) **cell** Specialised epidermal cell of a plant that are found adjacent to the guard cells and may help in opening and closing of stomata.

accessory chromosome (B-chromosome; supernumerary chromosome) An additional chromosome to the regular karyotype of a species, e.g. extrachromosome that partly duplicates chromosome 22 in people with cat eye syndrome.

accessory molecules Cell surface proteins that promote or suppress the response of immune cells. These include many Cluster of Differentiation (CD) antigens.

accessory nerve (cranial nerve XI) One of the pair of nerves that arises from the posterior region of the medulla oblongata in the brain of higher vertebrates.

accessory pigment A pigment (light-absorbing compound) such as a carotenoid or chlorophyll-*b*, *c* and *d* that captures light energy and transfers it to chlorophyll-*a* (an accessory pigment).

acclimatisations The process of an organism adjusting to a gradual change in its environment (such as a change in temperature, humidity, photoperiod or pH) allowing it to maintain performance across a range of environmental conditions; acclimation. It occurs in a short period of time (days to weeks) and within the organism's lifetime.

accomodation The adjustment of an eye to focus on object at varying distances.

acellular

1. Containing no cells, not made of cells.
2. Of or related to unicellular organisms that are complete within a single unit.

acentric A chromosome or fragment of a chromosome that lacks a centromere (centre).

acephalus Organism without head.

acetabulum The concave surface of the pelvis. The head of the femur fits into the pelvis at the acetabulum, forming the hip joint. There are three bones of the os coxae (hip bone) that come together to form the acetabulum. Two-fifths of the structure is contributed by the ischium, which provides lower and side boundaries to the acetabulum. The ilium forms the upper boundary, providing a little less than two-fifths of the structure of the acetabulum. The rest is formed by the pubis, near the midline.

acetate An ester of acetic acid.

acetic acid A carboxylic acid, CH_3COOH, that is used as a carbon source by certain green algae.

acetoacetic acid A metabolic acid formed from oxidation of fatty acids.

acetyl coenzyme-A (acetyl Co-A) Activated acetic acid, with an energy-rich sulphur bond. It is an important molecule in metabolism, used in many biochemical reactions. One of the end products of oxidative decarboxy- lation of pyruvic acid, produced near the beginning of Krebs' cycle.

acetylation The introduction of an acetyl group (CH_3CO—) into a compound.

acetylcholine (ACh) The acetyl ester of choline. It is secreted at the ends of many nerve fibres (cholinergic fibres), when nerve impulses (action potentials) arrive there. It is important in the nervous system of many animals, where it transmits impulses between synapses of nerves, i.e. acts as a neurotransmitter. After transmission, it is broken down by the enzyme cholinesterase.

acetylcholinesterase An enzyme that breaks down acetylcholine in the synapse.

achene The dry, indehiscent, one-seeded fruit formed from a single carpel, with seed distinct from the fruit wall.

acheulean tools The large stone hand axes made and used by *Homo ergaster* and *Homo erectus* 1.76 million years ago and by *Homo hiedelbergensis* 5,00,000 years ago.

achlamydeous Flowers lacking sepals and petals.

achlorhydria The absence of hydrochloric acid in gastric secretions of the stomach.

acholia The absence or lack of bile secretion or a condition that prevents bile from entering the duodenum.

achromasia

1. Lack of normal skin pigmentation.
2. Inability of tissues or cells to be stained.

achromatin The weakly staining substance of cell nucleus.

aciclovir (acyclovir, acycloguanosine) A synthetic purine nucleoside analog $C_8H_{10}N_5O_3$, used to treat herpes and varicella-zoster virus infections.
It is an analog of guanosine which selectively interferes with viral DNA synthesis.

acicular Needle-shaped, e.g. slender and pointed leaves of pine trees.

acid-base balance The maintenance of the pH of body fluids by regulating the concentration of acid and bases. This usually requires a buffer system, e.g. mammalian blood pH must be maintained at 7.4. The optimum pH for higher plants is around 6.7 and they tend to be more tolerant to pH changes.

acid growth theory The theory that describes how auxins stimulate cell expansion in certain plant tissues.

acid protease A protein- digesting enzyme that exhibits maximum activity and stability in acid conditions and is inactivated at pH values above 6.

acid rain The deposition of acids by natural precipitation (rain, snow and fog) leading to acid pollution. It is formed by reaction of gaseous waste products (sulphur dioxide, nitrogen dioxide, carbon monoxide) with moisture present in the air. Unpolluted rain is slightly acidic having a pH of 5.0-5.6; damage occur to sensitive ecosystems when the pH falls below 4.5.

acid soil The soil with a pH value less than 6.6.

acidosis A body condition in which there is increased acidity in the blood and other body tissues (opposite of alkalosis).

acinus (pl. acini) The smallest sac-like dilation of a compound gland such as pancreas.

acne (acne vulgaris) A disease characterised by the presence of pimples on face, neck and upper trunk.

acoelomate Animals lacking a coelom in the body between gut and body wall, e.g. phylum– Coelenterata, Platyhelminthes and Nematoda.

acquired characteristics The changes in the structure or function of an organ or organ system during the life of an organism, caused by the use or disuse of that organ or system or by environmental influences, e.g. sportsmen may develop strong muscles and plants growing near coasts show adaptations to the drying effects of sea air. These characteristics are not inherited.

acquired immunity (adaptive immunity) Immune responses directed at specific invaders and mediated by antibodies.

acquired immunodefeciency syndrome See AIDS

acraniata The chordate animal with a notochord and lacking a brain and skull. Acraniates include the hemichordates, urochordates and cephalochordates.

acridine A planar heterocyclic compound group with a formula $C_{13}H_9N$ that is capable of causing frameshift mutations in DNA sequences.

acrocentric A chromosome with centromere near one end. Humans normally have five pairs of acrocentric chromosomes.

acromegaly A chronic disease causing abnormal growth of cartilage and soft tissues due to excess growth hormone secretion by pituitary in an adult.

acromion The point of attachment of clavicle to scapula in mammals.

acropetal The development of organs in succession towards apex, the oldest at base, youngest at tip. Also used in reference to direction of transport of substances within a plant.

acrosome A cap-like covering over the nucleus in the head of a male gamete (sperm). It is formed by the modification of Golgi apparatus during spermatogenesis and contains enzymes to digest the egg membrane during fertilisation.

acrosome reaction

1. The stepwise reactions occuring when sperm comes in contact with ova releasing enzymes (sperm lysins).
2. The reaction allows the spermatozoa to penetrate zona pellucida membrane of oocyte.

ACTH (Adreno corticotrophic hormone; Corticotrophin) An anterior pituitary hormone that regulate secretion of cortisol from adrenal cortex.

actin One of the two contractile proteins present in muscles that in conjugation with myosin provide the contractile mechanism in muscles and in other cells.

actin binding proteins Proteins that bind to monomeric actin or actin filaments and alter the function of the actin.

actinobacteria (actinomycetes, actinomycota) The phylum of Gram-positive non-motile bacteria with high guanine and cytosine content in their DNA and are mostly anaerobic.

actinomorphic Regular and capable of dividing vertically into two or more planes having similar halves, also known as radially symmetrical. e.g. Daisy flower and starfish

actinomycin-D (Dactinomycin) An antibiotic that combines with DNA and prevents transcription.

action potential The rapid change in the electrical potential of a plasma membrane. It occurs when the cell has been activated by a stimulus. It is caused by a selective closing and opening of voltage-sensitive ion channels.

action spectrum A graphical plot/representation of amount of photosynthesis for different wavelengths of light.

activated sludge process The sewage and wastewater treatment method that involves enhancing microbial oxidation.

activator The substance which enhances the activity of an enzyme, e.g. calcium.

active immunity The immunity acquired due to the body's response to a foreign antigen.

active site A specific site on the surface of an 'enzyme molecule' that binds and acts on the substrate molecule. The properties of an active site are determined by the three-dimensional arrangement of the polypeptide chains of the enzyme and their constituent amino acids.

active transport The movement of molecules across a membrane against concentration gradient; accomplished by of use of energy.

activin The peptide hormone from the gonads that stimulates FSH secretion.

activity space The entire range of climatic and environmental conditions suitable to normal functions, process and activities of a living organism.

actomyosin A complex of two proteins, i.e. actin and myosin which is the major constituent of muscles.

acuity The ability of an eye to distinguish between objects present close together.

acute disease A disease that develops rapidly but lasts for a relatively short time.

acute-phase response An innate immunity mechanism in which liver proteins (acute phase proteins) act as opsonins and enhance the inflammatory response.

Adam's apple Larynx of respiratory tract is covered by pieces of cartilage forming a voice box called Adam's apple.

adaptation A genetic mechanism of an organism to survive, thrive and reproduce by constantly enhancing itself, by altering its structure or function, in order to become better suited to the changing environment.

a

adaptive enzyme (inducible enzyme) An enzyme that is produced by a cell only in the presence of its substrate.

adaptive radiation The evolutionary diversification from an ancestral group of organisms, into a number of new, more specialised forms, each suited to live in new habitats.

adaptive zone A set of ecological niches that may be occupied by various distinct types of organisms exploiting the identical resources in a like manner.

adaptor proteins In signal transduction, proteins that promote formation of signalling complexes by binding signalling proteins in a signalling pathway.

adaxial

1. Toward or facing toward the axis or center line.
2. On the upper side (e.g. of a leaf).

Addison's disease A deficiency disease of adrenocorticosteroid hormone produced by adrenal cortex.

adductor (depressor) The type of muscle whose function is to pull a limb inwards, towards the body of an animal.

adenine A purine base found in nucleic acids. In DNA, it is one member of the base pair AT (adenine-thymine).

adenohypophysis (anterior pituitary gland) An endocrine gland in the brain that secretes multiple hormones.

adenosine A nucleoside composed by adenine and ribose proteins.

adenosine deaminase An enzyme converting adenosine to inosine.

Adenosine Diphosphate (ADP) The diphosphate ester of adenosine. Closely-linked with adenosine triphosphate in the transfer of energy during respiration.

adenosine monophosphate cyclic (cyclic AMP/cAMP) A nucleotide that participates in the transfer of signals between external environment and the cell.

Adenosine Triphosphate (ATP) A nucleotide with multiple functions. In its phosphate bonds, it stores energy produced by photosynthesis and cellular respiration. This energy is consumed by various metabolic processes when ATP is hydrolysed into ADP.

adenovirus One of the group of DNA containing viruses, about 80 nm in diameter, causing respiratory diseases and tumours in human and animals.

adenylate cyclase An enzyme that catalyses the formation of cyclic AMP. It is bound in the inner surface of the plasma membrane. The hormone binds to a receptor on the outer surface of the plasma membrane, which then activates adenylate cyclase on the inner surface.

adermin AB-vitamin essential for metabolism of amino acids and starch.

adherens junction (Zonula adherens) A cell-cell junction that links actin myofilaments in adjacent cells together with the help of cadherins.

adhesion The attraction of dissimilar molecules. In a flower, fusion of members of distinct whorls, as in the case of union of calyx with the ovary wall.

adipocyte (fat cells) The cells specialised for storing triacylglycerols and for releasing them in blood in the form of fatty acids and glycerol as required.

adipose tissue A body tissue having cells containing fat and oil. It is mainly found below the skin and around major organs acting as an energy reserve, providing insulation, protection and generating heat.

adjuvant A non-antigenic substance that increases the antigenic response in combination with an antigen.

A-DNA Right-handed double-helix similar to B-DNA whose base pairs are not perpendicular to the helical axis as in B-DNA.

adnation One organ attached to another of different kind by its whole length.

adolescence The period in human development that occurs during the teenage years, between the end of childhood and the start of adulthood. It is characterised by various physical and emotional changes associated with the development of the reproductive system.

adoptive immunity It is the transfer of immunity from an immune competent organism to another non-immune organism.

adoral Located toward or near the mouth.

adrenal gland A pair of endocrine glands situated immediately above the kidneys. The outer layer of it is called **adrenal cortex** (produces steroid hormones). The inner portion is called **adrenal medulla** that produces catecholamines.

adrenaline (epinephrine) A hormone secreted by the adrenal medulla in times of fear, anxiety or tension. It increases the body's efficiency to face threat or danger.

adrenergic The type of nerve fibre that releases epinephrine or norepinephrine from its ending when stimulated by a nerve impulse. Vertebrate sympathetic motor nerve fibres are adrenergic.

adsorption The binding of molecules or particles on to a surface of a solid or of a liquid.

adventitious Arising in unusual position, e.g. roots developing from stem or leaf cuttings; buds developing from parts other than axils of leaf.

adventive embryony (sporophytic apomixis) Origin of embryo directly from the cells of the nucleus or integuments.

aeciospore (aecidiospore) An asexual spore formed in the rust fungi from the fusion of cells. The nuclei from the two cells do not fuse and the spore is binucleate.

aeolian soil A type of soil that is transported from one place to another by wind.

aerenchyma A plant tissue with thin-walled cells surrounding large air spaces. Found in hydrophytes.

aerial hypha The hypha of a fungus which is outside the material in which the fungus is growing.

aerial root The root above the ground, mostly arising adventitiously from stem.

aerobe An organism that require presence of free oxygen, respiration and life.

aerobic respiration A type of respiration in which food is completely oxidised to carbon dioxide and water with the release of chemical energy in presence of oxygen.

aerosol The dispersion of solid or liquid particles in a gas eaus form, e.g. smoke.

aerotaxis The movement of a cell or a microorganism in response to oxygen concentration gradient. For instance, motile aerobic bacteria are positively aerotactic, whereas motile obligate anaerobic bacteria are negatively aerotactic.

a

aerotropism The growth curvature in relation to oxygen concentration.

aestivation

1. Arrangement of the parts in a flower bud, especially of the sepals and petals (in botany)
2. Dormancy during summer or dry season, e.g. lungfish respond to the drying up of water by burying themselves in the mud bottom. They re-emerge at the start of the rainy season (in zoology).

aetiology The study of causes, especially the causes of medical conditions.

afferent Refers to bringing inward to a central part of a body or organ, e.g. afferent arteriole transport blood to glomeruli and afferent neuron carries information to the CNS from the receptors.

aflatoxin A toxin compound made by grain-inhabiting fungi, especially *Aspergillus flavus*.

after birth The placenta and associated fetal membranes expelled from the uterus after a child is born.

after ripening A dormant period after the formation of seed, needed by the embryo to undergo certain chemical and physical changes before it can grow.

agameon (agamospecies) An organism which reproduces only by asexual means, e.g. bacteria.

agamospermy The reproduction in which the embryo develops without fertilisation, e.g. parthenogenesis.

agar A gelatin-like product of certain red seaweeds that is solidified and is used as a culture medium.

Agaricus A large group of fleshy fungi with gills. They include many familiar fungi such as the field mushroom, the chanterelle, the red and white spotted fly agarics and the notorious death cap.

age polyethism The phenomenon where an animal shows different forms of behaviour at different ages.

age structure The relative proportion of individuals in each age group in a population.

ageing

1. The process of growing older.
2. Assessing the age of an animal.

agglutination The clumping together of red blood cells or bacteria. Agglutination may occur in transfusion, if blood of incompatible group is given. The surface of the donor's red blood cell contains antigen molecules which reacts with antibodies in the serum of the recipient. It causes the red cells to clump together. These clumps may block capillaries, causing fatal damage to heart or brain. Agglutination of bacteria by antibodies causes them to disintegrate.

agglutinogen Antigens that are present on the outer surface of red blood cells. There are more than 100 different agglutinogens and they form the basis for identifying the different blood groups.

aggregated distribution The distribution of organisms in which individuals are close to each other rather than being evenly distributed.

aggregated fruit Fruit formed from a single flower in which gynoecium is apocarpous. Each carpel forms a fruitlet. Thus, the fruit consists of number of small fruitlets. The aggregates of simple fruitlets are called eatrios.

aggressin A toxic substance that is secreted by certain parasitic microorganisms which inhibits the natural defence mechanisms of a host organism.

aglycon The non-sugar part of a glycoside.

Agnatha A class of most primitive vertebrates without true jaws. These are in contrast to Gnathostomata which includes Pisces, Amphibia, Reptilia, Aves and Mammalia.

agonist A substance such as a drug or hormone, that binds to a cell's receptors and mimics a response.

agonistic behaviour (agonism) A type of animal behaviour exhibiting features of both aggression and avoidance, e.g. the establishment of a territory involves both attack and escape behaviour, if two neighbours meet at the boundary of their territories.

agranulocyte A white blood cell (leukocyte) that does not contain granules in its cytoplasm. These are of two types, i.e. lymphocytes and monocytes. Both have large nuclei and a small amount of clear cytoplasm.

agrestal Growing in a cultivated land, but not itself cultivated, e.g. a weed.

agriculture The study and practice of cultivating land for growing a wide range of crops (e.g. cereals, pulses, vegetables, fruits, fodder, etc) and rearing of livestock.

agrin A large proteoglycan released by motor neurons and muscle cells that is important for normal development or regeneration of neuromuscular junction.

Agrobacterium tumefaciens A soil bacterium that causes crown gall disease in species of eudicots. It is of interest as its cells contain a Ti-**plasmid** (see *Ti* plasmid). Its DNA transmission capabilities have been explored in biotechnology as a means of inserting foreign gene into plants.

agronomy The science of soil management and the production of field crops.

agrostology The study of grasses.

agynous A flower in which gynoecium is not developed or absent.

AIDS (Acquired Immuno Deficiency Syndrome) A set of diseases (of humans) characterised by defective cell-mediated immunity and increased susceptibility to infections. It is caused by Human Immunodeficiency Virus (HIV), a member of a group of virus called retrovirus. These viruses have envelope enclosing the RNA genome. Transmission occurs by (i) sexual contact with HIV infected person, (ii) by transfusion of contaminated blood and blood products, (iii) by sharing infected needles, (iv) from infected mother to child through placenta.

air Mixture of gases that forms the earth's atmosphere. Its composition varies from place to place. It contains nitrogen (78%), oxygen (20.9%), argon (0.9%) and other gases (0.1%).

air bladder (swim bladder) An air-filled structure present in many bony fishes. Its main function is hydrostatic, i.e. to make the body light for swimming and maintain equilibrium in water.
In some fishes, it also helps in transmission of water-borne sound waves to the internal ear. In case of lung fishes, air bladder is also related to respiration.

air pollution (atmospheric pollution) Introduction of harmful chemicals particulates or gases into the atmosphere.

a

air sacs

1. A thin-walled sac in birds that are connected to the lungs and increase the efficiency of ventilation. Some of the air sacs penetrate the internal cavities of bones.
2. The structural extension of the trachea in insects which increases the surface area available for the exchange of oxygen and carbon dioxide in respiration.

akinesis The motionless state of an insect resembling sleep, particularly after loss of or damage to sensory organs such as antennae.

akinete Thick-walled non-motile spore formed in algal cells for the purpose of perennation.

alanine A non-essential amino acid.

alary muscles A series of triangular wing-like muscles inserted on to the membrane that separates the perivisceral and pericardial cavity of an insect. Contraction of these muscles causes blood to flow into the pericardial cavity and then through small opening or ostia into the long dorsal tubular heart.

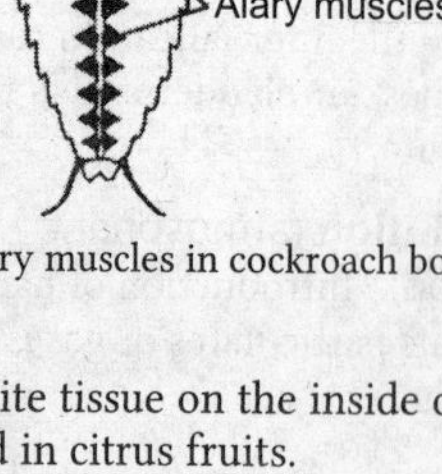

Alary muscles in cockroach body

albedo

1. White tissue on the inside of the rind in citrus fruits.
2. White area of the retina resulting from edema or infarction.

albinism The absence of natural pigment in skin, hair or feather. In mammals, it is due to an autosomal recessive gene. Such persons are called albinos.

albumin (albumen) One of the major proteins in blood plasma. It is without a prosthetic group and is soluble in water and dilute salt solution.

albuminous cell A vertically elongated parenchyma cell, found in groups in the rays of the secondary phloem in gymnosperms.

albuminous seed A seed that contains endosperm at maturity.

alburnum An obsolete term for sap wood.

alcohol Any organic compound with a hydroxyl group (– OH) bound to a carbon atom of an alkyl or substituted alkyl group. The general formula for a simple alcohol is $C_nH_{2n+1}OH$.

alcohol dehydrogenases A group of dehydrogenases that interconvert alcohols and aldehydes or ketones with the reduction of NAD^+. In animals and humans, they breakdown toxic alcohols. In yeast and many other microorganisms, they catalyse the reverse process, producing alcohols *via* fermentation.

alcoholic fermentation The anaerobic formation of ethanol and CO_2 by pyruvate. It is used in the production of alcoholic beverages and ethanol fuel and in rising of bread dough.

alcoholism The addiction to alcoholic drinks. It affects the person's mental functions and leads to deterioration of physical skills. The heart, liver and nerves may also be affected.

aldose Sugar, containing an aldehyde (—CHO) aldehyde group e.g. glucose.

aldosterone A hormone produced by the cortex of the adrenal gland which affects the rate of carbohydrate metabolism. It also helps to control the electrolyte balance of the body by allowing the retention of sodium ions and the excretion of potassium ions.

aleurone layer The outermost protein-rich layer of the endosperm of cereal grains. At germination, the embryo produces gibberellins, which stimulates the aleurone layer or synthesises enzymes, especially amylase. The latter causes hydrolysis of the starch in the endosperm.

aleuroplast (proteinoplasts, aleuronaplasts and proteoplasts) A colourless plastid (leucoplast) storing protein.

alfalfa (lucerne) Perennial forage legume. It can restore nitrogen in the soil by way of its root nodules.

alga (pl. algae) A unicellular or multicellular simple organism which contain a variety of photosynthetic pigments and are present in moist and aquatic habitats.

algal bloom The rapid increase in the growth of algae and other phytoplanktons, especially cyanobacteria in the inland water sources like ponds, lakes, wells, tanks, rivers, etc. It is one of the most common source of water pollution.

algesia Hypersensitivity to pain.

alginic acid (algin) An anionic polysaccharide occuring widely in cell walls of brown algae.

alien (exotic) The species introduced in a region deliberately or accidentally by humans. These are not native to a locality (in ecology).

alignment A process of matching up base sequences or amino acid sequences to reveal similarities or differences between them (in bioinformatics).

alimentary canal (digestive tract; gut) A tube (8-9 m long in human) through which food is passed for digestion and absorption into the blood-stream of an animal.

alkali A substance that is either a soluble base or a solution of a base. Alkalis have a pH of more than 7 (pH = 7 is neutral, a pH of less than 7 is acidic) and react with acids to produce salts and water.

alkaliphiles The organisms that grows best under alkaline conditions, i.e. pH 8.5-11.

alkaloid A group of nitrogen-containing organic compounds that are found in some plants. Many are toxic or medicinal. e.g. heroin, atropine, digitalis, morphine, quinine, etc.

alkalosis Body condition in which pH of the blood and body fluids rises above normal level, i.e. above 7.35. It may result from loss of acidic digestive juices through vomiting or from an excess intake of alkali.

alkaptonuria (alcaptonuria, black urine disease) An inherited rare genetic disorder of phenylalanine and tyrosine metabolism. It is due to defect in the enzyme homogentisate 1, 2-dioxygenase.

allantois One of the four membranes enclosing an embryo in reptiles, birds and mammals. It helps in exchange of gases and throws out the nitrogenous wastes produced by the embryo.

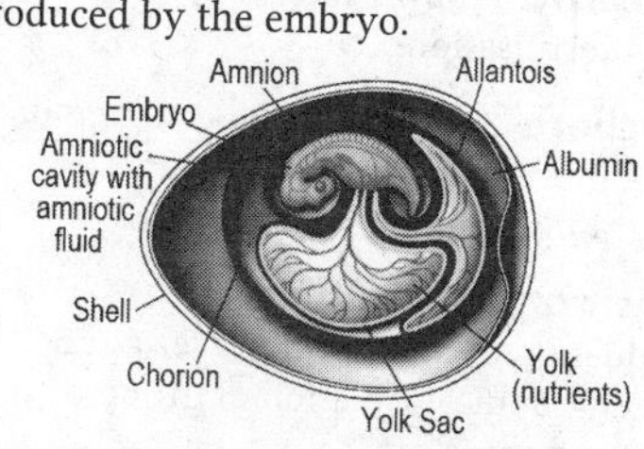

Inside structure of an egg

a

allele (allelomorph) Alternative form of a gene occurring at the same locus.

allele frequency (gene frequency) The relative proportion of a particular allele among individuals of a population.

allelic exclusion The process by which only allele of a gene is expressed while the other allele is silenced. It has been observed most often in genes for cell surface receptors and has been extensively studied in *B*-lymphocytes.

allelochemical (Allomone) A substance emitted by an organism of a species that influences the physiology of organism of different species. e.g. herbicides and insecticides are synthetic allelochemicals.

allelopathy A inhibition of the germination, growth or reproduction of an organism effected by a chemical substance released from another organism. This is a common anti-competition mechanism in plants.

Allen's rule A generalisation that mammals tend to have shorter extremities in colder than in warmer climates.

allergen Refers to any antigenic substance that can stimulate strong immune response (allergic reaction) in the body of animals.

allergy Excessive sensitivity of the immune system towards certain substances which do not affect majority of people. It is not a disease.

allocarpy Fruiting after cross-fertilisation.

allochore Any organism occurring in two different habitats in the same geographic region.

allochronic speciation The production of new species over a time, by geographical isolation of group of common ancestors.

allochthonous An organism that is not indigenous (in ecology).

allogamy Cross-fertilisation in plants.

allogenic It means having genetic variation between the members of a population of same species.

allogenic succession Changes in the species that make up a community as a result of changes in the external environment.

allograft (homograft) A graft of tissue obtained from donor and implanted in recepient of same species but with a different genetic make up, e.g. transplantation of kidney.

allometric growth The growth of different parts of the body of an organism at different rates or at different times. In humans, e.g. brain growth stops at about the age of five years while other parts of the body continue to grow.

allopatric The species or populations in separate geographical locations.

allopatric speciation An evolutionary change between populations initiated by geographical separation or because intervening geographical populations become extinct.

allophore A chromatophore that contains a red pigment, that is soluble in alcohol. It is found in the skin of fishes, amphibians and reptiles.

allopolyploid An organism that combines two distinct chromosome sets from different sources. Allopolyploids are produced by hybridisation and chromosome doubling.

all-or-none rule (all or nothing) A response that either occurs maximally or does not occur at all, depending on the strength of stimulus, e.g. transmission of a nerve impulse.

allosome Sex chromosome that is different from autosome in form, size and behaviour.

allosteric control It is a regulation of the activity of an enzyme by a molecule that interacts at a site other than active site.

allosteric enzyme An enzyme having both the active and inactive forms. It has two binding sites, one for the substrate and another for binding the regulatory substance.

allosteric site The binding on protein molecule or an enzyme other than the active site where the regulatory susbtance binds.

allosyndesis

1. Pairing in a cross of two polyploids or of chromosomes derived from opposite parents.
2. Pairing in an allopolyploid, between chromosomes derived from ultimate diploid ancestors, opposite to autosyndesis.

allotetraploid An allopolyploid is derived by doubling the set of chromosomes resulting from fusion of haploid gametes.

allotopic Closely related sympatric populations, whose distributions are such that both occupy the same geographical range, but each occurs in a different habitat within that range.

allotype Genetic variant within a locus of a given species population, such as allelic forms with in a blood group system.

allozymes Enzymes produced at the same locus, but differing in molecular structure. They are coded by distinct alleles.

almond A species of deciduous tree bearing pink flowers and leathery fruits. The stones, commonly but mistakenly called nuts, contain edible kernels.

α-amantin A compound inhibiting transcription by RNA polymerases. It is obtained from the poisonous mushroom *Amanita phalloides*.

alpha cells A group of cells present in islets or Langerhans of the pancreas which synthesise and secrete glucagon hormone.

alpha diversity (alpha richness) The diversity of species within a particular habitat or community.

alpha helix

1. The spiral arrangement of DNA molecule.
2. The secondary structure of proteins.

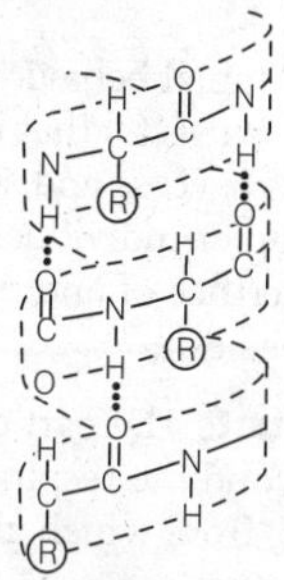

--- Hydrogen bond, R=Amino acid side chain

alpine vegetation The plant life comprising herbs, shrubs etc. found above timber-line mountains. Timber-line is the limit of tree growth on a mountain.

alternate host One of the two hosts required for the completion of the life cycle of a parasitic organism.

alternate splicing The mechanism by which single genes produce different functional proteins by rearrangement of axons.

alternation of generation The occurrence of two generations during the life cycle of an organism. It occurs mostly in the life cycle of organisms, such as mosses, ferns and many coelenterates.

A generation that reproduces a sexually alternates with a generation that reproduces sexually. As a result, life cycle is divided into haploid and diploid phases.

alternative respiratory pathway A cellular respiration pathway that permits electron transport and reduction of oxygen to water in the presence of cyanide and other substances that completely inhibit respiration in animals. It acts as an alternative to electron transport pathways common to both plants and animals.

altricial The mammals or birds born or hatched in a state requiring parental care.

altruism Animal behaviour (common among social insects) that benefits the reproductive success and survival chances of other individuals because of potential sacrifice of another member of the same species.

alveolar ducts The part of respiratory passage beyond the respiratory bronchioles, from which alveolar sacs and alveoli arise.

alveolates A eukaryote subgroup that share a honeycomb system of flattened membrane bound sacs beneath cell membrane.

alveolus (pl. alveoli)

1. A minute air-filled sac in vertebrate lungs, thin-walled and surrounded by blood capillaries. Exchange of oxygen and carbon dioxide takes place here, during respiration (pulmonary).
2. A socket in the jaw, in which the roots of teeth are embedded (dental).

Alzheimer's disease A type of dementia seen in elderly, probably resulting from a deficiency of acetylcholine. It leads to degeneration and death of neurons.

amacrine cell A retinal nerve cell without axon. They are inhibitory neurons.

amber (codon) The UAG coding for termination.

ambilateral Occurring on both sides.

ambiparous bud A bud containing both the young vegetative leaves and young flowers.

amblyopia (lazy eye) Impaired vision due to poor transmission by the optic nerve; usually occurs in only one eye.

amblypygi An order of arachnids composed of whip spiders and tailless whip scorpions.

ambulacra Radial bands of tube feet present in echinodermata.

ameloblast A type of epithelial cell involved in forming dental enamel.

amenorrhea The absence of menstruation in a pre-menopausal woman.

amensalism Type of biological interaction in which one organism is inhibited or damaged by other that remains unaffected.

ametabolous Absence or no metamorphosis during development of an insect. Immature larval stages appear very similar to adults, except they lack genetalia, e.g. silverfish.

amino acid One of 20 organic molecules serving as the monomers of proteins. An amino acid has both carboxyl ($-COOH$) and amino groups ($-NH_2$). The various amino acids differ only in what is attached to the 4th bond of the carbon. This variable attachment is generally indicated by the letter *R*. The *R* group is also called the "side chain." The properties of the side chain determine the unique properties of an amino acid.

aminotelism The excretion of excess amino acids as such. Common in some molluscs and echinoderms.

amitosis The direct division of nucleus by constriction. Spindle and chromosomes are not formed during this cell division. It occurs in meganuclei of *Ciliophora.*

ammonification An important stage in nitrogen cycle. It is carried out by decay causing organisms.

ammonotelism The process of excreting ammonia as waste., e.g. many bony fishes, aquatic amphibians and aquatic insects are ammonotelic.

amniocentesis A test in which a sample of amniotic fluid is taken from the pregnant woman. The cells from amniotic fluid are tested particularly for determining any hereditary disorders in unborn baby.

amnion The innermost of the four membranes enclosing the embryo of a reptile, bird or mammal.

amniote Vertebrates having an amnion during embryonic development. It includes reptiles, birds and mammals. The development of the amnion, which permits gas exchange was the evolutionary step that first enabled eggs to be laid on dry land.

amniotic egg Egg of an amniote. The presence of an amnion prevents drying and allows amniotic eggs to be laid on land.

Amoeba A single-celled organism, belongs to Protozoa, that moves and feeds with the help of pseudopodia.

amoebiasis (amoebic dysentery) *Entamoeba hystolytica* that is a protozoan parasite of the large intestine causes this disease. Symptoms are abdominal pain and cramps, stools with excess mucous and blood clots. Houseflies act as mechanical carriers.

amoebocyte A freely wandering animal cell i.e without a fixed position. It resembles *Amoeba* in its movement. e.g. certain leucocytes in blood of mammals.

amoeboid movement *Amoeba*-like movement of a cell by the help of protrusions called pseudopodia.

amphetamine A drug that is similar in molecular structure to adrenaline and noradrenaline and shares their stimulatory properties. Stimulates sympathetic nervous system.

amphiaster Two asters and the spindles connecting them, produced during mitotic cell division.

amphibians Animals that can survive and live on land as well as in water. These are cold-blooded vertebrates.

amphiblastic Segmenting unequally, said like that of telolecithal ova with complete segmentation.

amphibolic pathway A biochemical pathway that serves both anabolic and catabolic processes, e.g. Krebs' cycle.

amphibolic siphonostele A central vascular bundle in which the tissues are arranged concentrically and an outer ring of phloem completely surrounds a central core of xylem.

amphicarpic Production of two kind of fruits that differ either in shape or in time of ripening.

amphicribral A type of vascular arrangement in which phloem surrounds the xylem.

amphidiploid An organism or cell having a diploid set of chromosomes from two different species, e.g. *Raphanobrassica.*

amphimixis Union of egg and sperm nuclei during fertilisation to form a zygote. It is the true sexual reproduction.

a

amphineura A class of mollusca, including the chitons having reduced head.

amphiphloic Type of vascular arrangement where phloem is on both sides of the xylem.

amphisbaenian A long reptile (worm-like lizard) of genus *Amphisbaena* with a short tail and ring-shaped scales that has well adapted itself to burrowing.

amphistylic jaw suspension A type of jaw suspension in which the upper jaw is held against the cranium and is also supported by the hyomandibular, e.g. in certain sharks.

amphithecium Outer layer of the cells of the spore containing capsule of a bryophytes (moss) derived by the periclinal division of the cell.

amphitrichous bacteria A bacteria having a single flagellum at each opposite end of the cell.

amphitropous ovule An inverted type of ovule. Ovule is bent transversely and is at right angle to the funicle.

amphoteric A compound that can act both as an acid and a base, e.g. amino acids.

amphoterin A binding protein essential for normal neurite growth.

amplexus Mating position of the frogs and toads, in which the male clasps the female about the back. The female sheds the eggs into the water and the male fertilises it. Fertilisation takes place outside the female's body.

amplification Increase in the copy number of a DNA fragment. Amplification can occur either *in vivo* or *in vitro.*

ampulla A small, rounded, sac-like dilation of a duct, canal or a tubular structure, e.g. semicircular canal of ear.

amygdala The two almond-shaped groups of nuclei located in brain of vertebrates.

amygdalitis (tonsillitis) Inflammation of tonsils.

amylase A group of enzymes that breaks starch into sugar. Amylase is present in saliva, digestive juices and microorganisms.

amyloid A tissue that consists of protein fibrils which may accumulate between cells in various animal tissues. The build up of this tissue is feature of certain diseases such as Alzheimer's.

amylopectin The water-insoluble fraction of starch.

amyloplast The colourless plastid (leucoplast) which stores starch in plants, e.g. potato tubers.

amylose The water-soluble fraction of starch, that reacts with iodine to give a characteristic blue colour.

anabiosis The state of complete dormancy or suspended animation in organisms triggered by desiccation and ends, when water is present again.

anabolic pathways (anabolic reaction) The metabolic processes that consume energy and build up complex molecules from simple ones.

anabolic steroids Steroid compounds that promotes tissue growth, especially of muscles. Androgens are its naturally occurring form.

anabolism The sum of all biosynthetic reactions. Anabolic reactions are concerned with building up of complex compounds like starch, glycogen, fats and proteins. These reactions generally require energy as ATP, produced by catabolism.

anadromous Animals which ascend rivers and streams to breed, e.g. lampreys and salmon.

anaemia The reduction in circulating erythrocytes, haemoglobin or both.

anaerobic An absence of molecular oxygen, referring to an organism, environment or cellular process that lacks oxygen and may be poisoned by it.

anaerobic respiration Process by which organisms obtain energy from the breakdown of food molecules in the absence of oxygen. In yeast and some bacteria, as fermentation, in which sugar is broken down to alcohol. In animals, muscle cells respire anaerobically to form lactic acid. Both processes yield less energy than aerobic respiration.

anaesthetic Compound that can render an animal unconscious of painful stimuli.

anagenesis An evolutionary change in a population without separation of the population into different forms. Also called phyletic evolution.

anaima Group of animals with no red blood cells. Includes all non-chordates.

anal canal The lowermost portion of the large intestine, leading from the rectum to the anus. About 3 cm in length, its aperture is controlled by an internal and external sphincter.

analgesic A substance that reduces pain. There are several categories of analgesic drugs, including morphine and its derivatives, which produce analgesia by acting on the central nervous system; non-steroidal, anti-inflammatory drugs and local anaesthetics.

analogous structures Structures that serve the same function in different organisms, but differ in structure and embryological development, for example, the wings of insects and birds.

analogous traits Traits that are presumed to be similar due to convergence. These are not inherited from a common ancestor.

analogy Similarity of structure between two species that are not closely related. It attributes to convergent evolution.

anamniota A group of vertebrate including Agnatha, fishes and amphibians. They do not have amnion during development in contrast to Amniota which includes reptiles, birds and mammals.

anamorph The asexual form in the life cycle of a fungus, when asexual spores, such as conidia or no spores are produced.

anandrous The flowers lacking stamen.

anaphase Stage in mitosis and meiosis (cell division) in which chromosomes migrate to opposite poles of the cell by means of the spindle.

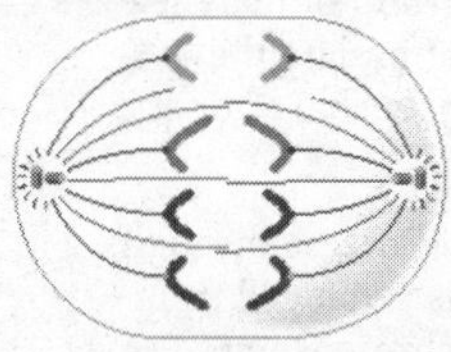

Anaphase stage of cell division

anaphylaxis An extreme sensitivity of the body to a particular foreign substance. In a mild anaphylactic reaction, there may be fever accompanied by itching and redness of the skin. Severe cases may involve constriction of the air passages (trachea and bronchi) resulting in unconsciousness and even death.

anaplerotic A metabolic pathway that replenishes an intermediate in one of the main metabolic pathways.

a

anapsid An extinct sub-class of reptiles except for the turtles, that have no opening in the temporal region of the skull.

anastomosis Interconnection between blood vessels or other tubular structures such as the veins of a leaf, fibres or other parts of a network.

anatomy Study of structure of organisms and their parts.

anatropous Position of the ovule in the ovary when the developing ovule has turned through 180°, so that micropylar end has folded over and lies close to the base of the funicle. Most common in flowering plants.

androcyte A cell in bryophytes that modifies to become a sperm cell.

androdioecious Species in which male and hermaphrodite flowers are borne on separate plants.

androecium A whorl containing male parts in a flower.

androgen A group of male sex hormones that stimulate the production of male characteristic features.

androgenesis Process through which embryo is produced from sperm-derived chromosomes.

androphore An elongation of the receptacle of a flower carrying the androecium and the gynoecium.

androsterone A steroid hormone formed in the liver from the metabolism of testosterone. It has weak androgenic activity.

anemochore A plant with seeds and/or spores suited to distribution (dispersal) by the wind.

anemophily Wind-pollination.

anesthesia

1. Total or partial loss of sensation, usually by damage to a nerve or receptor.
2. Loss of the ability to feel pain caused by administration of a drug etc.

aneuploid Condition due to non-disjunction of homologous chromosomes at meiosis, in which one or more chromosomes are missing from or added to the normal somatic chromosome number.

aneuploidy The presence of an abnormal number of chromosomes in a cell.

aneurysm An excessive localised swelling of the wall of an artery.

angina pectoris It is the chest pain associated with inadequate blood flow to the heart muscles, during heart attack.

angiocarpous

1. A fruit partially or wholly enclosed in a shell, involucre or husk.
2. (of a fungus or lichen) Having fruiting body enclosed in thallus.

angiography X-ray study of blood vessels.

angioplasty The procedure of repairing the damaged blood vessels.

angiosperms (anthophyta; angiospermophyta) A member of a major group of flowering plants. Their characteristics are the possession of flowers and production of seeds present in a fruit. Unlike gymnosperms, angiosperms form their seeds within the protective capsules called ovaries.

angiotensin Any of the several vasoconstrictor peptide hormones that cause narrowing of blood vessels.

animal husbandry The agricultural practice of breeding and raising livestock.

animal pole A point on surface of egg, nearest to its nucleus and contains less yolk. The other end is the vegetal pole.

Animalia A taxonomic kingdom of animals that includes organisms that are multicellular, eukaryotic (having cells with membrane-bound nuclei) and heterotrophic (require organic compounds for nourishment).

anisogamy A common type of reproduction in which the two gametes that join to produce the zygote differ greatly in size and sometimes in form. The smaller of the two gametes is mobile and seeking out the larger, which is immobile.

ankle

1. The joint between the tarsals and the tibia.
2. The slender section of the leg immediately above the foot.

ankylose An immobilisation of a joint by pathological or surgical process.

annelid A member of the phylum–Annelida.

Annelida An invertebrate phylum that comprises segmented worms, with cylindrical soft bodies, e.g. earthworm.

annual A plant that completes its life cycle within one season and then dies, e.g. cereals, grasses.

annual ring A cylinder of secondary xylem cut by the cambium in a dicot stem in one year. It comprises early wood and late wood. Occurs as a result of distinct variation in seasonal activity of the cambium. Successive annual rings are seen in cross-section of stem.

annular thickening An internal thickening of a xylem vessel or tracheid wall.

annulus

1. Ring of tissue surrounding the stalk of the mature fruiting body of the basidiomycete fungi. Also called velum (in botany).
2. A special arc or ring of cells in the sporangia of ferns for spore dispersal.
3. A ring-shaped structure of an external segment of annelid worm (in zoology).

anoestrus A period between breeding seasons in female mammals when oestrous cycles are absent and the species are sexually inactive.

anorexia nervosa A psychiatric eating disorder characterised by fear of gaining weight. Affected individuals are underweight, but diet or exercise excessively to lose weight. Often, they take diet pills. Most of them are adolescent females and about 10% are males.

anoxemia Lack of proper amount of oxygen in the blood flowing through the arteries.

antagonism

1. Inhibition of the growth of one kind of organism by another. This may be due to production of antibiotics, competition for food, etc.
2. Ability of one toxic substance to reduce or remove the toxic effect of another (antagonist).
3. (Of muscles) Opposing actions of muscles, e.g. flexor and extensor muscles are antagonistic in their action.

antenna (antennule) A sensory apparatus found on the heads of insects and most arthropods. It is usually in pairs.

antennal gland (green gland) A duct that occur in pairs in the third segment of a crustacean and opens to outside at the base of the second antenna. They function as osmoregulatory organs.

a

anterior Directed towards the head or front.

anterograde transport In vesicle transport, it is the movement of material from the ER through the Golgi apparatus and onto their final compartments (e.g. plasma membrane).

anther The upper two lobed part at the top of a stamen in the flower, with each lobe containing two pollen sacs filled with pollen grains.

antheridium Male sex organ of Thallophytes, Bryophytes and Pteridophytes; it produces ciliated antherozoids.

antherozoid (spermatozoid) The male gamete of algae, fungi, bryophytes, pteridophytes and some of the gymnosperms. It is motile and is produced in antheridium, except in certain gymnosperms (*Cycas, Gingko*) where they develop from the generative cells of the pollen tube.

anthesis

1. An opening of a flower bud.
2. A duration in life of a flower from the opening of the bud, to the setting of a fruit.

anthocyanins A group of water soluble pigments (flavonoids) occurring in solution in vacuoles, flowers, fruits, stem and leaves. Can change colour depending on the acidity or basicity of solution. Responsible for most red, purple and blue colours in plants, especially in flowers.

anthotaxy Arrangement of the flower on a stem or parts on a flower.

Anthozoa These are cnidarians of the class-Anthozoa including the corals and sea anemones.

anthropoid apes Any ape of the family Pongidae and Hylobatidae, resembling humans anatomically, i.e. gibbon, chimpanzee, gorilla, etc.

anthropology Study of the origin, behaviour and the physical, social and cultural development of humans.

antibiotics A substance produced by some microorganisms, that can kill bacteria or inhibit their growth, e.g. Penicillin.

antibody A protein produced by plasma cells in response to foreign substances (antigens). An antibody couples specifically with its antigen (and with molecules sufficiently similar to its antigen). Each plasma cell produces one and only one type of antibody or immunoglobulin.

anticholinesterase The substance that inhibits the enzyme cholinesterase. Often cause death of vertebrates by paralysing the respiratory muscles.

anticlinal A line formed due to cell division at right angles to the surface of an organ or part (in botany).

anticoagulant A chemical substance that prevents blood from clotting, e.g. coumarin, heparin (natural anticoagulant), warfarin.

anticoding strand A strand in DNA duplex that is used as a template for *m*RNA assembly during transcription. It is complementary to the coding strand of DNA. Also called non-coding strand or template strand.

anticodon A nucleotide triplet in transfer RNA. It is complementary to and bonds with the corresponding codon sequence in messenger RNA on the ribosomes.

anticodon loop The loop conventionally drawn at the bottom of *t*RNA molecule that contains the anticodon.

antidiuretic hormone (ADH; vasopressin) Hormone secreted from the posterior pituitary and synthesised in the hypothalamus. In mammals, it stimulates the reabsorption of water in the kidney and thus, diminishes the volume of urine produced.

antienzyme An inhibitory enzyme or any other agent that retards, inhibits or destroys enzymic activity.

antigen A foreign substance that elicits an immune response, such as the production of an antibody.

antigen-antibody complex Formed when an antibody binds to an antigen.

antigen presenting cell (APC) A cell such as a macrophage, displaying foreign antigen complex with MHC (Major Histocompatibility Complex) on its surface. T-cells may recognize this complex using their T-cell receptor (TCR).

antigenic variability A change in an antigen present on the surface of a microorganism. In influenza viruses, antigenic variation results in new strains fo the virus.

antigibberellins Organic compound which cause plants to grow with short and thick stems.

antihistamine A chemical or drug used in the treatment of allergic reactions such as urticaria and hay fever. Antihistamines bind to histamine receptors and block the site and action of histamine.

antihypertensive A drug used in medicine to treat high blood pressure. Antihypertensives lower blood pressure either by reducing the volume of the blood or by dilating blood vessels.

antioxidants A compound that delay the oxidation of substances such as food by molecular oxygen. Most antioxidants are organic compounds. Natural antioxidants are present in vegetable oil and in some fruits.

antiparallel strands of DNA The relative polarities of two strands in a DNA double helix, if one strand goes 5′—3′, top to bottom, the other goes 3′—5′.

antipetalous Opposite to the petals related to flower whorls.

antipodal cells Three haploid cells found in the embryo sac of seed-bearing plants. These migrate to the chalazal end of the sac opposite to the micropyle. They are eventually absorbed by the developing embryo and their function is uncertain.

antiport A membrane carrier protein that transports two different ions or small molecules across a membrane in opposite directions, either simultaneously or sequentially.

antipyretic A drug that reduces fever by lowering body temperature, e.g. paracetamol and aspirin.

antisense DNA A single-stranded DNA molecule that can bind to a complementary base sequence in a m-RNA molecule. It prevents protein synthesis encoded by mRNA.

antisense RNA The RNA molecule with complementary base sequences to that of the RNA transcript of a gene.

antiseptic Mild chemical substance which when applied on the body, kill or inhibit the growth of germs. e.g. carbolic acids, iodine, benzoic acid, boric acid, etc.

antiserum Blood serum containing antibodies, used in vaccines to treat or prevent a disease or to combat animal venom.

antitoxin An antibody produced in response to a bacterial toxin.

antiviral A drug or any other substance that kills or inhibits viruses.

antler One pair of bony, and branched horn-like structue, found on the head of a deer, moose, elk, etc.

a

antrum
1. The cavity in a bone, e.g. sinus.
2. The part of the stomach next to the pylorus.

anucleate A cell that lacks a nucleus e.g. mature RBC's in man.

anus An opening at the lower end of the digestive tract through which indigestible material is eliminated from the body.

aorta The major blood vessel in higher vertebrates through which oxygenated blood leaves the heart from the left ventricle. Through it, branches off numerous arteries to supply oxygen and nutrients to all living cells in the body.

aortic arches Six pairs of blood vessels present in all vertebrate embryos. These connect the ventral aorta leaving the heart with the dorsal aorta.

aortic body Area in the wall of the aorta that contains chemoreceptors. These receptors respond to changes in the concentration of carbon dioxide in the blood.

aortic valve A semilunar valve between the left ventricle and the aorta. It prevents blood from flowing back into the heart from the aorta.

apetalous Flower without petals. Having no corolla at all.

aphid A small insect that feeds by sucking the sap from the phloem of plants.

aphotic zone Refers to the region of a lake or sea, where light cannot penetrate. This zone is situated beneath the euphotic zone. Inhabitants include carnivorous animals or organisms that feed on sediment or detritus. Also called the bathypelagic zone.

apical dominance A phenomenon in higher plants, in which growing apical bud inhibits the growth of the lateral (axillary) buds. It is brought about by action of auxins and ABA. Removal of shoot tips usually results in the growth of lateral buds. Mostly applied in tea plantation, hedge-making.

apical growth An alternative name for primary growth. Growth promoted by the apical meristem region of plants at the root and shoot tips.

apical meristem Mass of meristematic cells at the very tip of a shoot or root. These cells are continuously dividing to form new stem and root tissue. The new tissues thus produced are collectively known as the primary tissues of the plant.

apical surface The surface of epithelial cells that faces the lumen.

apicomplexa (sporozoa) A phylum of protists containing numerous organisms that are important causes of disease. In particular, these are the blood parasites of the apicomplexan genera *Babesia, Plasmodium* and *Theileria*, and the coccidian parasites that infect the intestinal tracts of animals.

apiculture Cultivation of bees on a commercial scale for the production of honey.

apiology The scientific study of bees and honey making.

aplanospore A non-motile asexual spore, characteristic of the pin moulds and green algae. It is formed in a-sporangium and is usually thick walled.

apnea Temporary absence or cessation of breathing.

apocarpous Gynoecium in which carpels are free.

apodeme A chitinous ingrowth of the exoskeleton to which muscles are attached. These ingrowths serve as tendons in insects, crustaceans and other arthropods.

apoenzyme The protein part of an enzyme. Together with the prosthetic group, it gives rise to the complete enzyme (holoenzyme).

apogamy In pteridophyte, the development of the sporophyte directly from a gametophyte, without fusion of gametes. The term also describes the development of an unfertilised female gamete into the sporophyte–a phenomenon described as parthenogenesis.

apolysis Separation of the cuticula from epidermis in arthropods and related groups.

apomixis The development of an embryo without fertilisation especially in plants. Includes parthenogenesis, apospory and apogamy.

apomorphic character A new specialized trait in an evolving organism which is completely different from its ancestral line.

apoplast Part of a plant that do not consist of living tissue. These are cell walls, xylem and spaces between the cells.

apoplastic pathway Movement of water and dissolved ions, etc, through the cell walls. It is the main route by which water travels across the root cortex to endodermis.

apoptosis Cell death that is genetically programmed. During embryonic development it plays a vital role in determining the final size and form of tissues and organs.

aporepressor A proteinous substance synthesised by regulator gene, forms a constitution of repressor for blocking the working of operator gene.

aposematic colouration Colour changing characteristics in animals either as a warning signal or as a self-defence mechanism, as in the skunk.

apospory Development of gametophyte directly from the cell of a sporophyte thus bypassing meiosis and spore production. Found in some bryophytes and pteridophytes. See also apogamy, apomixis.

apothecium A cup-shaped ascocarp of some lichens and the fungi Ascomycetes.

appendage Any structure attached to or an outgrowth of a major part, e.g., a limb of the trunk, eyelids and eyelashes ot the eye, hair and the nails of the skin and fallopian tubes of the uterus.

appendicular skeleton The component of vertebrate skeleton that are attached to the main supporting skeleton. It comprises 126 bones as pectoral girdle (4); Forearm 60; Pelvic girdle : 2 Hip bones, Leg : 60.

appendix A small diverticulum of caecum at the junction of small and large intestine of man and apes. It contains lymphoid tissue. It is a vestigial organ in human beings and performs no digestive functions.

apposition Deposition of successive layers of cellulose on the inner wall of a plant cell, resulting in an increase in thickness of the wall.

apposition eye A compound eye characteristic of diurnal insects, light enter the retina of each ommatidium as a single spot and the image is a composite of all the spots.

apricot

1. A tree producing a golden fruit with a stone, native to Asia or,
2. The downy yellow juicy edible fruit of this tree.

a

apterous Without wings, e.g. silver fish (*Lepisma*).

aquaculture The cultivation of aquatic animals and plants.

aquaporins A protein embedded in the cell membrane that regulate the flow of water.

aquarium A pond, tank, bowl or the lake in which living aquatic animals or plants are kept.

aqueous humour The fluid that fills the space between cornea and the lens of vertebrate eye. Iris and lens lie in it. Continuously secreted by ciliary body every 4 hrs.

arabidopsis A genus of flowering plants of the family Cruciferae. Its species *A. thaliana* is widely used in molecular genetics because of its simple genome. Its full genome sequence was published in year 2000.

arabinose A pentose sugar produced by a wide variety of organisms. The L-arabinose operon of *E. coli* has been extensively studied in molecular genetics.

arachnida A class of phylum–Arthropoda including ticks, mites, spiders, scorpions and king crab.

arachnoid membrane One of the three membranes that surround the brain and spinal cord of vertebrates.

arboriculture The cultivation of ornamental trees and shrubs is known as arboriculture.

Arbuscular Mycorrhiza (AM, endomycorrhiza) It is a symbiotic association between a non-pathogenic or weakly pathogenic fungus and the roots of plants in which fungal hyphae invade cortical cells of the root and produce vesicles and arbuscles.

arbuscule (of vesicular-arbuscular mycorrhizae) A much-branched, microscopic haustorial structure of the fungal symbiont that forms within living cortical cells of the root.

archaea A prokaryotic organism containing the *archaeobacteria.*

archaeabacteria A group of microorganisms that thrive in O_2 poor environments. It includes methanogens and halobacteria.

archaeocyte A type of amoebocyte in sponges capable of developing into other types.

Archaeopteryx A reptilian fossil bird from late Jurassic period having teeth, a feathered tail, long neck, wings and beak.

archegonium Female sex organ of liverworts, mosses, ferns and related plants of gymnosperms. It is multicellular and flask-shaped.

archenteron (gastrocoel) Cavity within early embryo of many animals, opening exterior by blastopore. Formed by invagination of mesoderm and endoderm cells during gastrulation. It ultimately becomes gut cavity.

archesporium Single cell or group of plant cells in the sporophyte from which spores may eventually develop in a sporangium in fungi.

areola

1. The word means a small circular area coloured differently from the surrounding tissue on the breast around a nipple or,
2. a small space in a network of leaf veins.

areolar tissue A type of connective tissue having mucin matrix, containing yellow and white fibres.

a

arginine The colourless, crystalline essential amino acid of the alpha-ketoglutaric acid family.

aril A fleshy, coloured covering on the seed. It arises as an upgrowth of the base of the ovule and may be a tuft of hairs.

arms

1. Forelimbs of a primate.
2. Locomotive limbs of certain invertebrates, such as the rays of a starfish.
3. Two portions of a chromosome lying on either side of the centromere; the shorter of the two is called the *p* arm, the longer the *q* arm.

arolium A pad-like lobe projecting between the claws of an insect's foot.

arousal A general level of alertness in an animal, resulting from the activity of a particular part of the brain.

arrayed library Individual recombinant clones placed in two dimensional arrays in microtiter dishes. The clones are hosted in cosmid, bacteriophage, YAC or other cloning vectors. Each clone in the array is identified by its plate and its location on that plate. Arrayed libraries are used for many purposes, in particular, for screening a specific gene or genomic region of interest.

arrector pili A small muscle attached to the hair follicle in the skin (in mammals). Contraction of these muscles in response to cold etc pulls the hair in vertical position (goose bumps).

arrhenotoky A reproduction phenomenon in many insect orders, particularly common in the hymenoptera (bees and wasps). The unfertilised eggs produce the males and fertilised eggs gives rise to females.

arrhythmia A disorder of the heart rate (pulse). It occurs due to problem with the heart's electrical conduction system (See tachycardia and bradycardia).

arteriogram An imaging test that uses X-rays and a special dye to see inside the arteries. The procedure is often called angiography.

arteriography A radiographic method of observing the arteries after injection of a radio-opaque solution.

arteriole A small-diameter blood vessel that extends and branches from an artery and leads to the capillaries.

arteriovenous anastomosis A small muscular blood vessel that carries blood directly from the arterioles to the venules and bypasses the capillary network. By dilating or contracting, it can regulate the amount of blood flowing through a particular capillary network at any given time. It is stimulated by sympathetic nerves.

ateriovenous shunt A direct connection between an artery and a vein that bypasses the capillary system. May be congenital, caused by trauma or result from a breakdown in adjacent vessel walls. Arteriovenous shunts are also in some cases created intentionally as part of a surgical procedure.

artery It is a blood vessel through which blood flows away from the heart, i.e. towards tissues.

arthritis The inflammation of one or more joints. It causes joint pain, swelling, stiffness and limited movement.

arthrodesis Surgical immobilisation of a joint by fusion of the adjacent bones *via* surgery. Also called artificial ankylosis or syndesis.

arthropod A group of invertebrate animals such as the insects, crustaceans, arachnids, centipedes, etc. that are characterised by chitinous exoskeleton and a segmented body with jointed appendages.

arthrospore A type of spore produced in some fungi (e.g. Actinomycetes) by the segmentation of hypha.

articular bone A small bone of the lower jaw in bony fishes, amphibians and reptiles that forms a hinge joint with the quadrate bone of the upper jaw. The articular is derived from the ossification of Meckel's cartilage.

articulation The joining of two bones usually with the help of a joint.

artificial chromosomes These are laboratory constructs that contain DNA sequences and perform critical functions of natural chromosomes. e.g. BACs, YACs. They are used as cloning vectors to introduce and control new DNA in a cell, to study chromosome functions and to map genes in a genome.

artificial classification Classification system based on organism's form and shape, i.e. morphology.

artificial fertiliser Mixture of inorganic substances that provides the nutrients necessary for the growth of crops.

Artificial Insemination (AI) Artificial introduction of semen into female genital tract. Technique used in animal breeding.

artificial kidney An apparatus that performs the work of the kidney in purifying blood in case of renal failure (Also see haemodialysis).

artificial selection The process of selective breeding of organisms by humans to change one or more characters over time.

artiodactyla Even-toed ungulates. An order of mammals with two equally functional digits, the third and fourth as in cattle, sheep, goat, antelope, camel, giraffe, pig and hippo. These are terrestrial, herbivorous and cursorial (fast runners).

asbestosis A chronic lung disease caused due to prolonged inhalation of asbestos. Can lead to cancer.

ascariasis Disease caused by an intestinal parasite *Ascaris.* Symptoms include internal bleeding, muscular pain, fever, anaemia and blockage of intestinal passage.

ascending aestivation An aestivation in which each petal overlaps the edge of the petal posterior to it.

ascending limb (loop of Henle) Portion of the nephron where dilute fluid is produced. It leads to the distal convoluted tubule in kidney.

ascending tract Spinal neurons that carry signals to the brain. Also called afferent tract.

ascent of sap Upward movement of absorbed water along with the dissolved salts through the xylem tissue of roots to aerial parts in plant.

Aschelminthes Phylum of the invertebrate comprising of unsegmented worms (roundworms, rotifers, etc). They are distinguished by lack of coelom and blood vessels.

ascidian Sea-squirt (*Herdmania*), A member of the sub-phylum–Urochordata (Tunicata) of lower chordates (Protochordata). It is a sedentary tunicate, without notochord, nerve cord and tail in the adult but the tadpole larva has all these in the life-cycle, which undergo retrogressive metamorphosis to become adult.

ascites Accumulation of excess fluid in the abdominal (peritoneal) cavity.

ascocarp A fruiting body of an ascomycete, usually developed from one fertilised ascogonium and consists of asci, paraphyses and an outer wall.

ascogenous hyphae The hyphae in fungi of Ascomycota that are made up of binucleate cells containing one nucleus derived from male antheridium and other from female ascogonium.

ascogonium Female sex organ of certain fungi Ascomycota, which after fertilisation produces the ascogenous hyphae (hyphae containing haploid 'female' nuclei).

Ascomycota Division of the kingdom–Fungi that form the subkingdom–Dikarya, together with Basidiomycota. Largest fungi phylum, commonly known as Ascomycetes.

ascorbic acid White crystalline water-soluble vitamin found in many plant materials, particularly fresh fruit and vegetables. It is a natural antioxidant. Also known as vitamin-C or simply ascorbate.

ascospore Spore within an ascus, the sexual spore of Ascomycetes.

ascus Cylindrical sac produced by Ascomycete fungi. It usually contains eight ascospores that are the products of one cell undergoing two meiotic divisions followed by one mitotic division.

asepalous Having no sepals in a flower.

asepsis Reducing or eliminating pathogens. Asepsis is important before and during surgery.

aseptic (sterile, germfree) Free from pathogenic microorganisms, particularly bacteria.

asexual reproduction Mode of reproduction occurring without fusion of gametes and involvement of a single parent, e.g. fission in protists.

A-site The site on a ribosome where charged amino acyl *t*RNA first binds to *m*RNA.

asparagine (ASn) A non-essential amino acid. Its codons are AAU and AAC.

asparagine synthase A mammalian enzyme. It uses glutamine as a nitrogen source. Also known as aspartate-ammonia ligase, it is a prokaryotic enzyme that catalyses the same reaction, but using ammonia as the source of nitrogen.

Aspartic Acid (ASP) A non-essential amino acid. Also called asparaginic acid.

aspect diversity Identifying different physical features that are found in a group of species living in a common habitat. These species are hunted by other animals that use visual hunting skills to identify and kill their prey.

aspirin (acetylsalicyclic acid) A drug that is commonly used as an analgesic, antipyretic and to reduce inflammation.

assimilation Process of incorporation of simple molecules of food (that has been digested and absorbed) into living cells of an animal and its conversion into complex constituents of an organism.

assimilatory power Two important end products of light phase of photosynthesis, *i.e.* ATP and $NADPH_2$ are used in dark reaction for the reduction of CO_2 to carbohydrates. ATP and NADPH are called assimilatory power.

assisted reproductive technology (ART) The reproductive technology used in infertility treatments, e.g. AI, *in vitro* fertilisation, etc.

assortative mating The reproduction between individuals on the basis of their phenotypic or genotypic similarities or differences.

a

astasia The inability to stand due to muscular incoordination.

aster

1. A star-shaped structure of microtubules that surround the centrosome during cell division (in cytology).
2. Any of various plants of the genus *Aster* having radiate flower head with white, pink or violet rays around a yellow disk (in botany).

Asteroidea The echinoderm class including the starfishes and sea stars.

asthenia Weakness, lack or loss of strength and debility.

asthma Condition in which the flow of air to the gas exchange surface of the lungs is restricted.

astigmatism The refractive error of the eye in which the various meridians of the cornea or lens do not have the same radius of curvature.

astragalus An adjacent bone of toes in higher vertebrates.

astral microtubules Microtubules that originate from the centrosome that only exists during and immediately before mitosis.

astringent The substance which reduces the discharge of mucus or blood by contracting body tissues.

astrobiology The study of the possible existence of life in outer space.

astrocyte A star-shaped glial cell of the central nervous system.

astrosclereid Irregularly branched sclereid found in the leaves of certain cotyledons in small groups or sligly.

asymmetrical Without symmetry, e.g. certain sponges like *Spongilla*.

asynapsis Failure of chromosomes to pair at pachytene or the absence of chiasmata formation.

atactostele A type of stele in which many vascular bundles are scattered in ground tissue. It is a characteristic of monocotyledons.

atavism Recurrence of a character in descendants which had been possessed by an ancestor, after an interval of several or many generations. It is usually due to genetic recombination.

atelopodia Abnormal or incomplete development of the foot.

atheroma Fatty deposits formed in the walls of arteries.

atherosclerosis Thickening and the loss of elasticity of the walls of arteries due to deposits of fatty substances, like cholesterol.

athlete's foot Common fungal infection of the skin of the feet; ringworm of the feet.

atlas First cervical vertebra which joins the skull to the spine and articulates with the axis allowing nodding movements of the head.

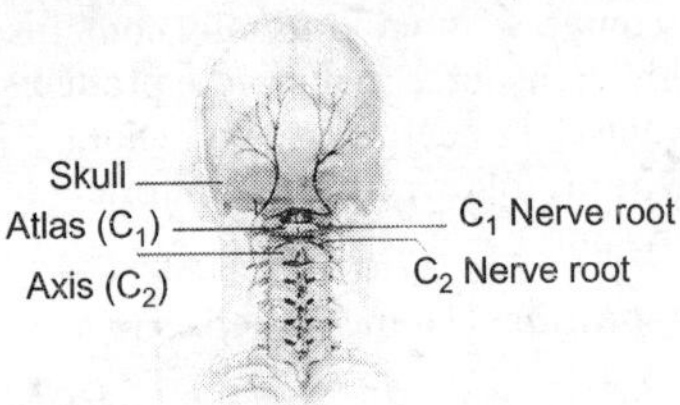

Atlas is the first cervical vertebra

Atomic Force Microscopy (AFM) A type of scanning probe microscopes, designed to measure local properties, such as height, friction, magnetism, with a probe.

ATP *See* adenosine triphosphate.

ATP synthetase Enzyme that couples proton movement across the inner mitochondrial membrane with ATP synthesis. Also called F_1-F_0 ATPase.

ATPase Enzyme that hydrolyses ATP to ADP and inorganic phosphate.

atresia A poptosis of ovarian follicles.

Atrial Natriuretic Peptide (ANP) A peptide hormone from atria of the heart that increases renal Na^+ and water excretion.

atrioventricular (AV) **node** Area of specialised cardiac muscles that receives the cardiac impulse from the sino atrial (SA) node and passes it to the ventricles. It is located in the septal wall between the left and right atria.

atrioventricular (AV) **valve** Valve between each auricle and ventricle of the heart. *See* biscuspid and tricuspid valve.

atrium Either of the thin-walled upper chambers of the heart that receives blood from the veins and forces it into a ventricle.

atrophy Shrinkage or disappearance of a part or organ.

atropine White crystalline poisonous alkaloid that occurs in deadly nightshade or belladona used to dilate pupils of the eyes and as an antispasmodic, antisudorific and anticholinergic.

attenuation Loss of virulence of pathogenic microorganisms.

attic Upper portion of the tympanic cavity of the ear.

atticitis Inflammation of the tympanic attic.

audibility State of being perceptible to hearing. The limits of audibility of the human ear are between 20 and 20000 hertz.

auditory canal Tube that leads from the outer ear to the ear drum (tympanum) along which sound waves travel.

auditory nerve (cochlear or acoustic nerve) One of two parts of 8th cranial nerves in amniotes. It transmits sensory information from the ear to the brain.

auditory ossicle Small bones of ear in terrestrial vertebrates.

auricle

1. Alternative name of atrium present in heart.
2. Outer projecting portion of the ear. Also called pinna.

auriculars Set of feathers that are found near the bird's ear openings.

auscultation Process of listening to the sounds of the body, generally of the heart or lungs for the purpose of diagnosis and evaluation.

australopithecines *(Australopithecus)* An early hominids discovered in Africa. A genus of fossil primates that walked erect and had teeth resembling those of modern humans, with brain capacity less than the modern human. Lived 4-2 million years ago, chiefly in East and South Africa.

autacoid An organic substance produced locally by one group of cells that exerts effects on other type of cells in same region, e.g. histamine, serotonin, etc.

autecic Pertaining to or being a parasite that throughout its entire life cycle, infects a single type of host.

autecology Study of interaction of an individual organism or a single species with the living and non-living components of its environment.

autism A kind of brain disorder that develops in infants and characterised by extreme learning difficulties and a lack of responsiveness to other people.

autoclave A steel vessel used for sterlising the equipments, media, etc to be used in an experiment with steam at high temperature and pressure.

autoecious Require only one host species to complete the various stages of their life cycle, e.g. rust fungi *Puccinia antirrhinum.*

autogamy Self-fertilisation in plants.

autogenic succession Change in species that make up a community as a result of effects of the organisms themselves.

autograft A type of graft involving transplantation of tissue or an organ from the part of an individual to another part of the same individual.

autoimmune disease Disease in which the immune system creates antibodies against body's own tissues.

autoimmunity A diseased state in which antibodies are formed and react against a normal component of the animal's own tissues. It is a contributory factor of a number of diseases such as rheumatoid arthritis and some forms of gout.

autolysis Self destruction of cells or tissues, etc. by action of enzyme activity of lysosomes. It is the final stage of cell senescence resulting in complete digestion of all cell components.

autonomic movements Movement of plants in response to internal stimuli and is independent of changes in external environment, e.g. cytoplasmic streaming, chromosome movement during nuclear division and growth itself.

Autonomic Nervous System (ANS) Part of vertebrate nervous system that regulates involuntary action of an animal's internal organs like the intestine, heart glands, etc. It is divided into sympathetic and parasympathetic nervous system.

autophagy A process involving the degradation of a cell's own components through the lysosomes. It speeds up during senescence.

autopolyploid A polyploid in which the multiple sets of chromosomes are derived from the same species.

autoradiography A technique used in analysing DNA fragments size. Typically, radioactive fragments are segregated according to length by allowing them to migrate through an agarose gel, subjected to a polarised electrical field (gel electrophoresis). The gel is then placed on X-ray film to visualise the radioactively labelled fragments.

autosomal dominant Gene that is both dominant and located on an autosome.

autosome Chromosome not involved in sex determination. The term is used for chromosomes other than sex chromosome in an organism. e.g. in man, out of 23 pairs of chromosomes, 22 pairs are autosomes.

autostylic jaw suspension A type of jaw suspension in which the upper jaw is directly connected to the cranium. Seen in lungfish and tetrapods.

autosyndesis Pairing of chromosomes from the same polyploid parents or remote polyploid ancestors.

autotetraploid A tetraploid with four similar sets of chromosomes.

autotomy Self-amputation of part of the body. Lost part is regenerated.

autotroph An organism capable of acquiring nourishment from its surrounding environment using photosynthesis or chemosynthesis.

autotrophic nutrition Preparation of organic food from inorganic materials using energy from an outside source. Most green plants are completely autotrophic.

autotropism Tendency of a plant part to grow in a straight line, on absence of an external stimuli.

auxanometer An instrument designed to measure the increase in length or rate of growth of a plant part.

auxin Plant hormone with multiple effects, such as cell elongation, apical dominance, inhibition of abscission and growth toward light (phototrophic response). Also, with gibberellins, it stimulates fruit production. Naturally occurring auxins are synthesised by plants in actively growing regions, e.g. IAA (Indole Acetic Acid). Synthetic auxins include NAA, 2, 4 D that are used as weedicides.

auxocyte A non-specific term for a spore or gamete-forming cell such as oocyte, sporocyte, etc. during its growth phase.

auxospore Spore produced by diatom.

auxotroph A mutant strain of microorganism which needs growth factors not needed by its wild type.

auxotrophic mutant An organism having a growth requirement of specific nutrients which are not necessary in the parental strain.

averbach's plexus (myenteric plesux) The plexus of fine unmyelinated fibres and postganglionic autonomic cell bodies lying in muscular coat of oesophagus, stomach and intestine.

Aves Class of vertebrates comprising the entire bird family. Birds are characterised by feathers, a beak, oviparous, high metabolic rate, a 4-chambered heart and a light-strong skeleton.

aviculture Breeding and keeping of birds.

avidin A glycoprotein that binds to biotin and inactivates it. It is produced in the oviducts of birds, reptiles and amphibians and deposited in their egg white.

avidity The total binding strength of antigen-antibody binding based on the rate at which the complex is formed (in immunology).

avitaminosis Any disease caused due to lack of one or more vitamins.

awn (arista) Stiff, bristle-like appendages occurring frequently on the flowering glumes of grasses and cereals.

axenic culture A culture medium in which only a single species, variety or strain of organism is present and entirely free of all other contaminating organisms. Such cultures are used in microbiology to determine growth requirements or effect of chemicals of a particular species.

axial

1. Pertaining to an axis.
2. Around an axis.
3. Along or parallel to an axis.

axial skeleton Bones constituting the head and trunk of a vertebrate body. In humans, it consists of 80 bones that comprises skull bones (22), ossicles of middle ear (6), hyoid bone (1), rib cage (24), sternum (1) and vertebral column (26).

axil The angle between a branch or leaf and the stem that bears it.

axile placentation Condition in which ovules arise on the axis of an ovary with several locules.

axillary bud During the formation of leaves and elongation of stem, some cells 'left behind' from shoot apical meristem form axillary bud. These buds are present in the axils of leaves and are capable of forming a branch or a flower.

axis The main stem or central part about which organs or plant parts such as branches are arranged (in botany). Second cervical vertebra supporting the head. Also called epistrophous. Its body elongates and forms the pivot upon which atlas can turn (in anatomy).

axolotl An aquatic salamanders of the North American genus *Ambystoma* in which larval stage is remarkable as it does not metamorphose. It breeds while being in its larval form (the phenomenon of neoteny).

axon A long, slender projection of the neuron that carries nerve impulse away from neuron's cell body.

axon hillock A specialised part of neuron's soma that connects to the axon.

axoneme The central core of microtubules found in cilia and flagella of eukaryotes. It consists of nine pairs of outer microtubules, surrouding the two single central microtubules.

axoplasm The protoplasm of axon.

axostyle A rod-like structure of protoplasm in many ciliated animals, which also functions as internal skeleton.

Azotobacter A genus of free-living aerobic nitrogen-fixing bacteria found in limestone soils and water. Due to their ability to fix molecular oxygen, these species are widely used as biofertilisers in agriculture.

azygospore A sexually formed zoospore in fungi. It is formed parthenogenetically and is a resting spore. It is haploid and does not undergo meiosis at the time of spore formation.

azygous vein It is the vein present in reptiles, birds and mammals, which collects blood from ribs.

B

baboon A ground-living primate of Africa which lives in highly organised family groups called troops belonging to the genus *Papio*. They forage for a wide variety of food during the day and sleep in the safety of trees at night. Lion is their chief enemy which preys on them.

Bacillariophyta A class of algae comprising the diatoms.

bacillemia Presence of bacteria in the blood.

Bacillus (pl. bacilli) Any rod-shaped bacterium. These are Gram-positive, spore forming and motile.

Bacillus anthracis A bacilli form bacterium; causative agent of anthrax.

Bacillus thuringiensis (Bt) A gram-positive soil dwelling bacterium used as a biological pesticide. It has been used to develop genetically modified crops with insect tolerance by expressing the *cry* genes. Its advantages are specific toxicity against target insects, lack of polluting residues and safety to non-target organisms.

back bone (vertebral column) It is formed during development by the replacement of notochord. It is dorsal in position and encloses and protects the spinal cord. It is made from individual bones called vertebrae.

back cross A cross between F_1-hybrid individual (obtained by crossing two parental types) and a second individual belonging to one of the parental types (Also see test cross).

bacteria A minute prokaryotic single-celled organism that cannot be seen with naked eye. Bacteria may be rod-shaped (bacilli), spherical (cocci), spiral (spirilli), comma-shaped (vibrio), etc.

Bacterial Artificial Chromosome (BAC) It is a DNA construct used for transforming and cloning in bacteria. This property is based on a functional fertility plasmid (F-plasmid).

bactericide A class of antibiotics that kills bacterial cells.

bacteriochlorophyll Photosynthetic pigments that occur in phototrophic bacteria, such as the purple bacteria. These are seven forms, designated as bacteriochlorophyll-*a*-*g*. All are structurally similar to chlorophyll-*a* of plants.

bacteroid Any of various bacteria that carry out nitrogen-fixation in the root nodules of leguminous plants and change in shape and size after they enter the roots, e.g. *Rhizobium*.

bacteriology Study of bacteria, including their identification form, function, reproduction and classification.

bacteriophage (phage) Virus that infect bacterial cells; composed only of DNA and protein. They infect their DNA into the bacterial cells, where it is selectively replicated and expressed. It is used as a cloning vector in genetic engineering.

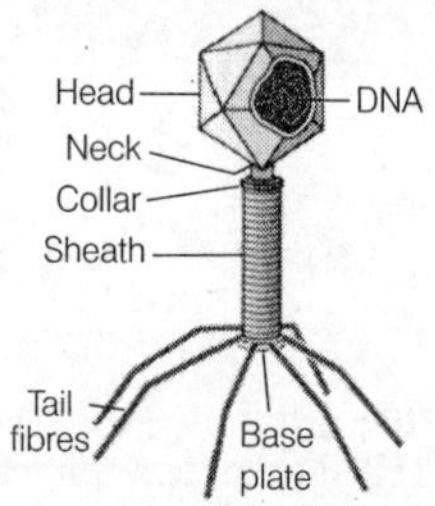

Structure of a bacteriophage

bacteriorhodopsin A pigmented membrane protein found in halophilic archaebacterium. It converts light energy to chemical energy (ATP) without involvement of chlorophyll.

bacteriostatic A compound that prevents or slows the growth and reproduction of bacteria, but does not kill them.

bagging A process wherein emasculated flowers are enclosed in bags to prevent random cross- pollination.

baker's yeast A species of *Saccharomyces cerevisiae* used in baking and bread products. It converts the fermentable sugars present in the dough into CO_2 and ethanol by anaerobic respiration (fermentation).

balanced genetic load The decrease in overall fitness of a population caused by alleles that persist in the population because they confer selective advantage in other genotypic combinations.

balanced polymorphism A situation where more than one allele (gene) are maintained in a population, which is the outcome of the heterozygote being superior to both homozygotes.

balancer (halter) A structure on the side of a dipterous fly that helps it to stabilise during flight (or swimming). In *Drosophila*, the balancers sit posterior to the wings and represent a modified pair of wings.

balausta Type of fruit that develops from multilocular, syncarpous, inferior ovary. Seeds are many and irregularly arranged in the fruit. Pericarp is rough and leathery, e.g. pomegranate.

Balbiani rings Enlarged bands of giant (polytene) chromosomes of the salivary glands in *Drosophila melanogaster*. Normally, the chromosomes are invisible in the interphase. Polytene chromosomes were first observed by Balbiani.

Baldwin effect (ontogenic evolution) An idea proposed by James Mark Baldwin in 1896, that describes the effect of learned behaviour on evolution. Though appear similar to Lamarckism, it proposes change in genetic make up of species through natural selection.

baleen (whale bone) Type of whale that filter planktons for ingestion out of large quantities of water. Baleen is the fibrous structure in their mouths, usually referred to as baleen plates, that enable them to feed in this way. They do not have teeth.

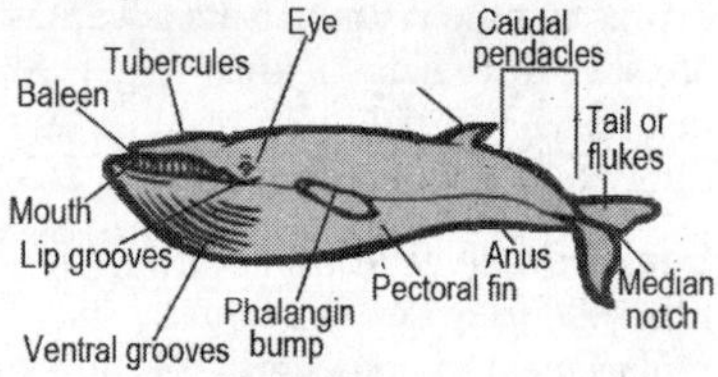

Baleen whale

ball and socket joint (spheroidal joint) Type of synovial joint that allows movement forward, backward, and from side to side. In this type, a ball like structure of one bone fits into a cup-like depression of another bone, e.g. hip joint between femur and the acetabulum of pelvis.

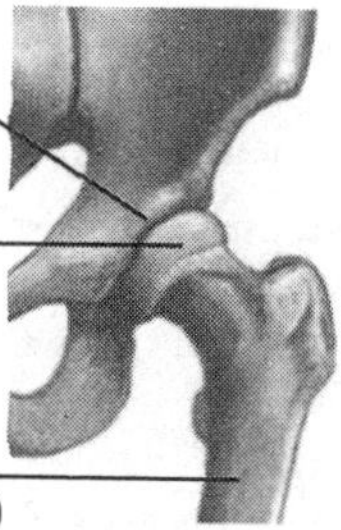

Ball and socket joint

banding (medical) A medical procedure that uses elastic bands for constructing, e.g. to tie off blood vessels, to stop bleeding. It is also done on animals for different purposes.

barbel A slim, whisker like tactile organ located near the mouth of certain species of fish. In some fishes, they contain the taste buds, and assist the fish to locate food in murky water.

barbicels Tiny curved structures on barbules, that connect continuous barbules to form the firm, mesh like structure of the feather vane.

barbiturate Any of a group of barbituric acid derivatives (such as phenobarbital) that acts as a depressant on the central nervous system. Originally used in medicines as a sedative, hypnotic but their clinical use has been limited due to their toxic side effects. Prolonged use can lead to addiction.

barbs

1. The processes attached to the main shaft of a feather, and make up its vane (in zoology)
2. A short sharp reflexed bristle or hair-like projection (in botany)
3. Several species of freshwater fish.

barbules Tiny filamentous structures that emerge from the barbs of a bird's feather forming a row on each side. They interlock, overlap and knit together, making the feather solid and stiff.

Barfoed's test A test to detect monosaccharide (reducing) sugars in solution. A positive test is indicated by red precipitate, on boiling the solution with Barfoed's reagent (ethanoic acid + copper (II) acetate).

bark Outer covering of the stems, branches and roots of trees and woody plants; composed of an outer layer of dead cells (cork) and an inner layer of phloem tissue.

baroreceptor A receptor sensitive to pressure, e.g. the carotid sinus baroreceptors respond to mean arterial pressure and pulse pressure.

Barr body An inactivated X-chromosome in a female somatic cell. Under the microscope, it is seen as a dense body lying near the nuclear envelope. In humans, with more than one X-chromosome, the number of barr bodies is (X-1), e.g. Klinefelter man (XXY) have one barr body (2-1).

barrier method A contraception based on putting a physical or chemical barrier between egg and sperm, e.g. condoms, spermicidal creams, etc.

Bartholian's glands A pair of glands occurring in some animals which lie on either side of the upper end of the vagina. They secrete mucus to lubricate the vagina and are homologous to bulbourethral glands in males.

basal body (kinetosome) An organelle formed from a centriole and a short cylindrical array of microtubules. It serves as a nucleation site for the growth of axoneme microtubules.

basal ganglia Nuclei surrounding the thalamus that connect the cerebrum with other parts of the nervous system. They help to control and coordinate voluntary movements.

b

basal lamina (basement membrane) A thin delicate layer of extracellular matrix that lies beneath the epithelium. It holds the epithelial cells to underlying cell layers.

basal metabolic rate (BMR) An individual's lowest metabolic rate. It is measured in terms of heat produced per unit time.

basal metabolism Minimum quantity of energy required to maintain vital functions in an organism when at complete rest.

base pair (bp) A pair of hydrogen-bonded nitrogenous bases that join the complementary strands of the DNA or RNA. A base pair is composed of a purine and a pyrimidine (guanine and adenine or cytosine and thymine).

base substitution Substitution of one base for another in a nucleotide of DNA. It constitutes the simplest type of mutation.

basement membrane Thin sheet of fibrous proteins that supports the cells of an overlying epithelium or endothelium, separating this from underlying connective tissue.

basidia (sing. basidium) Specialised club-shaped cells on the underside of club fungi (Basidiomycetes) within which basidiospores form.

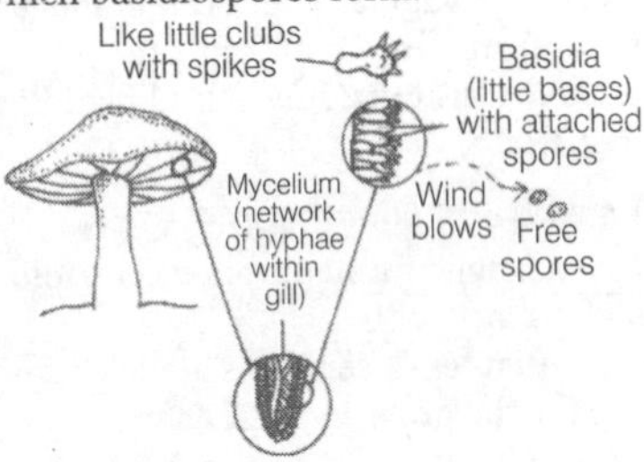

A fungus (mushroom) showing basidia

Basidiomycetes (Basidiomycota) It is club fungi, a major phylum of fungi that reproduces sexually by means of basidiospores, e.g. mushrooms and toadstools.

basidiospores Spores formed on the basidia of club fungi (Basidiomycetes).

basilar membrane Membrane that separates the cochlear duct from the tympanic duct. It supports the organ of Corti.

basipetal Downward from the apex towards the base of a shoot; referring to development in the direction of the base, so that the apical part is oldest, successive from apex to base.

basket cells An inhibitory GA Bergic interneurons present in several brain regions i.e. cerebellum, hippocampus and cortex.

basophil Leukocyte that stains dark purple with basic dyes. They have a lobed nucleus with granular cytoplasm. Least abundant (0.8-1%) among WBCs. Involved in many types of inflammations, particularly allergic reactions. Secrete histamine, serotonin, heparin, etc.

bast Old name for phloem.

bast fibre (skin fibre) Sclerenchymatous fibre in phloem or bast surrounding stem of dicot plants. Provide strength and support the conductive cells of phloem.

bat A mammal of the order–Chiroptera. The wings of a bat are elaborate folds of skin supported by its elongated fingers and its arms and legs. They are nocturnal species.

batch culture A technique used to grow microorganisms or cells. A limited supply of nutrients for growth is provided, when these are used up or some other factor becomes limiting, the culture declines.

Batesian mimicry Type of mimicry where a harmless organism looks similar to other species that is poisonous or harmful to its predators.

bathypelagic Pertaining to ocean levels that are deep, but above the abyssal zone (deep water).

B-cell (B-lymphocytes) Type of lymphocyte that differentiates in the bone marrow in mammals and in Bursa of Fabricus in birds (the B in B-cell originally stood for Bursa since these cells were discovered in birds. Principal functions are to make antibodies against antigens, to develop into memory B-cells and perform the role of APCs.

B-cell receptor An assembly of proteins, attached to the plasma membrane of a B-cell, that binds specific extracellular antigen.

BCG Bacillus Calmette Gurein vaccine (BCG) was first introduced in France in 1908 by Calmette and Gurein. It produces significant immunity against the tuberculosis and treatment of some bladder cancers.

B-DNA The naturally occurring form of DNA duplexes *in vivo*, that is same as the model proposed by Watson and Crick in 1953.

bead theory A disproved hypothesis that states genes are arranged on the chromosome just like beads on a necklace, indivisible into smaller units by mutation and recombination.

beak (bill or rostrum) Protruding part of mouth of several groups of vertebrates, including some cetaceans. Birds use them not only to eat, but also to groom, kill prey, manipulate objects, in courtship and to feed the young.

behavioural genetics Study of genes influencing behaviour.

benign Harmless, non-malignant usually used for tumours that lacks the ability to invade neighbouring tissue.

benthic An ecological region that encompasses the bottom most level of any body, such as a river, lake or ocean. When used in relation to a living organism, it refers to bottom-dwelling.

benthos Plants and animals living on the bottom of sea or lake from high water mark down to the deepest levels. The benthos is divided into littoral organisms (down to 200 meters deep) and deep water organisms.

Bergmann's rule The generalisation that animals living in colder climates tend to be larger than those of the same group living in warmer climates.

beri-beri A disease caused by the deficiency of thiamine (vitamin-B_1). Symptoms are weakness, swelling and pain in legs, loss of appetite, enlarged heart and shortness of breath and paralysis.

berry A fleshy indehiscent fruit with many seeds. It is developed from mono or poly carpellary, superior or inferior ovary with axile or parietal placentation. Epicarp forms the outer dry cover, mesocarp is fleshy and endocarp is membraneous. Examples include brinjal, banana, grape and tomato.

beta-adrenergic receptor (β-receptor) Membrane receptor that bind and transduce signals from epinephrine and nor-epinephrine.

beta-amyloid proteins An amyloid that circulates in blood and CSF of humans and is deposited into plaques in Alzheimers patients.

beta blocker (beta-adrenoceptor antagonist) Any of a group of drugs that interfere with the ability of adrenaline to stimulate the beta-receptors of the heart.

beta-carotene An orange pigment composed of 8 isoprene units. It occurs in most plants as an accessory pigment to photosynthesis. It is a precursor of vitamin-A.

beta cells Cells present in endocrine part of pancreas (islets of Langerhans), which secrete a hormone-insulin. *See* insulin.

beta-cyanin (Beta xanthin) A group of red pigments found in plants or order-Chenopodiales (goose foot, cactus, etc). They are nitrogen containing glycosylated compounds responsible for red colour of beet-roots.

beta-diversity (beta richness) The diversity in the variety of organisms in a region. It is influenced by the turnover of species among habitats.

beta-galactosidase An enzyme which hydrolyses the β-galactosides into monosaccharides.

beta-oxidation A sequence of biochemical reactions that oxidise fatty acids into a series of 2–C compounds that are converted to acetyl Co-A.

beta sheet (β-pleated sheet) A form of secondary structure in proteins, that occur in many globular proteins. In this, the extended polypeptide chains lie parallel to each other and are linked by hydrogen bonds between the N—H and C = D groups.

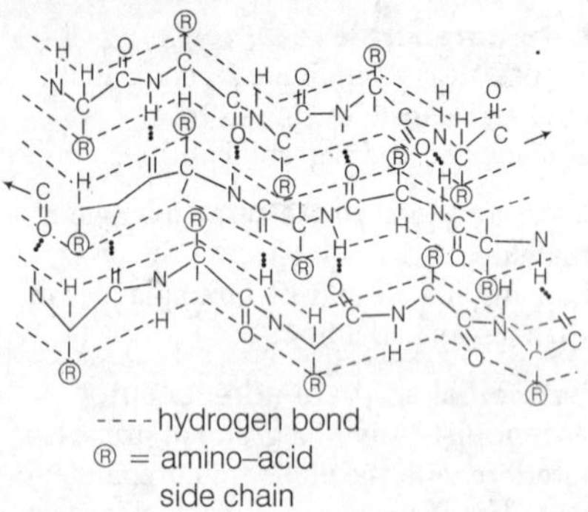

bet-hedging Evolutionary strategy in which an individual adopts two alternative tactics, so as to maximise probability of survival or reproduction.

biceps (biceps brachii) Two-headed muscle that lies on the upper arm and is connected to the radius at one end and the scapula at the other. It works antagonistically with the triceps.

bicollateral bundle Arrangement of tissues in the vascular bundles in which the phloem is situated on both sides of the xylem. This is thought to aid transport of nutrients in plants with long trailing stems, such as the marrow.

bicuspid valve (left AV valve; nutral valve) A heart valve, located between the left atrium and left ventricle and composed of two triangular flaps. It regulates blood flow between these chambers.

biennial Plant that completes its life cycle in two years. In the first year, it produces foliage and only photosynthesise. In the second year, the stored food is used to produce flowers, fruits and seeds.

Big Bang theory A model for the evolution of the universe that holds all matter and energy in the universe were concentrated at one point, which suddenly exploded. Subsequently, matter condensed to form atoms, elements, and eventually galaxies and stars.

bilateral symmetry This type of symmetry is exhibited by most animals. Body can be divided into equal and symmetrical halves only along one plane.

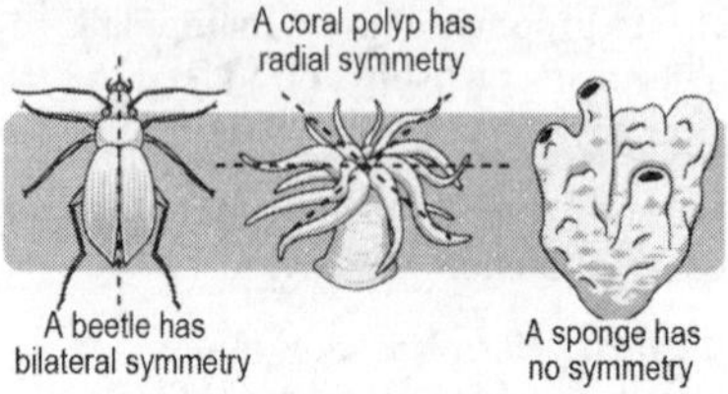

Bilateral symmetry

bile (gall) Bitter, alkaline greenish-yellow fluid produced by the liver, stored in the gall bladder and secreted into the duodenum. It contains bile salts (sodium glycocholate and sodium taurocholate), bile pigments and cholesterol. It aids in the absorption of fats by facilitating their digestion and emulsification.

bile duct The tube through which bile passes from the liver or (when present) the gall bladder to the duodenum.

biliary system The system that creates, moves, stores and releases bile into the duodenum. It consists of the liver, gall bladder and associated ducts (bile duct).

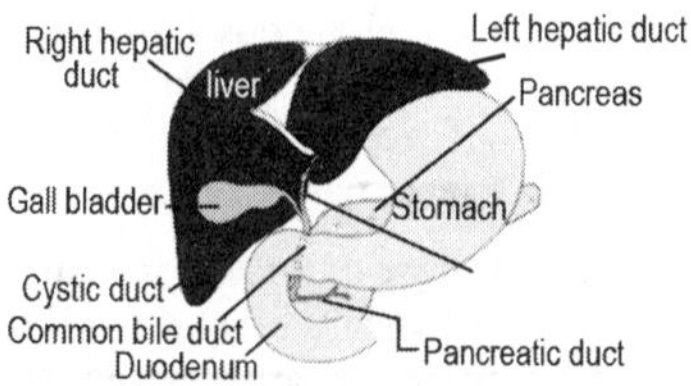

Biliary system

bilirubin An orangish-yellow pigment in bile. A catabolite of haemoglobin breakdown, bilirubin is carried to the liver where it is chemically converted and excreted as the bile into the duodenum.

biliverdin A green bile pigment formed by catabolism of haemoglobin and converted to bilirubin in the liver. It may also arise by oxidation of bilirubin.

binary fission Method of reproduction used by prokaryotes, in which a single parent cell divides into two daughter cells, each of which receives a single, complete copy of the parental chromosome, e.g. in *Amoeba, Paramecium.*

binding sites The region on a protein, DNA or RNA to which specific molecules or ions bind. The binding initiates a chemical or physiologic action.

binocular vision The ability, found in animals with forward facing eyes, to produce a focussed image of the same object simultaneously on the retinas of both eyes. This permits three-dimensional vision and contributes to distance judgement.

binomial nomenclature A system of taxonomy developed by Linnaeus in the early eighteenth century. In this system, each species is given a two part name, the first term is the genus, and the second is the species, e.g. *Mangifera indica*-scientific name of mango.

bioaccumulation Increase in concentration of chemicals (such as mercury, DDT) in organisms. These chemicals are non-biodegradable, so their rate of absorption and storage is greater than rate of excretion and thus gets accumulated in organisms. (*Also see* biomagnification)

bioactivation The metabolic activation of xenobiotic (foreign) compounds into reactive toxic compounds.

bioassay A technique in which the presence of a chemical substance is quantified by using a part of or an entire living organism.

bioaugmentation Increasing the activity of bacteria by addition of bacterial cultures to speed up the degradation of a pollutant, a technique used in bioremediation.

biochemical evolution (molecular evolution) The change that occur at the molecular level in organisms over a period of time, particularly in the molecular constitution of the genome.

biochemical fuel cell A system that exploits biological reactions for the conversion of biomass to electricity.

biochemical genetics Study of relationships between genes, protein and metabolism.

b

Biochemical Oxygen Demand (BOD) The amount of oxygen required for aerobic breakdown of organic matter by microorganism in water at certain temperature over a specific time period. A high BOD indicates a high level of pollution.

biochemistry Study of living organisms at the chemical level, in particular the structure and function of proteins, carbohydrates and nucleic acids.

biochip A collection of miniaturised test sites (microarrays) arranged on a solid substrate that permits many tests to be performed at the same time. It allows rapid screening of gene profile to detect genetic defects like sickle cell anaemia, cystic fibrosis, types of cancer. Biochip contains all the human genes.

bioclimatology Study of the influence of climate on ecosystems and structure and function of living organisms.

biocoenosis A group of interdependent organisms living in a particular ecosystem under uniform environmental conditions.

biodegradable Organic compounds that are susceptible of decomposition by bacteria and other microorganisms, such as the constituents of sewage.

biodiversity (biological diversity) The diversity of organisms in a habitat or ecosystem. Biodiversity can be measured based on the number of species or genetic variation that exists within an ecosystem or region.

biodiversity gradient The gradual reduction in biomass and species number with increasing latitude.

bioelement An element that is found in A living organisms, e.g. oxygen, carbon, nitrogen, hydrogen are the most common bioelements in human body.

bioenergetics A field of biochemistry that deals with energy relationships of living organisms.

bioengineering A hybrid field of scientific research that uses the principles of biology and the techniques of engineering to produce useful products, e.g. artificial replacements or organs, such as artificial limbs, heart pacemaker, etc.

bioenrichment Addition of nutrients or oxygen to increase microbial breakdown of pollutants.

biofeedback A technique whereby a person can consciously control physiological responses that are normally controlled unconsciously by the autonomic nervous system. It is used to control heart rate, blood pressure, migraine and to relax spastic muscles.

biofilm A colony of microorganisms adhering to the surface of an organic or inorganic structure together with their slimy matrix, e.g. dental plaque.

biofortification Process of breeding crops to increase their nutritional value, either through conventional breeding or genetic engineering.

biofuel Any gas, liquid or solid that can be converted from raw biological material to fuels useful for industry, transport, etc e.g. biogas.

biogas A mixture of gases (mainly methane) produced by anaerobic decomposition of organic matter by methanogens. It is about 50% methane and can be used as fuel.

biogenesis A theory which states that living things originate only from other living things as opposed to non-living matter.

biogenetic law It is the theory proposed by Ernst Haeckel that stages in the embryonic development of an individual (ontogeny) recapitulates (repeats) the evolutionary history (phylogeny) of clades to which the organism belongs. Also called Haeckel's law, law of recapitulation, recapitulation theory.

biogenic amine Any amine that is produced by living organisms, especially the physiologically active amines that serve as neurotransmitters in animals.

biogeochemical cycle The flow of an element through the living organisms and physical environment of an ecosystem, e.g. the carbon, hydrogen, oxygen, nitrogen, sulphur, and phosphorus cycles.

biogeography Study of the geographic distribution of living organisms.

bioinformatics Application of computer technology to manage and analyze biological data.

biolistics or gene gun Method of introducing alien DNA into living cells. Cells are bombarded with high velocity microparticles of gold or tungsten coated with DNA, using a special gun.

biological clock Internal mechanism of many organisms that regulates circadian rhythms and various other periodic cycles in behaviour or physiology.

biological control Reduction in population size of an unwanted organism by means of another organisms, e.g. *Bt* used to control butterfly catterpillars.

biological rhythm (biorhythm) Regular cyclic fluctuation in the physiological process or behavioural activity of an organism, e.g. circadian rhythm.

Biological Species Concept (BSC) The thesis that the primary criterion for separating one species from other is reproductive isolation.

biological warfare Use of living organisms, particularly microbes or their products to induce disease or death in humans, livestock and crop plants. Most often used during wars.

biologics Agents, such as vaccines, those give immunity to diseases or harmful biotic stresses.

biology Scientific study of living organisms.

bioluminescence Production and emission of light by living organisms.

biomagnification Increase in the concentration of toxicants at successive trophic levels.

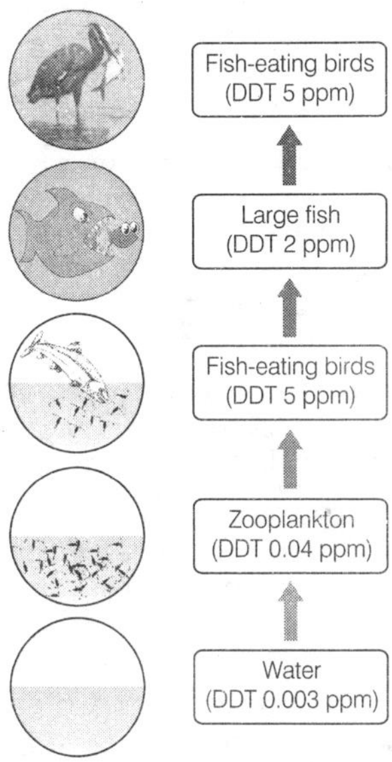

Biomagnification of DDT in an aquatic food chain

biomarkers (biological markers) A normal metabolite whose abnormal concentrations in body fluids indicate presence of a particular disease, etc., e.g. PSA for prostate cancer.

biomass Total dry weight of all organisms in a particular sample, population or area.

b

biomathematics An interdisciplinary field of study that attempts to model biological processes using mathematical techniques.

biome A region defined, as based on its climate and geography, having ecologically similar communities of plants, animals, and soil organisms. The similarity is based on plant structures (such as trees, grasses and shrubs), plant spacing (forest, savanna and woodland), leaf types (such as needle leaf and broadleaf) and climate, e.g. tundra, desert, etc.

biomechanics Application of the principles of mechanics to living systems, particularly systems with coordinated movements, e.g. study of muscles involved in leg movement in humans.

biomolecule Organic molecules associated with living organisms and are involved in their maintenance and metabolic processes. These include carbohydrate, lipid, protein, nucleic acids and water molecules.

biophysics An interdisciplinary science applying the theories and methods of physics to biological phenomenon.

biopiracy Refers to the use of bio-resources by multinational companies and other organisations without proper authorisation from the countries and people concerned, without compensatory payment.

biopoiesis Origin of organisms from self replicating organisms that are non-living. Biopoiesis is a cornerstone of abiogenesis. DNA is the best example of a self-replicating molecule and is found in the chromosomes of all higher organisms.

biopsy The medical removal of tissue from a living subject to determine the presence or extent of a disease. The biopsied tissue is examined under a microscope by a pathologist.

bioreactor Vessel in which large quantities of cultures can be processed. A bioreactor provides optimal conditions (temperature, pH, substrate, salts, vitamins and oxygen) for achieving the desired product.

biosensor An analytical device which converts a biological response into an electrical signal. These are used in diagnostic tests to allow quick sensitive and specific analysis of a wide range of products.

biosphere Whole region of earth's surface including the sea, the land and the atmosphere in which living organisms exist.

biostratigraphy (Biozone) The characterisation of rock strata on the basis of the fossils they contain, e.g. numerous ammonite species provides an important means of zoning rocks of Mesozoic era.

biosynthesis Chemical reactions in which a living cell builds up its necessary molecules from other molecules present.

biosystematics Area of systematics in which experimental taxonomic techniques are applied to investigate the relationships between taxa.

biota Flora (plants) and fauna (animals) of an era, geographic region or biome.

biotechnology Practical application of biological science; particularly applied to research and product development that uses recombinant DNA technology, cell fusion, and various other modern bioprocessing techniques.

biotic These are the living components that shape an ecosystem.

biotic factor Any living component that affects other living organisms in the environment.

biotic potential Denoted by symbol '*r*', it is also called the intrinsic rate of increase. It is the maximum reproductive capacity of an organisms, under optimum environmental conditions.

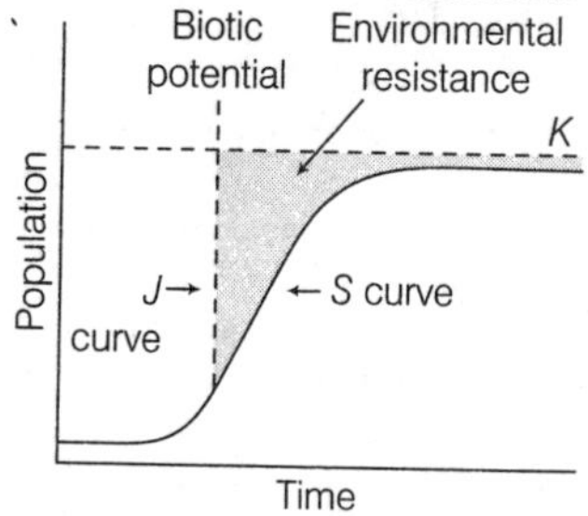

Graph showing biotic potential

biotic stress Adverse effect on plants, by other living organisms.

biotin A member of the vitamin-B complex and it is a cofactor for various enzymes. It also helps in the metabolism of fatty acid and amino acid. Rich sources are raw egg white, cereals, vegetables, milk and liver. Also known as vitamin-H or coenzyme R or vitamin-B_7.

biotope A geographical area uniform in environmental conditions and types of organisms. It is the smallest ecosystem.

biotype

1. A group of individuals having the same genotype.
2. A physiological differentiable strains of species.

biparous

1. Producing two offsprings per birth (vertebrate in zoology).
2. Bearing branches on dichotomous axes (in botany).

bipedalism The exclusive use of hindlimbs for locomotion, as seen in humans. It was an important factor in hominid evolution.

bipinnaria The first stage larva characteristic of starfish. It bears lobes that carry ciliated bands used for feeding and locomotion.

bipolar neuron A type of neuron with two processes, an axon and a dendron extending in different directions. These are specialised sensory neurons for transmission of special senses, e.g. bipolar cells of retina.

biramous appendage Characteristics of arthropods of the phylum N-Crustacea. It consists of a basal protopodite that gives out two branches outer exopodite and inner endopodite, each with one or two segments.

bird *See* Aves

birth Production of new individuals or release of fertilised eggs by a female.

birth control (contraception) Use of methods to intentionally avoid pregnancy without hindering sexual activity. These methods can be natural or artificial (*See* condoms, IUDS, pills, vasectomy, tubectomy).

birth rate (natality) Total number of young ones produced within a specific period of time. It is calculated per 1000 individuals of a population per year and is an important population attribute.

bisexual

1. Pertaining to a type of organism having a sexual orientation to person of other sexar
2. Pertaining to an individual having both male and female sexual organs (hermaphrodite)
3. Denoting a single flower that contains functional staminate and pistillate structures.

bivalent A pair of homologous, synapsed chromosomes; this configuration occurs during meiotic prophase-I. Also known as a 'tetrad.'

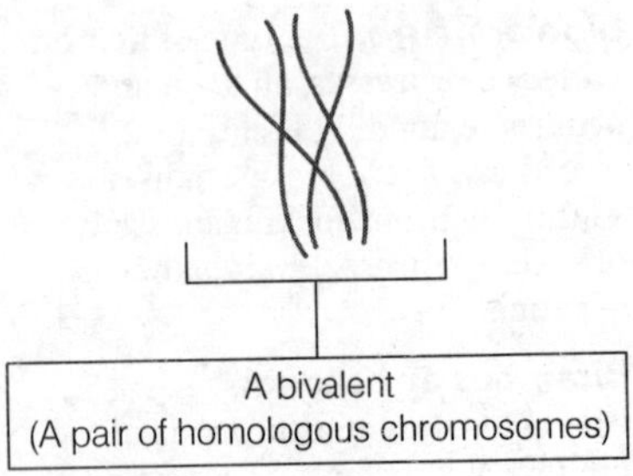

bivalvia (pelecypod; lamellebranch) Any marine or freshwater mollusc of class—Pelecypoda. They are characterised by a laterally compressed body, a shell with two hinged valve and gills for respiration. Includes clams, cockles, oysters and mussels.

blackhead

1. Sebaceous glands of the face become enlarged due to accumulated sebum. Melanin and oxidised oil give it a black colour (Comedo, in pathology)
2. Infectitious fatal disease of turkeys, fowls, etc, by a parasitic Protozoa (in veterinary science)
3. Any of various birds (gills or ducks) with black plumage on the head.

bladder

1. A hollow, distensible organ with muscular walls that stores urine and expels it through the urethra. Also known as urinary bladder (in anatomy)
2. Any of various sac-like organs in animals for storage of liquid or gas, e.g. swim bladder (in zoology)
3. Any of various hollow or inflated sac-like organs or structures, e.g. floats of certain sea weeds or specialised traps of bladderworts.

blade

1. Broad, expanded part of a leaf
2. Long narrow leaf of a grass or related plant.

blanching Treatment with heat for a short time to destroy cellular enzymes.

BLAST Stands for Basic Local Alignment Search Tool. It is a computer program that finds regions of local similarity between sequences. The program compares nucleotide or protein sequences to sequences databases and calculates the stastical significance of matches. It can be used to infer functional and evolutionary relationships between sequences and also helps to identify members of gene families.

blastema (pl. blastemas or blastemata) Group of cells giving rise to a new individual (in asexual reproduction) or to an organ or part (in either normal development or in regeneration).

blastocoel Cavity within a blastula during animal embryonic development.

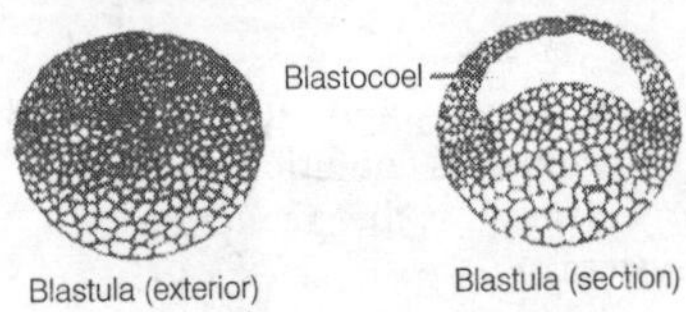

Position of a blastocoel

blastocyst Embryo at the time of implantation in the wall of the uterus. It is the modified blastula stage of mammalian embryos. It consists of an inner cell mass and a thin trophoblast layer enclosing the blastocoel.

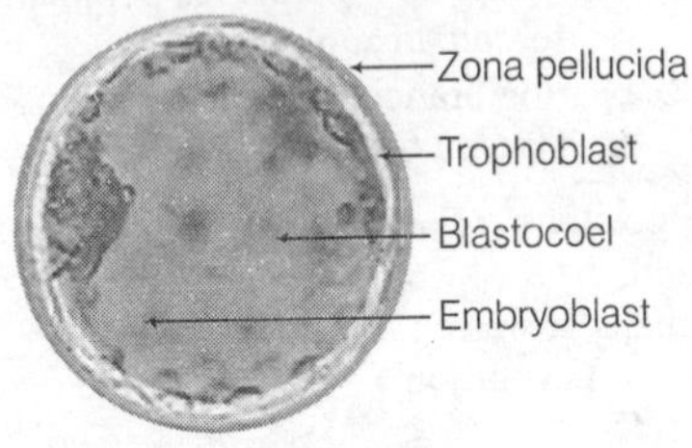

Human blastocyst

blastoderm Single layer of cells that surrounds the blastocoel in blastula. In yolky eggs, it forms a layer covering yolk as in insects.

blastomere One of the cleavage products of an animal zygote. The zygote usually divides into two, then four, then eight blastomers and so on until the normal nuclear/cytoplasmic ratio of the embryo's cells is achieved.

blastopore An opening formed by the invagination of cells in the embryonic gastrula, connecting its cavity (archenteron) to the outside. In protostome phyla the blastopore is the site of the future mouth. In deuterostomes the blastopore becomes the anus and the mouth forms else-where.

blastula A hollow ball of cells that occurs early in embryonic development of animals.

blending inheritance An abandoned theory that offspring inherit a blend of parental traits rather than the genes that determine those traits.

blepharoplast Basal granule of a flagellate.

blind spot A small area lying below the yellow spot on the retina which contains openings through which the axons of the ganglion neurons exit as the optic nerve. Since it does not contain rods and cones, an image cannot be transmitted from it.

blood A type of fluid (connective tissue) that acts as a transport medium through animals, carrying nutrients and oxygen to the body cells, and removing waste products and CO_2. It is composed of fluid matrix called plasma in which blood cells are suspended.

Plasma 55%

Constituent	Major functions
Water	Solvent for carrying other substances
Ions Sodium Potassium Calcium Magnesium Chloride Bicarbonate	Osmotic balance, pH buffering and regulation of membrane permeability
Plasma proteins Albumin Fibrinogen Immunoglobulins (antibodies)	Osmotic balance, pH buffering clotting defense
Substances transported by blood Nutrients (e.g., glucose, fatty acids, vitamins Waste products of metabolism respiratory gases (O_2 and CO_2) hormones	

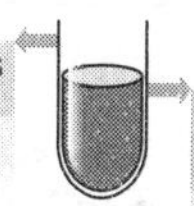
(Blood)

Cellular elements 45%

Cell type	Number (per mm³ of blood)	Functions
Erythrocytes (red blood cells)	5-6 million	Transport oxygen and help transport carbon dioxide
Leukocytes (white blood cells)	5000-10,000	Defense and immunity
Platelets	250,000–400,000	Blood clotting

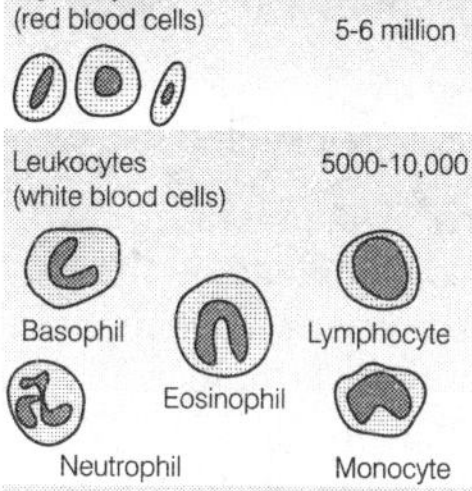

blood-brain barrier A specialised capillary arrangement in the brain that restricts the passage of most substances into the cerebro spinal fluid, thereby preventing dramatic fluctuations in the environment of CNS.

blood capillary *See* capillary

blood cells (blood corpuscle) Any of the cells found in the blood plasma. They constitute about 45% of the blood volume. These include RBCs, WBCs and platelets.

blood clotting (blood coagulation) Conversion of blood from a liquid to a semi-solid state which occurs when an injury to the blood vessel exposes blood to air. The clot closes the wound and prevents further blood loss. Blood clotting occurs by action of clotting factors and platelets.

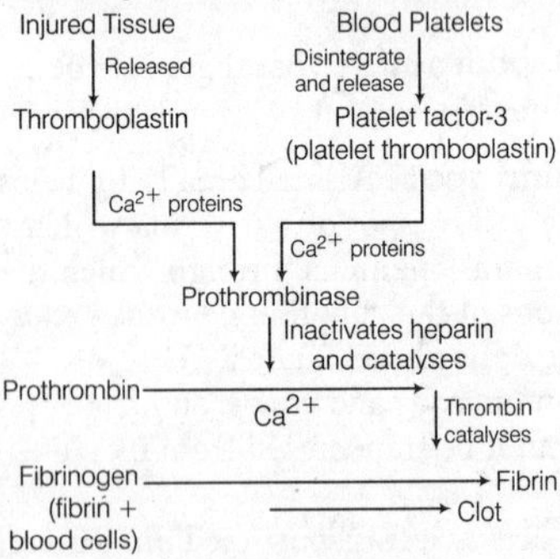

Clotting of blood

blood plasma Blood without its cellular constituents. It consists of water containing a large number of dissolved substances including proteins, salts, food materials, hormones, vitamins and excretory materials. It constitutes about 55% of blood volume.

blood platelets *See* platelets.

blood pressure The pressure exerted by the flow of blood on the elastic walls of the arteries. The normal blood pressure in humans is measured as 120/80 mmHg (millimetres of mercury pressure), in which 120 mmHg is the systolic pressure (pumping pressure) while, the 80 mmHg is the diastolic pressure (or resting pressure). Persistent increase in blood pressure above 140 mmHg (systolic) and 90 mmHg (diastolic) is termed as hypertension.

blood serum Blood plasma from which fibrin and clotting factors have been removed by centrifugation or any other method. Serum containing a specific antibody or antitoxin may be used in the treatment or prevention of certain infections.

blood vascular system The network of anatomic structures, including the heart and blood vessels that circulate blood throughout the body.

blood vessel A tubular structure through which the blood of an animal flows (see artery, arteriole, venule, vein, capillary).

blow hole Opening on the top of a cetacean's head, from which air is inhaled and exhaled, e.g. whale.

blue-green algae *See* cyanobacteria.

B-memory cells Long-lived B-cells that are produced after an initial exposure to an antigen and play an important role in secondary immunity. They remain in the body and facilitate a more rapid response, if the antigen is encountered again.

BMR *See* basal metabolic rate

body cavity Internal cavity of the body of an animal which is present in most invertebrates and all vertebrates and contains the major organs. Also *see* coelom

body fluid Any of the fluids found within animals, including blood, lymph, tissue fluid, urine, bile, sweat, and synovial fluids. These are generally involved with the processes of transport, excretion or lubrication.

body plan The 'blueprint' according to which an organism develops a predetermined number, arrangement, and size of body components. The body plan is embodied and implemented by the organism's genes (e.g. homeotic genes), but can be influenced by environmental factors, such as malnutrition or exposure to toxic agents (e.g. teratogens).

bog *See* hydrosere.

bolus A ball of food that has been chewed and mixed with the saliva.

bone Hard connective tissue of vertebrates. It consists of non-pliable ground substance rich in calcium salts and collagen fibres providing strength to the bone. The cells of bone are found in a calcified matrix made up of **ossein**. The bone cells known as osteocytes lodged in the spaces called **lacunae**. They also interact with skeletal muscles attached to them to bring about movements. The bone consists of four parts, i.e. periosteum, matrix, endosteum and bone marrow.

bone marrow A highly vascular, cellular substance in the central cavity of some bones. It is the site of synthesis of erythrocytes, some types of leucocytes and platelets.

Bone marrow is of two types:

- **Yellow marrow** Rich in fat cells called adipocytes.
- **Red marrow** Blood cells are formed in this marrow.

bony fish (osteichthyes) A term applied collectively to all groups of fish with bony (as opposed to cartilaginous) skeletons.

bony labyrinth *See* labyrinth.

book lung A saccular organ used for respiration in arachnids, such as spiders and scorpions. The book lung is located inside the ventral abdominal cavity and consists of several parallel membranous folds arranged like pages in a book.

Bordeaux mixture A fungicide made up of copper sulphate and quick lime.

boreal Forest of or relating to the forest areas (such as taiga) found at high latitudes and dominated by evergreen conifers (spruce, firs and pine).

Borrelia A genus of spirochetes, that are pathogens of humans, other mammals and birds, e.g. causative agents of Lyme disease (*Borrelia burgdorferi*), louse-borne relapsing fever (*Borrelia recurrentis*), and tick-borne relapsing fever (*Borrelia hermsii*).

botany Study of plants.

botryoidal tissue In *Hirudinea,* parenchyma and connective tissue invade the coelom. It consists of grape like masses of cells with brown pigment and serve an excretory function.

bottleneck Drastic short-term reductions in population size caused by natural disasters, disease or predators. It can also lead to random changes in the population's gene pool.

botulinum toxin A nerve toxin produced by *Clostridium botulinum* which cause fatal food poisoning.

bouton *See* synapse.

Bovine Spongiform Encephalopathy (BSE) Degenera-V disease of the brain that effects cattle and is caused by an abnormal form of a cellular protein.

bow riding An activity carried out by cetaceans (most commonly dolphins), in which they swim or drift along the crests of waves in the ocean.

Bowel Refers to the part of alimentary canal below the stomach, that is the intestine.

Bowman's capsule (renal capsule) Funnel or cup-shaped structure in vertebrate nephron in kidney, which receives the glomerular capillaries. It is initial, expanded segment of the nephron where filtrate enters from the blood. Its epithelium contains podocytes which facilitates this filtration.

b

box A small portion of a gene or protein that appears in many genes or proteins that are related in structure. The box usually has some specific functions sometimes called a 'motif', like binding DNA or interacting with specific proteins or other molecules.

bp Symbol for base pair(s), used as a unit at the molecular level for measuring distance along a duplex polynucleotide and corresponding to the number of paired bases in a particular segment of DNA (or duplex RNA).

Brachiopoda A phylum of marine invertebrates that have left an excellent fossil record; brachiopods live on or in the ocean floor.

brachydactyly Human genetic disorder that causes shortness of fingers and toes (digits). It is an autosomal dominant inherited trait.

brachydont Having short crowns and well-developed roots (said of teeth).

bract Small, leaf-like structure at the base of a flower or inflorescence.

bracteole A small leaf that arises on a flower stalk.

bradycardia Slowing of heart rate, usually less than 60 times per minute.

bradykinin Non-peptide hormone of submaxillary salivary gland, involved in dilation and increased permeability of blood vessels and smooth muscle contractions; released by parasympathetic stimulation; implicated in allergic reactions.

bradymetabolism Characteristics of 'cold-blooded' animals. Metabolism that is sustained at a relatively low rate.

bradytelic Pertaining to evolution that occurs at a slower rate than usual standard rate.

brain

1. Enlarged anterior, highly developed portion of the central nervous system of the vertebrates, encased within the cranium of the skull.
2. Concentration of nerve ganglia at the anterior end of an invertebrate animal.

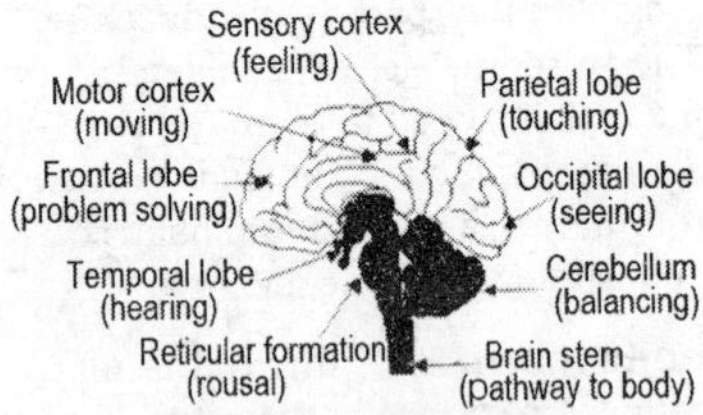

Structure of human brain

brain death Permanent absence of vital functions of the brain, which is marked by cessation of breathing and other reflexes controlled by the brainstem and by a zero reading on an electroencephalogram.

brain stem Portion of the brain that is continuous with the spinal cord and consists of the medulla oblongata, pons of the hindbrain and the midbrain.

branchial Of or pertaining to gills. Most fish possess five pairs of branchial arches.

branchiate Having gills.

brassin (Brassinosteroid; BR) Any of a group of steroid derivatives that occur at very low concentrations in plant tissues and are potent plant hormones.

breathing Physical process of inhaling and exhaling respiratory gases.

breed A domesticated variety of an animal or rarely a cultivated variety of plant.

breeding The process of sexual reproduction and bearing offspring.

breeding season (mating season) Specific season of the year in which many animals, such as birds mate and ensures that offsprings are produced only at certain times of the year.

breeding system A breeding system includes all the different breeding behaviours of a population and the methods in which the members of the population adapt to them.

brilles A transparent, immovable layer of scale/skin that covers the eyes of some creatures, (such as snakes and lizards), and provides protection.

bristles Long, stiff strands of hair or feathers. In birds, they are situated near the mouth or eyes. Their function may be to assist the bird in eating and give protection to the eyes.

Broca's area An area in the frontal lobe of the left cerebral hemisphere of the human brain associated with the motor control of the speech.

bronchi (bronchial tube) Tubes that carry air from the trachea to the lungs (sing bronchus).

Bronchial-Associated Lymphoid Tissue (BALT) It is the part of the network of mucosal associated lymphoid tissue. Found in the walls of the airways, it is an organised collection of lymphocytes and lymphoid tissue, where it protects the respiratory epithelium. It comprises B and T-cells.

bronchioles Small tubes in the lungs that are formed by the branching of the bronchi; in reptiles, birds and mammals terminate in number of alveoli in reptiles and mammals.

bronchitis A respiratory disorder characterised by excess mucus production and swelling of the bronchioles; caused by long-term exposure to irritants such as cigarette smoke and air pollutants.

brood parasite A brood parasite (usually a bird) manipulates another individual of the same or different species to raise its offsprings. A strategy known as brood parasitism, e.g. cuckoo.

brood patch Located on the lower abdomen of hen birds. It develops by the shedding of feathers in this area, and the consequent thickening of the skin, after which it becomes densely populated with blood vessels. The brood patch is used to incubate the eggs and keep the young ones warm.

brood reduction A reproductive strategy in which a female bird produces more eggs than she would normally be capable of raising. The clutch of eggs hatch sequentially, if there is inadequate food supply. The weakest chick or chicks, being deprived of food either fail to survive out of starvation, or are devoured by their stronger siblings.

brooding A pattern of behaviour of certain egg laying animals especially birds, marked by cessation of egg laying. It is the parent birds continued to provide warmth to their young, till the time, they are unable to maintain their own body temperatures.

brown algae Multicellular protistans placed in the division–Phaeophyta, includes kelps and wracks. They contain a brown pigment in addition to chlorophyll.

brown fat A special fat layer found between neck and shoulders of some mammals such as bats and squirrels. It enables production of large amounts of heat.

browser An animal (herbivore) that feeds on the twigs, shoots, leaves and fruits of bushes and trees.

b

brucellosis

1. Spontaneous abortion in cattle and sheep. A highly infectious and economically important disease caused by the bacterium of genus–*Brucella.* Also called Bang's disease.
2. An infectious disease of humans caused by bacteria of genus *Brucella* and is transmitted by contact with infected animals or raw milk products. Characterised by fever, malaise, severe headache and joint pain (Also called Gibraltar fever, Malta fever).

Brunner's glands (duodenal glands) Glands present in the submucosa of the duodenum that secrete mucus and an alkaline fluid that neutralises the acidic chyme leaving the stomach.

brush border Collection of microvilli forming a border on the intestinal side of the epithelial cells in the small intestine and renal tubule.

bryology Study of bryophytes.

bryophytes A group of non-vascular plants comprising the mosses, hornworts and liverworts. Also known as amphibians of the plant kingdom.

Bryozoa (Ectoprocta) Aquatic (mainly marine) invertebrate phylum comprising the sea mats and moss animals.

***Bt* cotton** Genetically modified variety of cotton. Bt refers to the bacterium *Bacillus thuringiensis* whose toxic gene has been introduced into cotton plants. Due to this, *Bt* cotton plant produces a chemical substance that kills the harmful insects (bollworm) which eats the cotton plant. The use of *Bt* cotton has decreased the use of pesticides on crops.

buccal Of or pertaining to the mouth or cheeks.

buccal capsule Structure connecting the oral cavity with the anterior portion of the oesophagus in nematodes. Also called the stoma.

buccal cavity (oral cavity) The beginning of alimentary canal which leads to the pharynx and (in vertebrates) to the oesophagus. In mammals, it contains the tongue and teeth.

buccogingival Of or pertaining to the cheeks and gums.

bucnemia It is the excess, diffuse, tense inflammation of the leg.

bud

1. Small projection on a plant from which a shoot, cluster of leaves or flowers can develop.
2. An outgrowth from a parent organism that breaks away and develops into new individual.

bud scale Specialised protective leaf of a bud.

budding

1. An asexual reproduction of new organisms; (usually found in yeasts) from a bud.
2. The process by which HIV and similar viruses leave the cell (other than by lysing).
3. Grafting method in which a bud of the scion is inserted onto the stock, beneath the bark usually.

Bufo A genus of toads in amphibia.

bug Insect member of the order–Hemiptera. A large order including both winged and wingless species. The most unique feature is their long proboscis adapted for piercing and sucking.

bulb An underground round part of a stem surrounded by fleshy modified leaves that contain stored food for the shoot within, e.g. onion bulb, tulip bulb.

bulbil A small bulb-like organ of vegetative reproduction that may form in the leaf axil, an inflorescence, or at a stem base. Modified axillary vegetative or floral bud.

bulbourethral (Cowper's) **gland** Paired pea-sized glands that secrete a mucus-like substance that is added to semen and provides lubrication during intercourse.

bulk flow Mass movement of water or air due to a difference in pressure between two locations (pressure gradient).

bulla

1. A rounded bony projection.
2. A large blister or vesicle.

bulliform cells (motor cell) Large, empty, colourless cells in grasses in which certain adaxial epidermal cells along the veins modify themselves.

bundle of His Specialised cardiac muscle fibres in the mammalian heart that receive electrical stimuli from the atrioventricular node and transmit them throughout the network of Purkinje fibres. The fibres are named after Swiss anatomist Wilhelm His.

bundle scar Scar from a vascular bundle remaining on a leaf, stalk, after abscission.

bundle sheath cells A layer of cells in plant leaves and stems that forms a sheath surrounding the vascular bundles. These may be parenchymatous or sclerenchymatous. Serve as site of Calvin cycle in C_4-plants.

bunodont Having molar teeth with rounded cusps as in mammals.

Burgess shale fossils A 545 year-old assemblage of superbly preserved fossils located in British Columbia, Canada, which record the Cambrian Explosion of animal body forms.

burkitt lymphoma Type of cancer of the lymphatic system, particularly B-lymphocytes manifested by tumours in the jaw.

bursa (pl. bursae) A cushion-like sac of fibrous connective tissue. Bursae are lined with synovial membrane and contain synovial fluid, a lubricant easing friction between tendons and bone. They are found between bones and other tissues, such as skin, ligaments, tendons and muscles.

burst size The number of phage particles released from a lysed bacterium.

buttress root (Stilt or prop roots) Tree root that extends above ground as a plate-like outgrowth of the trunk supporting the tree.

Buzz pollination Pollinating technique in which the vibrations of a buzzing bee shake dry pollen from anthers to the bee.

C

caching (hoarding) Storing of food for later use, when food is not available or is short in supply, e.g. rodents such as hamsters and squirrels cache food.

cactus A flowering plant adapted to desert conditions of plant family –Cactaceae. Many cacti have vast spreading root system covering large areas close to the surface of the ground, and are thus able to absorb great volumes of water quickly when the rain comes. They have succulent stems and branches with scales or spines instead of leaves.

caecum A blind pouch, open only at one end. In vertebrates, it is present at the junction of small and large intestine, very important in digestion process of herbivores. Vestigial in humans.

Caenohabditis elegans A nematode worm which was the first animal whose genome was fully sequenced in 1998. It is extensively used as a model organism in genetics and developmental biology due to its small genome size and transparent skin.

caffeine A white crystalline alkaloid that occurs in tea-leaves, coffee beans and other plant materials. Has a stimulatory action on the CNS.

calcareous Calcium or calcium carbonate containing parts, such as shells, bones and exoskeletons, which protects an animal.

calcicole Plants that thrive on neutral to alkaline chalk, carboniferous or limestone soils, such as marl.

calciferol (vitamin-D) Formed by the action of ultraviolet radiation on ergosterol. Controls the deposition of calcium compounds in the body; its deficiency causes rickets.

calcifuge Plants that grows best on acid soils, but not in lime rich soils.

calcineurin (CN) (protein phosphatase 3) An enzyme that activates T-cells of the immune system. It is inhibited by drugs as cyclosporin.

calcitonin (thyrocalcitonin) A 32-amino acid linear polypeptide hormone, secreted by the thyroid gland. It acts to reduce blood calcium levels; found in a broad range of vertebrates.

callose A carbohydrate which is deposited seasonally or permanently on sieve tubes, leading them to stop functioning. It is also found in the microspore mother cell walls and in the cells of some algae. It is digested by the enzyme callase.

callus

1. Mass of undifferentiated parenchyma cells formed in response to a cut or damaged plant surface. In tissue cultures, callus can be induced by various hormone treatments. Adventitious shoots and roots often differentiate from callus, a phenomenon exploited in the rooting of cuttings (in botany)
2. Thickened hard mass found in the outer layer of the skin formed as a result of continued pressure or friction (in pathology).

3. A tissue formed around the fragments of a broken bone which develops into a bone to heal the fracture (in physiology).

calmodulin An intracellular protein that binds calcium and regulates many physiologic processes, such as muscle contraction.

calnexin Integral protein of the Endoplasmic Reticulum (ER). It is a molecular chaperone that assist in protein folding and secretory pathway.

calorie (cal) Amount of heat needed to raise the temperature of one gram of water by 1°C (1K) at a pressure of one atmosphere.

calreticulin (Calregulin)
A multifunctional protein that acts as a major Ca^{2+} binding protein in the lumen of endoplasmic reticulum. It is also found in nucleus, proposing its role in regulation of transcription.

Calvin cycle (Calvin-Benson Bassham (IBB) cycle or dark reaction) A series of biochemical reactions taking place in the stroma of the chloroplasts of those plants using the C_3, Carbon fixation pathway. Rubisco (ribulose bisphosphate carboxylase/oxygenase) is the key enzyme of the Calvin cycle.

calyptra A layer of cells derived from lower cells of the archegonium. It covers the developing saprophyte. In bryophytes, it ruptures as the seta elongates, being taken up as a bond over the capsule in mosses. The presence of the calyptra is necessary for the proper development of the capsule in mosses and the embryo in ferns.

calyptrogen The apical meristem of the roots that contains a region which divides to produce the root cap (calyptra).

calyx Whorl of sepals that encloses the petals, stamens and carpels and protects the flower in a bud and subsequently embraces the base of an open flower.

cambium (lateral meristem)
A cylinder of undifferentiated tissue of actively dividing cells (meristem) that allows the secondary growth of roots and stems of a plant.

cambrian explosion The appearance of fossils of many different forms of animals in the cambrian period, interpreted as a rapid diversification of body plans.

cambrian period The most ancient (about 543 – 488 million years ago or mya) period of the Palaeozoic era. It was during the Cambrian period that the invertebrate phyla with mineralised shells first made their appearance.

camera eye The eye with a focusable lens. Light is received through a single opening. Found in vertebrates and cephalopods.

cAMP period Cyclic adenosine monophosphate (cyclic AMP or cyclic 3'– 5'-adenosine monophosphate). A diffusible signaling molecule important in many biological processes. It is synthesised from adenosine triphosphate (ATP) by adenylyl cyclase. cAMP is a secondary messenger that acts as a signaling molecule carrying signal of extracellular origin to some site within the cell. Generally, such signals originate from a hormone that itself is too large to enter the cell.

campylotropous ovule Refers to position of the ovule in the ovary, when the funicle appears to be attached half way between the chalaza and the micropyle. The micropylar end is turned through 90° relative to the orthotropous condition, so that the ovule is horizontal.

camouflage A feature, common to invertebrates which helps them blend with their surroundings using its skin colours or patterns. This helps them to avoid predators or remain unseen to their prey.

canaliculus (pl. canaliculi) Refers to a very small channel that occurs between the cells of the liver and bone. In the liver, the bile canaliculi carry bile to the bile ducts; in bone canaliculi connects lacunae, the cavities containing bone cells.

canalisation A developmental mechanism that limits variation of the phenotype within narrow bounds by repressing underlying genetic variation.

cancer A disease caused by the multiplication of an abnormal cell. Cells continue to multiply by uncontrolled cell divisions forming a ball of cells called tumour, e.g. sarcoma, carcinoma. Cancer cells lost the property of **contact inhibition** (contact with other cells inhibit the uncontrolled growth). Cancer cells move from tumour to new sites through blood for forming secondary tumours. This invasion of cancer from one part to other parts by the body fluids is called **metastasis**.

canidae The mammalian family that contains dogs, wolves, foxes, jackals and other dog like mammals.

canine

1. Pertaining to any animal of the family – Canidae, especially dogs.
2. A pointed tooth between the incisors and premolars; in mammalian predators, the canines are usually elongated, conical, bicuspid and function in seizing and piercing prey.

canopy The upper level of a forest composed of the leafy upper branches of the trees.

CAP (Catabolic Activator Protein) A protein which binds cAMP. Its presence is indispensible for the activation of operons whose activity is repressed by presence of catabolic repressiors (lactose, arabinose, galactose). Also known as cAMP receptor protein, CRP.

capacitation The biochemical changes that occur in sperm as a result of exposure to the female reproductive tract and precede the acrosome reaction.

capillary (blood capillary) Microscopic vessels connecting the arterial system with the venous system. These are the narrowest blood vessel in the vertebrate circulatory system with walls only one cell layer thick.

capitulum

1. A racemose inflorescence with compressed disc-like pedicel bearing sessile flowers – oldest at the margins and younger in the centre. Found in Compositae family.
2. A rounded articular protuberance of a bone, e.g. lower end of humerus.
3. The blood sucking probing mouthparts of a tick.

capsid The outer protein coat of a virus particle. Its chemical nature is important in stimulating immune response of body.

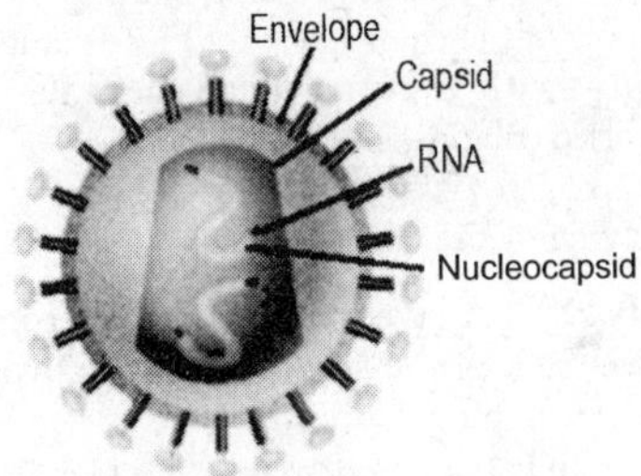

Human Immunodeficiency Virus (HIV)

capsomere (capsomer) Protein molecules that make up the outer protein coat (capsid) of a virus.

capsule

1. A dry dehiscent fruit that is formed from several fused carpels and releases its seeds when ripe. The numerous seeds may be released through pores, (e.g. poppy) or by splitting and separation of carpels (e.g. crocus).

2. The muciloginous rigid covering around the cell wall in some bacteria.
3. A protective sheath that surrounds an organ or part of the body, e.g. kidney
4. The ligamentous sheath of connective tissue that surrounds various skeletal joints.
5. In mosses and liverworts, part of sporophyte where haploid spores are produced.

carapace A hard exoskeleton which shields the dorsal side of an animal's body, e.g. crab (crustacean). It is used more specifically to refer to the dorsal part of a tortoise or turtle's shell.

carbohydrate Biochemical compounds with general formula [CHO] e.g. sugars, starch and chitin. They fall into three main classes, monosaccharide, disaccharides and polysaccharides. They are called carbohydrates because they contain carbon, hydrogen and oxygen in a two to one ratio.

carbon assimilation Incorporation of carbon from atmospheric carbon dioxide into organic molecules. It happens during photosynthesis.

carbon cycle The most common biogeochemical cycle composed of two primary processes, photosynthesis and respiration. Photosynthesis produces oxygen and glucose from carbon dioxide and water. Respiration reverses this by creating carbon dioxide and water from glucose and oxygen.

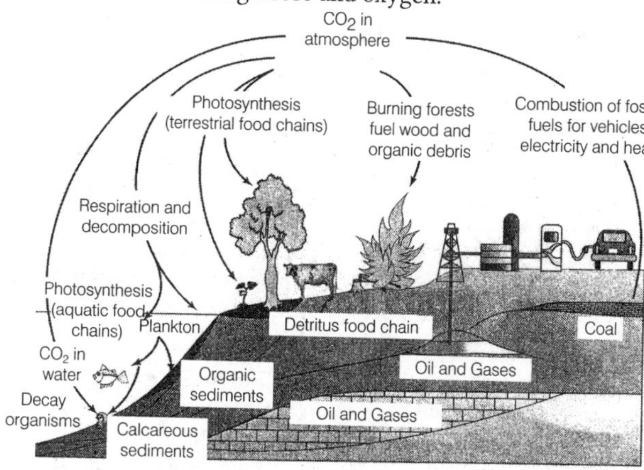

Simplified model of carbon cycle in the biosphere

carbon dating (radio carbon dating) A method of estimating the age of archaeological specimens of biological origin.

carbon fixation An enzymatic reaction in which CO_2 is attached to a receiver compound (e.g. ribulose biphosphate) thereby adding to the supply of organic carbon occurs chiefly in photosynthesis.

carbon nutrient balance hypothesis (environmental constraint hypothesis) A controversial hypothesis that seeks to explain the allocation of a plant's resources to defensive substances in terms of the relative availability of carbon and major nutrients, particularly nitrogen.

carbonic acid (H_2CO_3) Formed in the red blood cells by carbonic anhydrase when CO_2 diffuses into them from surrounding tissue cells.

$$CO_2 + H_2O \rightleftharpoons H_2CO_3$$

The acid remains in equilibrium with dissolved CO_2. It diffuses into $HCO_3^- + H^+$ and thus helps in the transport of CO_2.

carbonic anhydrase An enzyme present in RBCs and kidney cells. It catalyses the reaction between CO_2 and H_2O to form H_2CO_3.

carboniferrous period Fifth interval of the Palaeozoic era characterised by low sea levels, extensive glaciation and heavy forestation which resulted in the deposition of the coal deposits that give the period its name. Fishes continued to diversify during this period and amphibians became more common.

carboxyhaemoglobin Formed by the combining of carbon monoxide with haemoglobin. It is a highly stable product that has strong affinity of haemoglobin (250 times greater than oxygen) and competes with oxygen for it.

carboxylation The fixation of CO_2 into a stable organic intermediate. Most crucial step in Calvin cycle, where CO_2 is utilised for the carboxylation of RuBP (Ribulose 1, 5 biphosphate).

carboxysome An organelle containing the carbon fixing enzyme, ribulose bisphosphate carboxylase that is found in certain autotrophic bacteria.

carcerulus A type of capsular fruit that breaks up at maturity into one-seeded segments or nutlets. The carcerulus is typical of the Labiatae family, e.g. deadnettle.

carcinogen Agent causing cancer, e.g. smoke, X-rays, etc.

carcinoma A malignant tumour arising in the epithelial tissue of the skin or the living of internal organs.

cardiac arrest Sudden failure of the heart to pump blood.

cardiac cycle The coordinated and rhythmic series of muscular contractions of heart that occurs during one complete heartbeat. It comprises of the systolic phase and the diastolic phase. Systole occurs when the ventricles of the heart contract. Diastole is the period between ventricular contractions when the right and left ventricles relax and fill.

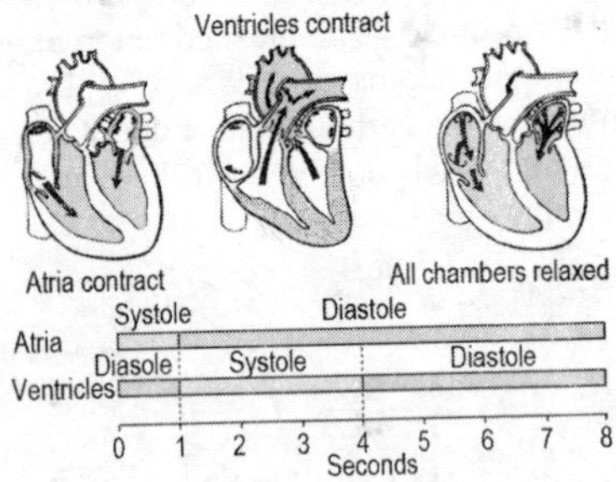

Cardiac cycle

cardiac muscle The contractile tissues present only in heart. These are striated and are involuntary with rich blood supply.

cardiac muscle cells (cardiomyocytes) Cells related to heart which continuously contract rhythmically and automatically without nervous stimulation.

cardiac output The volume of blood pumped out by each ventricle in one minute. It is equal to stroke volume X number of beats/min. Thus, average to approx 5 L in normal individual.

Cardiopulmonary Restrucitation (CPR) The emergency treatment aimed at restarting the heart and breathing after cardiac arrest.

cardiovascular centre An area in the brain responsible for modification of cardiovascular system based on autonomic nervous system.

cardiovascular system *See* circulatory system.

caries Decay and disintegration of tooth. It is associated with the inflammation and the formation of abscesses in the surrounding tissues. Commonly called as tooth decay.

carina A keel-shaped ridge or structure, e.g. in the flower of legumes. It has two fused lower petals which enclose the stamens and stigma and the ridge on the breastbone of a bird.

carnassial tooth Last upper premolar and first lower molar tooth, which is used to efficiently tear and slice meat of prey. This tooth is seen only in the carnivores.

carnation Refers to any cultivated plant descended from various kinds of wild pinks–mainly *Dainthus caryophyllus* which is a native of Southern Europe.

carnivore Mammal belonging to the order carnivora that mainly sustains by eating the flesh of other animals, e.g. wolves.

carnivorous plant (insectivorous plant) Any of the plant the derives its nutrient supply from consuming animals or protozoans mainly insects, e.g. Venus fly trap.

carotene An organic compound occurring as an orange-yellow and red pigment in many plants. It gives plants their characteristic colour, e.g. carrot (β-carotene), tomato (lycopene). In animal tissues, it is converted to vitamin-A by the liver.

carotenoids The group of yellow, orange, red or brown pigments occurring in plants and algae where they function as accessory pigments. These are chemically related to terpenes, e.g. xanthophylls (yellow).

carotid artery One of the pair of blood vessels that supplies oxygenated blood to the head and neck. Branches in the neck region into an internal and external carotid artery.

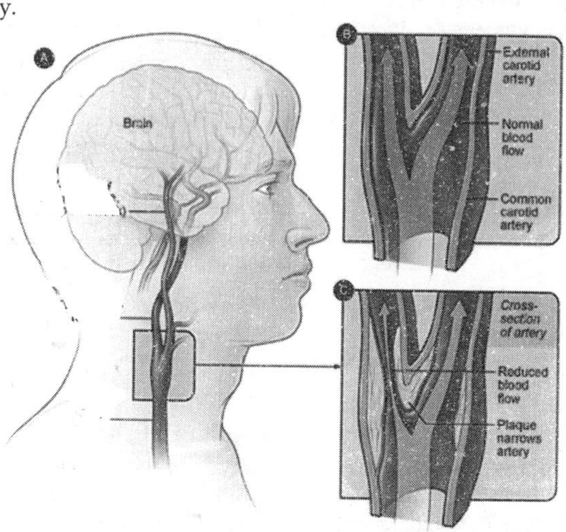

(a) Figure showing location of external, internal and right common carotid artery.
(b) Blood flow inside external and common carotid artery.
(c) Cross section of an artery.

carotid body A vascular structure at the base of the external carotid artery adjacent to the carotid sinus that contains chemoreceptors. These monitor carbon dioxide, oxygen concentrations and pH level of the blood.

carotid sinus The enlarged portion of the internal carotid artery, near its origin in the neck, containing sensory receptor, that monitor changes in blood pressure.

carpal bone (carpal) One of the bones that form the wrist in terrestrial vertebrates.

carpel Female reproductive organ of a flower; made up of a stigma, style and an ovary. Each flower may have one carpel (monocarpellary) or many (polycarpellary), either free (apocarpous) or fused together (syncarpous).

carpogonium Female gametangium of red algae found at the tips of the female gametophyte. These are flask-shaped structures from which trichogyne is produced. The male spermatia attaches to the trichogyne during sexual reproduction.

carpospore Spore of the red algae, produced in the fertilised carpogonium. After release, they develop into a sporophyte or gametophyte.

carpus Bones of the wrist taken as a whole in terrestrial vertebrates. The number of small bones (carpuls) varies with species, e.g. humans have 8 carpals.

carrageenan (carrageen) Linear sulphated polysaccharide isolated from red algae. It is used in desserts, shampoo, in biotechnology to immobolise cells, etc.

carrier

1. An individual who possess a mutant allele but does not express it because of the dominant allelic partner such individuals only pass on the defective allele to offspring, e.g. human females are carries of red-green colour blindness and haemophilia. The defective allele is carried on the X-chromosome (in genetics).
2. An individual who shelter a pathogen without ill-effects and transmit it to others (in medicine), e.g. mosquitoes are carriers of malaria.

carrier molecule An immunogenic molecule to which a hapten is coupled. This hapten-carrier conjugate can induce an immune response.

carrier protein (also transport protein) Protein that transports a specific enzyme across a biological membrane during facilitated diffusion. It has a binding site that is specific for the particular enzyme it transports.

carrying capacity (K) Maximum population density that can be sustained by a particular type of environment, without its damage.

cartilage Flexible but tough supportive tissue, such as that in the tip of the human nose. It is rich in collagenous fibres, which are suspended in a matrix of rubbery chondrin. It forms adult skeleton of cartilaginous fish (e.g. sharks). The cartilages are of three types, i.e. hylaine, fibrous and caltified.

cartilage bone (replacing bone) Bone formed by replacing the cartilage of an embryo skeleton by endochondral ossification.

cartilaginous fishes (Chondrichthyes) A class of vertebrate fishes with cartilaginous skeletons. Includes sharks, rays and skates. These are mostly marine, viviparous, with no operculum and swim bladder.

caruncle
1. A fleshy outgrowth (without feathers) seen on the neck and face of a bird. It is also referred to as fowl's comb and usually seen in the turkey family (in zoology).
2. A spongy outgrowth of the seed coat, especially in castor bean seed (in botany).

caryopsis A small single seeded, dry indehiscent fruit with the pericarp fused to the seed coat. Typical fruit of grasses and grains.

casein A milk phosphoprotein. It is precipitated from milk by the action of rennin. It forms the basis of cheese.

casparian strip Thickenings due to the deposition of lignin or suberin on the walls of endodermis. These are impervious to gases and liqids.

caspase Any of a group of proteases that mediate apoptosis.

caste A physically distinct individual or group of individuals specialised to perform a particular function. Found in eusocial insects, such as ants, bees, wasps and termites.

castration The surgical excision of one or both testicles or ovaries.

catabolic pathways Metabolic processes that release energy as complex molecules are broken down into simpler ones.

catabolism Metabolic reactions involved in the breakdown of complex molecules to simple compounds. The function of catabolic reactions is to provide energy which is used in the synthesis of new structures, for work, for transmission of nerve impulses and for the maintenance of functional efficiency, e.g. respiration.

catabolite A product of a catabolic process.

catalase Enzyme present in both plant and animal tissues that catalyses the breakdown of hydrogen peroxide, a toxic compound produced during metabolism, into oxygen and water.

catalysis Speeding of a chemical reaction by a catalyst.

catalyst A molecule that increases the speed of a chemical reaction without itself undergoing a change, e.g. enzymes are biological catalysts that are proteinaceous in nature.

catalytic RNA *See* ribozyme.

cataphyll Any rudimentary leaf preceding the true foliage leaves in a bud scale form.

cataract Any opacity which develops in the crystalline lens of the eye. This is mostly due to the yellowing of the lens with age.

catarrhine primates Old World monkeys such as the macaque and the apes. They have nostrils set close together and opening to front of face.

catecholamines A group of chemicals that occur in animals, especially vertebrates and act as neurohormones or neurotransmitters, e.g. norepinephrine, epinephrine and dopamine.

category *See* rank.

cathepsin Several protease enzymes present in lysosomes.

catheter A tube for introducing liquids into or extracting them from the body.

catkin (ament) A pendulous spike usually hanging with several unisexual flowers occurs in birch, hazel, etc.

caudal
1. Pertaining to the tail (in animals).
2. Inferior in position (in humans).

caudal vertebrae Bones of the tail which articulates with the sacral vertebrate. Their number varies with species. In humans, these vertebrae are fused to form a single bone, coccyx.

cauliflory Production of flowers on older, thickened, leafless branches or main trunk. Common among the angiosperm trees in tropical forests, e.g. cocoa.

cauline Belonging to the stem or arising from it.

cauterize Burn a tissue with hot iron, caustic, etc., in order to destroy or sterilise it (in surgery).

caveola (pl. caveolae). A small depression lined by caveolins (transmembrane proteins) and contain receptors for extracellular signalling molecules, present in the plasma membrane of a cell.

cavitation

1. Formation of a void or gas bubble in a fluid subject to reduced pressure. Occurs in plants in their xylem vessels and tracheids.
2. Formation of a cavity, as in lung due to the TB (in pathology).

C-banding A selective method for staining chromosomes showing locations of heterochromatin in the stained banded regions of the chromosomes. It uses Giemsa Stain. also called centromere banding stain.

C-cell (parafollicular cell) Group of cells invertebrate derived from terminal pairs of gill pouches. In mammals, these cells occur in the thyroid and parathyroid glands. In other vertebrates, they occur in lung and paired ultimobranchial bodies.

cDNA *See* complementary DNA.

cDNA clone A cDNA sequence carried in a cloning vector.

cDNA library A library of cloned cDNA sequences inserted into a collection of host cells. Such libraries are used to express eukaryotic genes in prokaryotes.

CDR *See* complementarity determining region.

cecidium *See* gall.

cecropin Produced by cecropia moths in response to bacterial infection, these are group of peptides with potent activity against Gram-negative bacteria.

cell The structural and functional unit of living organisms. Cell consists of cell wall, protoplasm, nucleus and a large vacuole in plants. In animal cells, cell wall is absent and it is bounded on the outer side by the plasma membrane only.

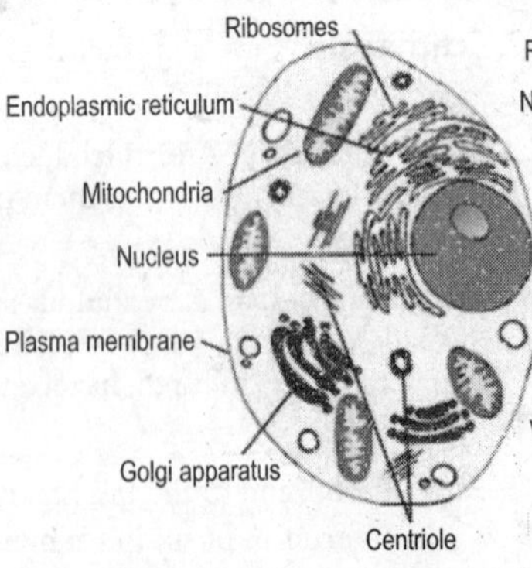

Animal cell

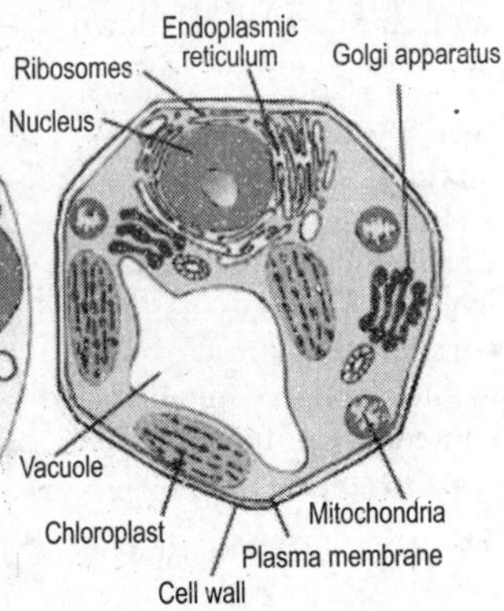

Plant cell

Cell Adhesion Molecule (CAM) Proteins present in the plasma membrane of animals cells that connects cells to the extracellular matrix or to each other, e.g. cadherins.

cell body (perikaryon; soma) Mass of cytoplasm in neuron which contains nucleus and Nissl granules. The branches of a nerve cell arise from it.

cell centre Alternative term for centrosome.

cell culture Maintenance and growth of cells, outside body in specially designed containers under precise conditions of temperature, pressure, etc.

cell cycle The orderly sequence of events by which a cell duplicates its genome, synthesises components and eventually divides into daughter cells. Can be divided into four main phases

(i) M-phase
(ii) G_1-phase
(iii) S-phase
(iv) G_2-phase

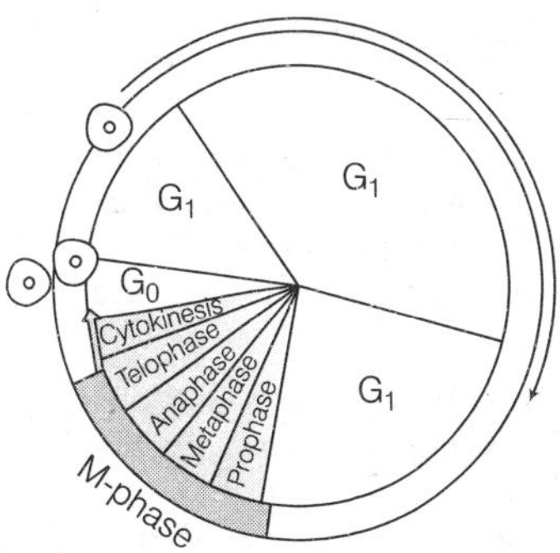

Diagrammatic view of a cell cycle indicating formation of two cells

cell division Process by which a single cell divides into two or more daughter cells. See mitosis and meiosis. The nucleus divides first followed by the division of the cytoplasm.

cell fractionation The separation of different constituents of the cell into homogenous fractions. This is achieved by breaking up the cells in a grinder and then centrifuging the resultant liquid.

cell fusion (somatic cell hybridisation) Technique of combining two cells from different species or tissues in a cell culture.

cell junction (intercellular junction) Kind of connection between cell, through specialisations of the cellular margins, e.g. desmosomes, zonula adherens, etc.

cell lineage A cell lineage is the developmental history of a differentiated cell as traced back to the cell from which it arises. The cell lineage of a structure traces the successive stages that the cells pass through from the time of their formation in the zygote to their appearance in the mature functional state.

cell-mediated immunity The specific immune response mediated by T-lymphocytes.

cell membrane (plasma membrane; plasmalemma). Extremely thin lipoprotein membrane which covers all cells. It is a three-layered structure and provides a selective barrier to the molecule and ions (selectively permeable).

cell plate Plate of materials, appearing across the equatorial plane of the spindle during telophase in plant cells. It separates the cytoplasm of two daughter protoplasts. Develops in middle lamella and primary wall of new cell wall.

cell sap A solution present in the vacuole of a plant cell, consist of water, amino acids, salts and waste substances (as tannins).

cell theory The theory was formulated by Matthias Schleiden (1838) and Theodore Schwann (1839) who postulated, respectively that plants and animals are made of cells. It states that

1. All organisms are composed of cell and cell products.
2. All cells arise from pre-existing cells.
3. Cells show similarity in chemical composition and metabolic activities.
4. Cells are the structural and functional unit of living organism.

cell wall The outer rigid covering present in plant, algal, fungal and bacterial cells. Its function is to protect and give shape to the cell. Its consists of plants and algae–cellulose, fungi– chitin bacteria-complex polymer of polysaccharides and amino acids.

cellular respiration A mechanism of breakdown of food materials within the cell to release energy and the trapping of this energy for synthesis of food.

cellulase An enzyme capable of digesting cellulose into simpler sugars. Associated with saprophytes and parasites and particularly important when produced by the microbes in the herbivore's large intestine.

cellulolytic Having the capacity to digest cellulose, e.g. bacteria in stomach of ruminants.

cellulose The unbranched polysaccharide composed of glucose molecules. Basic constituent of cell wall, especially of higher plants, many algae and some fungi.

cement (cementum) A thin layer of bony material that covers the dentine of roots of a tooth and fixes the tooth to the jaw.

Cenozoic (Cainozoic) The present era. This began, about 65 million years ago often called as age of mammals.

centipede Refers to the group of arthropods of the order-Chilopoda. They have a pair of walking legs on almost every segment.

central cell The central portion of an embryo sac consisting of the polar nuclei (or secondary nucleus if they have fused) and surrounding cytoplasm called the primary endosperm cell after fertilisation.

central dogma The doctrine that in all organisms, the genetic information flows from DNA to RNA and then to proteins. Proposed by Francis Crick. In some viruses, the flow of information is reversed, i.e. from RNA to DNA.

Central Nervous System (CNS) One of the two main division of the nervous system, includes brain and spinal cord. It is the site of information processing and control and acts as the 'command and control system'.

central placentation Ovules attached to the central axis of a non-loculed ovary.

centrifugal Developing from the center to the outwards, so the youngest structures are at the outer edge. Centrifugal xylem differentiates from the center outwards and a centrifugal inflorescence is one in which the progression of flower opening is from the center to the periphery.

centriole Present in animal cells and basal region of cilia and flagella. It consists of two short cylinders, oriented at right angles to each other and composed of microtubules.

centroblast A rapidly proliferating B cell (B-lymphocyte) found within the germinal centre of a lymph node follicle.

centrolecithal eggs Pertains to eggs with yolk in the centre as in arthropod, insects.

centromere (kinomere) The region of a chromosome where the sister chromatids attach to each other and to which spindle fibres attach during mitosis and meiosis.

centrosome (cell centre; centrosphere) In a eukaryotic cell except fungi, a small cytoplasmic organelle located near the nucleus that serve to organise microtubules. It contains a pair of centrioles.

centrum *See* vertebrae.

cephalic Pertaining to head.

cephalisation The evolutionary tendency for development of a head in which sense organs, brain and feeding organs are concentrated in animals.

Cephalochordata A chordate sub-phylum that contains lancelets. In these animals, the notochord persists throughout life, e.g. *Branchiostoma* (*Amphioxus*).

Cephalopoda A mollusc class that includes *Octopus*, squids, cuttlefishes, ammonites and the chambered nautilus.

cephalothorax A fused head and thorax found in Arachnida and many crustaceans. It is connected to the abdomen.

cercaria A tadpole like larva of flukes (trematode worms) that develops in body of secondary host (mollusc).

cerci A pair of filamentous, jointed appendages at the posterior tip of the abdomen on many insects and other arthropods, e.g. mayflies.

cerebellum A sub-division of brain which lies behind the forebrain and above the brainstem. It controls the muscle movement for smooth integrated body movement as well as maintain posture and balance. It is divided into two hemispheres with several types of neurons, the most important being Purkinje cells and granule cells.

cerebral cortex (pallium) The outer layer of gray matter in the cerebrum of many vertebrates.

cerebral hemisphere A pair of structures in vertebrate forebrain (cerebrum) that contain the centers concerned with major senses, voluntary muscle activities and higher brain functions, such as language and memory. In mammals, these are the largest part of brain and control the voluntary activities.

cerebroside Any of a class of glycolipids in which a single sugar unit is bound to a sphingolipid, e.g. galactocerebrosides.

Cerebrospinal Fluid (CSF) Liquid surrounding the brain and spinal cord, formed by the choroid plexus in the ventricles. Its function is to protect the CNS.

cerebrum *See* cerebral hemispheres. It is the dorsal anterior largest part of the vertebrae forebrain, consisting of two 'hemispherical' parts linked by corpus callosum. It plays an important role in the learning process.

cerumen A waxy secretion that collects in the external ear. Secreted by the ceruminous glands, to prevent the entry of dust particles, thus preventing damage of the tympanum.

cervical Pertaining to neck region. Most mammals (including humans) have seven vertebrae. Cervical vertebrae are the vertebrae of neck.

cervix A narrow posterior part of the uterus of mammals which leads into the vagina. During labour, it enlarges greatly to allow passage of fetus.

Cestoda A class of phylum–Platyhelminthes including mostly endoparasitic forms (no digestive tract but possess with hooks and suckers for attachment and other parasitic adaptations) and a complex life-cycle, e.g. *Taenia*.

Cetacea A marine mammalian order that includes whales, dolphins and porpoises. They have shape a fish and forelimbs modified to flippers with one or two blowholes for breathing.

cetotolite A fossil whale bone, usually an ear bone.

chaeta A slender chitinous structure on of the integument in many invertebrates like annelids (earthworm and *Neries*) for gripping. Also called seta or bristle.

Chaetognatha The invertebrate phylum that consists of arrow worms. Covered by a cuticle, their body is divided into a distinct head, trunk and tail. They lack organs for excretion, circulation and respiration and are hermaphrodite.

chalaza

1. The basal region of angiosperm ovule where nucellus and integuments are joined.
2. In a bird's egg, two twisted strand of fibrous albumen, attached to the vitelline membrane, at opposite poles of the yolk, lying in the long axis of the egg. They stabilise the position of the yolk and early embryo in the albumin.

chalazogamy A method of fertilisation in angiosperm in which the pollen tube enters the ovule through the chalaza, instead of the micropyle. Seen in certain trees and shrubs.

chalone The substance secreted by a tissue that inhibits mitosis in that tissue.

chamaephyte A perennial plant that is able to produce new growth from resting buds near the soil surface. Chaemophytes are usually small bushes.

chaperone *See* molecular chaperone.

chapparal A type of shrub found in temperature regions with little summer rainfall. They have yellow flowers and resinous foliage.

character (trait) A distinguishing feature that is inherited by an organism.

character displacement Adaptations of different sets of characteristics in two similar species, brought about by overlapping territories, resulting in competition.

Chargaff's rule According to this rule, concentration of adenine always equals the concentration of thymine and the concentration of cytosine always equals the concentration of guanine.

chasmophyte A plant which grows on rocks, rooted in debris in the crevices.

chela (pincer) The terminal segment in appendage found in lobsters, crabs and other arthropods.

chelicerae The first pair of appendages on the head of arachnids and other arthropods of the phylum Chelicerata. These appendages are modified for grasping or tearing food.

chelicerata A sub-phylum of arthropods containing the classes spiders and scorpions horseshoe crabs and sea spiders. The body of these animals is divided into an anterior prosoma and a posterior opisthosoma and bears six pairs of appendages, typically comprising chelicerae, pedipalps, and four pairs of walking legs.

chemical control The use of chemicals to kill pests (see pesticide). Compare biological control.

chemical dating The determination of relative or absolute age of minerals and other objects by measuring the chemical composition of a specimen, e.g. phosphate in buried bones is slowly replaced by fluoride ions from the ground water.

Measurement of the proportion of fluorine present gives a rough estimate of the time that the bones have been in the ground.

chemical fossil Any of various organic compounds found in ancient geological strata that appear to be biological in origin and are assumed to indicate that life existed when the rocks were formed.

chemical fusogen Chemical used for fusing of two cells or protoplasts, e.g. PEG (polyethylene glycol).

chemiosmotic theory This theory explains how ATP is generated in the mitochondria *via* the electron transport chain. It was postulated by Mitchell. He proposed that electron transport is accompanied by an accumulation of protons in the intermembrane space of the mitochondrian which in turn creates osmotic pressure. The protons driven back to the mitochondrian under this pressure generate ATP.

chemoautotroph A bacterium which obtain energy by oxidation of simple organic molecules, e.g. the oxidation of hydrogen sulphide to sulphur by *Thiobacillus* and some nitrifying bacteria like *Nitrobacter*.

chemokine A small protein produced by the cells and tissues at the beginning of an immune response to infection, etc.

chemolithoroph A bacterium that is able to obtain its energy by the oxidation of inorganic compounds, including those of iron, nitrogen and sulphur.

chemoorganotroph The organism, especially a microorganism, that obtains its energy by the oxidation of organic compounds.

chemoreceptors A sensory organ (e.g. afferent nerve endings or cells associated with them) which is capable of reacting to a chemical stimulus.

chemosynthesis An autotrophic nutrition in which organisms (called chemoautotrophs) synthesise organic materials using energy derived from the oxidation of inorganic chemicals, e.g. bacteria, including *Nitrosomonas*, which oxidises ammonium to nitrite; and *Thiobacillus* which oxidises sulphur to sulphate.

chemotaxis A movement in response to a chemical concentration gradient. The spermatozoids of primitive plants are often positively chemotactic, swimming towards the female organs in response to a chemical secreted by the latter.

chemotaxonomy Use of biochemistry for the classification of plants and microorganisms. It is based on similarities and differences in their natural products and the biochemical pathways involved in their manufacture.

chemotherapy Use of chemicals, especially drugs, in the treatment of disease. The term is often used specifically to denote drug therapy for cancer.

chemotropism The growth or movement of a plant or plant part in response to a chemical stimulus, e.g. growth of a pollen tube towards ovule, in response to the presence of sugars in the style.

Chengjiang fossils The assemblage of fossil organisms found in deposits of Chengjiang, China.

chiasma (pl. chiasmata) The attachment of two non-sister chromatids in a bivalent of diplotene in prophase-I of meiosis. Each chiasma results in the exchange of genetic material between the non-sister chromatids i.e. in crossing over.

Chilopoda A class of arthropods belonging to Myriapoda, including the centipedes. The body is divided into head and segmented trunk, with a pair of appendages on each segment.

chimaera An organism with combination of tissues of different genetic constitution in same part especially of the plant.

chiroptera The order of mammals highly specialised for flight. Includes bats. Their front limbs are modified as wings and are specialised for echolocation.

chitin A horny protective substance forming the chief part of the cuticle of arthropods; an amino polysaccharide or a polymer of acetylglucosamine analogous in chemical structure to cellulose. It is insoluble in water, alcohol, ether and other solvents and is resistant to acids and alkalis.

chitinase An enzyme that hydrolyses chitin. It is a majorly present in fungal cell walls and insects exoskeleton.

chlamydia A genus of microorganisms causing a variety of diseases in humans and other animals, e.g. *Chlamydia trachomatis* cause trachoma.

chlamydospore A thick-walled, unicellular resting spore formed from vegetative hyphae in fungi, particularly parasitic fungi.

chloranthy An abnormal flower in which all the parts are developed as leafy structures.

chlorenchyma A thin-walled, parenchymatous tissue containing chloroplasts. It is found in mesophyll tissue of plant leaves and also in stems of certain plants.

chloride secretory cells Present in the gills of marine bony fish that remove chloride from the blood and secrete it into the sea. thereby maintaining the osmotic composition of the body. This is accompanied by movement of sodium ions from the blood to the sea, thus keeping the electrochemical balance of the fish's body neutral.

chlorocruorin Greenish iron containing respiratory pigment that occurs in the blood of some polychaete worms.

Chlorofluorocarbons (CFCs) Compounds which are obtained by replacing the hydrogen atoms of hydrocarbons by chlorine and fluorine atoms. Their high stability to temperature makes them suitable for a variety of uses including aerosol propellants, oils, polymers, etc. They are often known as freons. They are responsible for ozone layer depletion.

chlorophyll Green coloured pigment found inside all photosynthetic organisms. It is the principle site of light absorption. In photosynthesis, it catalyses the combination of carbon dioxide with water to form carbohydrates.

Chlorophyta (green algae) A large phylum of kingdom–Protista containing marine or aquatic green algae. Contain photosynthetic pigment chlorophyll-*a* and chlorophyll-*b* as found in plants.

chloroplast An organelle (plastids) that performs photosynthesis in plant cells. It contain chlorophyll responsible for the green colour.

chlorosis The loss of chlorophyll from plants resulting in yellow leaves. It may be the result of the normal process of senescence, lack of key minerals that are for chlorophyll synthesis (iron and magnesium) or disease.

chloroxybacteria (grass-green bacteria, prochlorophytes) Green pigmented cyanobacteria that resemble the chloroplasts found as organelles in green plants and green algae.

choanae (internal nares) Paired openings that open into mouth cavity in air-breathing vertebrates. In mammals, they open posteriorly beyond the secondary palate.

choanocyte (collar cell) In sponges, a cell bearing a flagellum and surrounded at the base by a raised cylindrical collar.

cholecalciferol *See* vitamin-D.

cholecystokinin (CCK; pancreozymin) A hormone, produced by the duodenal region of the small intestine. CCK induces the gall bladder to contract and eject bile into the intestine and stimulates the pancreas to secrete its digestive enzymes. Cholecystokinin output is stimulated by contact with the contents of the stomach.

cholemesis Presence of bile in the vomit content.

cholera An acute infectious form of gastroenteritis caused by endotoxin-producing strains of the bacterium *Vibrio cholerae*. Spread throurh the ingestion of foods and liquids contaminated with the excreta of infected individuals.

cholesterol One of the most important sterol in animals, also occur in some plants and algae. It is a product of the liver and is essential to the functioning of the animal cell membranes.

choline An organic compound that is a constituent of the neurotransmitter acetylcholine and some fats.
A constituent of B-complex group of vitamins.

cholinergic Type of nerve fibre that releases acetylcholine from its ending when stimulated by a nerve impulse or is stimulated by it. In vertebrates, motor fibres to striated muscle, parasympathetic fibres to smooth muscle.

cholinesterase (acetyl-cholinesterase) An enzyme that hydrolyses the neurotransmitter acetylcholine to choline and acetate. It is secreted by nerve cells at synapses and by muscle cells at neuromuscular junctions.
Anticholinesterases inhibiting the action of cholinesterase, e.g. organophosphorus insecticides.

chondrichthyes *See* cartilaginous fish.

chondrin A gelatinous protein-carbohydrate matrix present in cartilage. It contains chondrocytes embedded in chondroitin sulphate.

chondriome (chondrioma) All the mitochondria of a cell taken together.

chondrocytes Cells that make up the cartilage matrix.

chondroitin sulphate The glycosaminoglycan (mucopolysaccharide) predominantly found in connective tissue, such as cartilage and in cornea.

chondrosarcoma The malignant tumour forming in bone cartilage. This is more common in people over the age of forty.

Chordata A phylum of animals in which characteristic features are solid dorsal notochord, a dorsal hollow nerve cord and pharyngeal gill slits. It is divided into three subphylum, i.e. Urochordata, Cephalochordata and Vertebrata.

chordotonal organs Sensory receptors that detect alterntions in the tension of insect muscles and crustanceans.

chorea A disorder of CNS characterised by involuntary, spasmodic movements of the face and extremities.

chorion The outermost of the extraembryonic membranes of amniotes. In mammals, a section of it becomes placenta part from embryo.

Chorionic Gonadotropin (CG) A hormone stimulating the production of progesterone and maintains corpus luteum. It is itself produced by the placenta in humans and is usually assayed for pregnancy tests.

Chorionic Villi Sampling (CVS) A method of prenatal testing in which foetal cells are extracted and analysed for chromosomal and biochemical defects.

choroid One of the eye's coats. It is composed of dark brown vascular tissue; it lies between the sclera and retina.

choroid plexus A membrane rich in blood vessels lining the ventricles of the brain. It secretes cerebrospinal fluid into the ventricles and also controls exchange of materials between the blood and cerebrospinal fluid.

chresard Total amount of water in the soil which is available to plants.

Chromaffin tissue Tissue which produces adrenaline or noradrenaline. Occurs in adrenal medulla and is widely distributed in vertebrate body.

chromatid One of the two identical nucleoprotein strands of a single replicated chromosome, which are joined at the centromere. They separate during cell division to become individual chromosomes.

chromatin Substance of which eukaryotic chromosomes are composed of DNA, RNA and proteins.
See euchromatin and heterochromatin.

chromatin silencing Inactivation of gene transcription as a result of condensation of chromosmal DNA into heterochromation.

chromatogram Record (pattern) of separated substances obtained by chromatography.

chromatography A method of analysing materials involving their separation by selective absorption of the various compounds as identifiable bands.

chromatophore

1. A pigment containing cell found in the skin of many lower vertebrates (e.g. chameleon) and in the integument of crustanceans e.g. melanophore, containing melanin.
2. A membrane-bound structure in photosynthetic bacteria that contains photosynthetic pigments.

chromomere One of the beads of locally coiled chromatin visible on a eukaryotic chromosome during meiotic prophase-I.

chromonema (pl. chromonemata) The chromosome thread, along which the chromomeres are arranged.

chromoplasts Pigment containing plastids in plant cells.

chromosomal mutations Changes in the karyotype. It may result from a translocation, an inversion, aneuploidy or polyploidy.

chromosome Physically discrete portion of the nucleus carrying many genes. A chromosome is composed of chromatin. Prokaryotic chromosomes are single and circular. Eukaryotic chromosomes are multiple and linear.

chromosome diminution The phenomenon seen in embryonic nematode worms in which the germ cells become separated from the somatic cells at an early stage of development. The chromosomes of the somatic cells tend to breakdown, with only a few remaining intact, whereas the integrity of the germ-cell chromosomes is maintained.

chromosome jumping
See chromosome walking.

chromosome map Diagram showing the order of genes along the length chromosome.

C

chronobiology The study of effect of time phenomena in living organisms.

chronospecies Different stages in the same evolving lineage that existed at different points in time.

chrysalis Third stage in the development of a butterfly or moth; a pupa.

Chrysophyta (golden-brown algae) The algae phylum containing carotenoid pigments as well as chlorophylls.

chyle Milk like substance contained within the lacteals and the lymphatic vessels of the intestine. A solution of the digestive products which consists primarily of fats.

chylomicron Any of the lipoprotein particles synthesised by intestinal epithelial cells and consisting mainly of triglycerides. Dietary fat is transported in the circulatory system in this form.

chyme Partially digested food leaving the stomach. Food mixes thoroughly with the acidic gastric juice of the stomach by the churning movements of its muscular wall and change it to chyme.

chymosin An enzyme found in gastric juices and responsible for the coagulation of milk. It acts by hydrolysing peptide links.

chymotrypsin (chymotrypsinogen) A proteolytic enzyme produced by pancreas which hydrolyses peptide bonds in alkaline medium. It is secreted as an inactive precursor, chymotrypsinogen which is activated by another pancreatic protease, trypsin.

cilia

1. Tiny hair-like processes (up to 10 μm long) covering certain internal surfaces of the body such as the epithelial surfaces of the respiratory tract or the internal lining of the inner ear.
2. Hair-like processes found on certain cells chiefly in protozoans.

ciliary body (ciliary muscle) The ring of ciliary muscles (which control the iris) and tissues that surrounds and supports the lens of vertebrate eye. The ciliary body produces the aqueous humour.

ciliary feeding The feeding method used by lancelets and many aquatic invertebrates. The movement of cilia causes a current of water inwards and also filters out the microorganisms in water.

ciliated epithelium Columnar or cuboidal cells of the epithelium bearing cilia on their free surface constitutes the cliated epithelium. They move particles or mucus in a specific direction over the epithelium. Occurs in respiratory tract, Fallopian tubes, etc.

Ciliophora Phylum containing ciliated protozoans such as *Paramecium*. Ciliates reproduce sexually by conjugation.

cilium A short minute hair-like structure present on the surface of many cells, e.g. protozoans and some types of vertebrate.

circadian rhythm (diural rhythm) Endogenous physiological or behavioural changes that occur in an organism in a period of about 24 hours, e.g. are leaf movement and growth in plants, sleep rhythms and running activity in animals, etc. Circadian rhythms are considered to be a basis of photoperiodism in plants.

circalunar rhythm A biorhythm corresponding with the lunar cycle (approx. 29.5 days), e.g. reproductive cycles of many marine organisms linked to changing levels of moon light and tidal cycle.

circinate Rolled from tip inward.

circulation The flow of liquid (e.g. blood or lymph) through the tissue and organs allowing transport and exchange of materials. See open and closed circulation.

circulatory system (also cardiovascular system) Heart, blood vessels, and lymphatic vessels, taken together as a circulatory system.

cirrus A small, slender and usually movable structure such as an appendage, barbel from a cell or body surface.

cisternae (cisterna) Flattened sac-like vesicles of endoplasmic reticulum and Golgi apparatus.

***cis-trans* test** A genetic test used for determining whether two mutations, affecting the same phenotypic function occur in a single gene or different genes. It is relevant for recessive traits. Also called complementation test.

cistron The functional unit of a DNA that controls protein manufacture.

citric acid cycle *See* Krebs' cycle.

clade A hypothesised group consisting of a single common ancestral form and its descendant form. This is usually thought of as being produced by the divergence.

cladistics (cladogram) A method of classification of animals and plants that holds that organisms should be classified on the basis of evolutionary relationship. A diagram called a **cladogram** consists of a system of dichotomous branching.

cladode (cladophyll) A modified internode of the stem that functions as a leaf. Its flattened and highly photosynthetic. It is a xerophytic adaptation and is seen in *Asparagus*. It is a phylloclade of single internode.

cladogenesis The branching of an evolutionary line into two or more separate lineages.

claspers

1. Any pair of processes on hind end of the abdomen of a male insect, serving to clasp female during mating.
2. A paired organ of male shark and related fish that assist in transfer of spermatozoa into female body.

class In taxonomy, a division of a phylum that itself contains one or more similar or closely related order e.g. mammalia.

classification A method to group and categorise organisms into categories, based on physiological, biochemical, anatomical or other relationships.

clathrin A major protein of the coated pits and vesicles formed during endocytosis.

clavicle The collar bone; articulates with the sternum and scapula and forms part of pectoral girdle.

clay The inorganic constituent of soil consisting chiefly of clay minerals in the form of particles less than 0.002 mm in diameter.

cleavage A mitotic division that divide the fertilised egg into smaller cells with equivalent nuclei. No growth occurs during cleavage.

cleidocostal Pertaining to both clavicle and rib.

cleidoic egg A shelled egg, limiting the exchange to gases and liquids, e.g. bird's egg.

cleistogamy The self-pollination that occurs before the flower opens.

cleistothecium A closed spherical ascocarp.

climate The characteristic pattern of weather elements in an area over a period.

climax community Steady, end stage in the ecological succession of a plant or animal species.

cline A graded series of differences exhibited within a population (usually along a geographic line or across a region of environmental transition).

clinistix A text strip used to test for glucose.

clinostat An apparatus used in tropism experiments to remove the unidirectional influence of a stimulus on a plant organ.

clitellum Thickened glandular non-segmented region in certain annelid worms. It secrete mucus which forms the cocoon where, fertilise eggs develop. In earthworms it is present in segments 14-16.

clitoris An organ in the female mammal corresponding to the penis of the male. It is a small cylindrical highly erectile body situated in the angle of the vulva.

cloaca The posterior part of the alimentary canal into which the urinary and reproductive ducts open, in birds, reptiles, amphibians and many fishes.

cloacal spur A claw in boas and pythons, which is an remnant of the pelvic girdle. It is used by the male snake, while courting.

clomiphene A synthetic chemical stimulant that causes ovaries to release eggs.

clonal selection theory A theory explaining how the cells of the immune system produce large quantities of the right antibody at the right time, i.e. when the appropriate antigen is encountered.

clone Two or more molecules, cells, or organisms identical to some ancestral molecule, cell or organism.

cloning

1. Producing multiple, exact copies (clones) of a single segment of DNA to obtain sufficient material for further study. This process is called DNA 'cloning.' The resulting collections of cloned DNA molecules are referred to as 'clone libraries'.
2. Producing by the natural process of cell division many copies of a cell. The collection of cloned cells identical to the original cell is called a cell line.
3. Producing complete, genetically identical organisms through mitosis.

cloning vector A virus, plasmid, bacteriophage, chromosome, or eukaryotic cell in gene cloning. It carry and replicate a DNA sequence of interest, so that more copies of that DNA can be produced; also can be used for producing the protein product of a particular DNA sequence.

clonotype

1. An attribute or specificity that is unique to a particular clone of cells, especially the structure of the surface receptors of a T-cell or of the secreted antibodies of a plasma cell in immunology.
2. A plant or line of plants, that has been propagated by cloning a type specimen (in botany).

clonus Involuntary, sudden and alternating contraction and relaxation of a muscle or group of muscles.

closed circulatory system Circulatory system of higher organisms in which blood flows in a system of closed vessels (arteries, veins, capillaries) in the body, e.g. humans.

clotting factors (coagulation factors) A group of substances present in blood plasma that undergo a series of chemical reactions leading to the conversion of blood from a liquid to a solid state (see blood clotting). (e.g. factor VIII, factor IX).

Lack of any of these factors in the blood results in the inability of the blood to clot. See haemophilia.

clove

1. An evergreen tropical tree which grows up to a height of 9 metres and has purple flowers (*Syzygium aromaticum*).
2. The dried unopened flower buds of this tree used as a spice.

clutch Collective eggs an organism lays at one attempt. If a bird lays five eggs, she has laid a clutch of five.

Cnidaria A phylum containing jellyfishes, sea anemonės and hydroids. The body of a cnidarian is made primarily of mesoglea contained in a bag of thin epithelium. The diagnostic feature of a cnidarian is the cnidocyte, a cell that fires a toxin projectile, e.g. *Hydra.*

cnidoblast *See* thread cell.

cnidocil A trigger or a sensory conical projection on the free surface of the cnidoblast to receive chemical as well as physical stimuli from the surroundings.

coacervate An aggregate of macromolecules, such as proteins, lipids, and nucleic acids, that form a stable colloid unit with properties that resemble living matter. Coacervate droplets arise spontaneously under appropriate conditions and hae been suggested as possible prebiological systems from which living organisms originated.

coadaptation The mutual, correlated adaptive changes in two interdependent organs.

cobalmin (Vitamin-B_{12}) A water soluble vitamin manufactured only by microorganism. Its natural sources are of animal origin, such as fish, meat, egg, etc. Plays an important role in normal functioning of brain and nervous system.

cocaine An alkaloid which occurs in the cocoa plant *Erythroxylum coca.* Used as a local anesthetic. It has a potent stimulating action on CNS, produces a sense of euphoria and increased energy.

coccus (pl. cocci) A spherical-shaped bacterium. Cocci may be found singly, in pairs (*Diplococcus*), in chains (*Streptococcus*), in regularly or irregularly packed clusters (*Staphy lococcus*).

coccyx Bone at the end of the vertebral column in tailless primates. It is usually comprised of 3-5 caudal vertebrae fused together.

cochlea Part of the membranous labyrinth (inner ear) in mammals, birds and some reptiles. It is concerned with the reception of sound and analysis of its pitch. A projection of the sacculus. In mammals, coiled in spiral, like a shell.

cocoon A protective covering for the egg or more usually for the pupa in many invertebrates.

codeine An analgesic drug derived from plant *Papaver somniferum.*

coding strand Refers to the DNA strand with the same sequence as the transcribed *m*RNA and containing the linear array of codons which interact with anticodons of *t*RNA during translation in order to give the primary sequence of a protein. Also called plus strand or sense strand.

codominance Situation in which two different alleles are equally dominant in a heterozygous organism. If they occur together, the resulting phenotype is intermediate between the two respective homozygotes.

codon A sequence of three nucleotide bases (triplet) that codes for a specific amino acid. Since four different bases are found in nucleic acids, there are 64 (4 x 4 x 4) possible triplet combinations.

coelacanth A primitive marine bony fish of the genus *Latimeria*, having pectoral fins and three lobed tail fin. Thought to be extinct until 1938, when *L.chalmnae* was discovered.

Coelenterate Any member of the phylum–Coelenterata comprising animals having a central body cavity with a single opening, the mouth, which is surrounded by a ring of tentacles and several arms with stinging cells. They include the jelly fishes, *Hydra*, corals and sea anemones.

coelom A fluid-filled cavity that forms the main body cavity of vertebrate and most invertebrate animals. It is formed by the splitting of the mesoderm.

coelomoducts Ciliated ducts that connect coelom to exterior. It helps in exist of waste products and gametes. These are specialised as oviducts, etc., in higher animals.

coenobium (pl. coenobia) The colony of algal cells, constant in number and arranged in a specific manner. It coordinates and behaves as a unit, e.g. *Volvox.*

coenocyte An area of cytoplasm containing many nuclei, typically found in certain fungi and algae.

coenzyme Non-protein organic molecule that plays an accessory role in enzyme catalysed reactions usually participate in the substratenzyme interaction by donating or accepting certain chemical groups. Many vitamins are precursors of coenzymes.

coenzyme A (Co-A) A complex organic compound that acts it linkage with enzymes involved in various biochemical reactions, e.g. the oxidation of pyruvate *via* the Krebs cycle. It comprises principally the B-vitamin pantothenic acid, the nucleotide adenine, and a ribosephosphate group.

coenzyme Q *See* ubiquinone.

coevolution The evolution of complementary adaptations in two species caused by the selection pressures that each exerts on the other.

cofactor Non-protein substance that helps an enzyme to carry out its activity. Unlike enzymes, they are in general, stable to heat. When a catalytically active enzyme forms a complex with a cofactor, a holoenzyme is formed. An enzyme without its cofactor is termed an apoenzyme.

cohesin A multiprotein complex that binds sister chromatids together during the initial stages of nuclear division in cells.

coitus The copulation of male and female. Term used generally in terms of mammals.

colchicine A drug obtained from the autumn crocus *Colchicum autumnale.* It is used to prevent spindle formation in mitosis or meiosis.

cold blooded animals (ectotherm) Animals that cannot maintain a regular body temperature. Reptiles and amphibians are cold-blooded animals.

Coleoptera The insect order, comprising of beetles and weevils. Forewings form elytra, mouthparts are biting type and metamorphosis occur in these species.

coleoptile The protective sheath surrounding the plumule (shoot tip) embryo of grass family.

coleorhiza The protective sheath surrounding radical in embryo in plants of grass faulty.

coliform bacteria A group of Gram-negative rod-shaped bacteria that are found in the vertebrate gastrointestinal tract; their presence in water is an indicator of faecal pollution, e.g. *E.coli* and *Salmonella.*

collagen The protein of fibrous connective tissues, present in bones, skin and tendon. It is the most abundant of all the proteins in higher vertebrates.

collateral bundle A kind of vascular bundle in which the phloem is external to the xylem and on the same radius.

collecting duct Any of the ducts in the mammalian. Kidney that drains into the renal pelvis. They are the main sites of water reabsorpiton from the glomerular filtrate. The cells of this region are relatively impermeable to water. However, the influence of antidiuretic hormone increases the permeability of the collecting ducts, allowing the reabsorpiton of water and controlling the final urine concentration according to the body state.

collenchyma A type of tissue in plants usually located just beneath the epidermis, that functions as supportive tissue. Most commonly found in stem cortex.

collobast *See* lasso cell.

colon A part of the large intestine, between the caecum and rectum. Its function is to absorb water and minerals from indigested food.

colony

1. A group of organisms of the same species living together and dependent on each other, e.g. Sponges (in zoology) Opposite of solitary.
2. A group of microorganism developed from a single parent.

Colony Stimulating Factor (CSF) Any of several cytokines that stimulate development of certain types of blood cells from progenitor cells in the bone marrow and in other tissues, e.g. GM-CSF.

colostrum The maternal milk of mammal formed during the first few days after the birth. Particularly rich in proteins, including antibodies in some animals, especially ungulates.

colour blindness A genetically controlled eye defect in which one cannot distinguish between colours. It is a recessive sex-linked trait, carried by X-chromosome. In female, it appears only when both the sex chromosome carry the gene (X^cX^c). The females function as carries in presence of a single recessive gene (XX^c). In males, the defect appears in presence of a single recessive gene (X^cY) because Y-chromosome does not carry any gene for colour vision, e.g. protanopia, deuteranopia, tritanopia.

colour vision The ability of the eye to detect different wavelengths of light and to distinguish between these different wavelengths and their corresponding colours. This is achieved by the cone cells, which are located in and around the retina in mammalian eye.

columnar epithelium The epithelium composed of a single layer of tall and slender cells, some of which produce (goblet cells) helps in secretion, absorption and protection, this epithelium occurs in lining of stomach, small intestine, etc.

commensalism A relationship between an organism and its host in which the host neither benefits nor suffers from the association, while the commensal benefits.

companion cells Small cells having dense cytoplasm and prominent nucleus lying side by side with sieve tube cells in phloem of flowering plants.

compensation point A stage in plants in which rate of synthesis of organic materials (photosynthesis) is equal to the rate of its breakdown by respiration.

C

competition Refers to the struggle between organisms of the same species (intraspecific) or between several different species (interspecific) for the same resources like food, water and space.

competitive exclusion principle The principle states that two species that occupy the same habitat cannot occupy the same ecological niche. Any two species that occupy the same niche will compete with each other to the detriment of one of the species, which will thus be excluded.

competitive inhibition Reduction in the rate of an enzyme catalysed reaction. In this, the inhibitior molecule resemble the substrate molecule and bind to the active site of enzyme.

complement receptor A protein found on the surface of certain cells of the immune system that binds specifically to complement components coating the surface of a bacterium or other pathogen, e.g. CR_1 receptor or macrophages.

complemental males The small males of certain animals that live in or on the females and are usually more or less degenerate apart from the reproductive organs. They occur in certain crustanceans (e.g. some barnacles).

complementarity Correspondence of DNA bases in the double helix in a way that adenine in one strand is opposite the thymine in the other strand and cytosine in one strand is opposite the guanine in the other.

Complementarity Determining Region (CDR) A region on the surface of an antibody molecule that contributes to the antigen-binding site.

complementary DNA (cDNA) The synthetic DNA transcribed from a particular RNA through the action of the enzyme reverse transcriptase has a base sequence that is complementary to that of the *m*RNA template used in gene cloning.

complementary genes Two or more genes that are interdependent, such that the dominant allele from either gene can only produce an effect on the phenotype of an organism if the dominant allele from the other gene is also present.

complex tissue A tissue made up of cells of the more than one type.

composite flower A plant made up of many tiny flowers called florets. There are over 20,000 species including daisy, sunflower and marigold.

composite fruit A type of pseudocarpic (false) fruit that incorporates the inflorescence.

compound epithelium Made up of more than one layer of cells and has a limited role in secretion and absorption. Their main function is to provide protection against chemical and mechanical stresses.

compound eye The type of eye found in crustaceans and insects. Consists of several thousands of units (ommatidia).

compound leaf A leaf whose petiole bears several leaflets.

conceptacle A flask-shaped cavity with a small opening found in the swollen tip of certain brown algae, such as *Fucus*. It contains the sex organs.

conception The fertilisation of a mammalian egg cell by a sperm cell, which occurs in the Fallopian tube. Conception is followed by implantation.

conchology The study of molluscan shells.

conditional response (conditioned reflex) A learned response that develops to any initially ineffective stimulus in classical conditioning.

conditioning A process by which animals learn about a relation between two events.

condyle A smooth round knob of bone that fits into a socket on an adjoining bone, forming a joint. Such a joint permits up-and-down or side-to-side movement but does not allow rotation, e.g. condyles present where the lower jawbone (mandible) is attached to the skull thus allowing chewing movements. See also occipital condyle.

cone

1. A reproductive structure occurring in gymnosperms, known as strobilus. It consists of sporophylls bearing the spore-producing sporangia (in botany).
2. A type of light sensitive receptor cell, found in the retinas of all diurnal vertebrates. These are specialised to identify colours (animal anatomy). Functions best in bright light.

conformer The organism whose internal environment is highly influenced by external factors, e.g. most aquatic organisms.

congenital The deformity or disease that exists at or before birth, e.g. Down's syndrome (conidiospore).

conidiospore *See* conidium.

conidium An asexual spore of certain fungi, e.g. *Pythium* and *Albugo*. They are cut off externally in chains at the apex of a specialised hypha, the conidiophore.

Coniferophyta A phylum of seed bearing plants comprising the conifers including the pines, firs and spruces.

coniferous forest A forest consisting of predominantly cone-bearing trees, with needle-shaped leaves. Their greatest extent is in the wide belt across Northern Canada and Northern Eurasia. Usually evergreen, some of these are deciduous, for example, the larch forests of Central Siberia.

conjuctive tissue Parenchymatous cells which lie between the xylem and the phloem tissue in plants.

conjugated protein A protein that on hydrolysis yields not only amino acids but also other organic and inorganic substances. They are simple proteins with non-protein groups.

conjugation

1. Sexual fusion of gametes, particularly isogametes.
2. A type of sexual reproduction found in some bacteria, most ciliates and certain algae involving the union of two individuals for the purpose of transfering genetic material.

conjunctiva A thin transparent layer of epidermis and connective tissue covering the cornea of the eye and lining the inner eyelid in vertebrates.

connective tissue A connecting or supporting tissue in animals, in which cells are distributed in a relatively extensive intercellular matrix. The matrix material may be rigid as in bone, flexible as in cartilage or fluid as in blood.

connexon A hexagonal array of protein subunits, typically 10 nm in diameter and arranged as a short hollow cylinder, that penetrates the plasma membrane of certain cells and forms part of a gap junction.

consensual Relating to movements that take place through the action of the nervous system independently of the will.

consensus sequence A sequence of nucleotides found in comparable regions of DNA or RNA, e.g. in the promoter regions (see operon) of different genes, in which certain bases occur with a frequency significantly greater than that expected by chance.

C

conservation The careful use of natural resources, including a slowing down of the use of non-renewable resources; development of renewable alternatives; recycling, reduction of pollution and caring for nature.

conservative replication In this model of DNA replication, the two parental DNA strands come together after replication. Thus, one daughter molecule contains both parental DNA strands and the other daughter molecule contains DNA strands of newly synthesised material.

conserved sequence A sequence of amino acids in a polypeptide or of nucleotide in DNA or RNA that is similar across multiple species.

consociation A climax plant community that is dominated by one particular species, e.g. a pine forest.

constipation A disorder of digestive system in which faeces are retained within the rectum as the bowel movements occur irregularly.

constitutive enzyme One of a group of enzymes that are always present in nearly constant amounts in a given cell. These enzymes are formed at constant rates and in constant amounts regardless of the metabolic state of the organism.

consumers Refers to the higher trophic levels in a food pyramid. They consist of primary consumers feeding on producers, and secondary consumers feeding on primary consumers.

continuous culture A technique used to grow microorganisms or cells continually in a particular phase of growth.

continuous replication During DNA replication, the synthesis on leading strand by continuous addition of nucleotides is known as continuous replication. The template of the leading strand is oriented in 3′-5′ direction.

continuous variation (quantitative variation) The range of differences that can be observed in many characteristics in a population.

contour feathers Feathers that are arranged in regular rows on a body of a bird giving streamlined shape. Each has a central horny shaft (the rachis) with a flatened vane on each side.

contraception *See* birth control.

contractile roots Specialised roots developed by certain bulb and corm forming plants that serve to pull the bulb or corm down to the appropriate depth in the soil. This counteracts the tendency for each new year's growth to be raised above the growth of the previous years.

contractile vacuole One or more membrane bound cavities in many protoctists that act as osmoregulators. They periodically expand as they fill with water by osmosis and contract to discharge their contents to the exterior

conus arteriosus A small thick walled chamber of the heart of vertebrate embryos. It receives blood from the single ventricle and leads to the ventral aorta. It is present as a distinct structure in adult fishes and (with modifications) amphibians, but in higher vertebrates it is incorporated into the roots of the aorta and pulmonary arteries.

convergent evolution (convergence) The development of similar structures in unrelated organisms as a result of living in similar ecological conditions. The wings of vertebrates and insects are an example of convergence.

convoluted tubule *See* distal convoluted tubule; proximal convoluted tubule; nephron.

coomassie blue A biological dye used for staining proteins.

coordination The process involved in the reception, integration and the subsequent response of the organism to sensory information controlled by regions of the brain that deal with specific functions, such as locomotion and breathing, and is carried out by the nervous system.

Copepoda A class of crustaceans, occurring in marine and freshwater habitats. Usually 0.5-2 mm long and lack carapace as well as compound eyes, e.g. *Cyclops*

coprophagy The ingestion of faeces as a means of obtaining nutrients. Coprophagous animals include dung beetles, which eat cow dung and rabbits, which ingest their own faeces.

copulation *See* sexual intercourse.

coral The marine sedentary invertebrates of phylum– Cnidaria belonging to class– Anthozoa. A coral colony consists of individual polyps within a protective skeleton that they secrete. The skeleton of stony, or true, corals consists of almost pure calcium carbonate and forms the coral reefs common in tropical seas.

coreceptor A receptor molecule on the surface of a cell that enhances the activity of another receptor following binding of its extracellular ligand, e.g. CD_4 and CD_8 proteins which lie alongside the T-cell receptor in plasma membrane.

corium

1. A deep inner portion of the skin beneath the epidermis, containing connective tissue, blood vessels and fat (derma or dermis).
2. The leathery basal part of the forewing of hemipterous insects.

cork (phellem) A protective waterproof plant tissue produced by the cork cambium. Its cells are dead with walls impregnated with suberin.

cork cambium (phellogen) Secondary formative tissue of plants, consisting of a single layer of cell. Together with its derivatives, it form periderm.

corm A rounded thick fleshy underground stem, bearing membranous or scaly leaves and buds. It acts as a vegetative reproductive structure. Occurs in plants such as *Crocus* and *Gladiolus.*

cornea Transparent part of the sclerotic at front surface of eye of vertebrates, overlying iris and lens. Mainly responsible in land vertebrates for refraction which results in focusing of image on retina.

cornification *See* keratinisation.

corolla A collective term for the petals of a flower. The corolla is denoted in the floral formula by the symbol 'C'. It encircles the stamens and carpels. The petals may either be free (polypetalous) or united to form a tube (gamopetalous).

corona

1. Any type of outgrowth from the petals or sepals, such as the trumpet of the daffodil flower in flowering plants.
2. A group of cells at the tip of the oogonium in *Chara.*

Coronary Artery Disease (CAD) *See* atherosclerosis.

coronary vessels Either of two pairs of blood vessels (coronary arteries and coronary veins) present in vertebrates that serve the heart. In mammals, the coronary arteries arise from the aorta and carry oxygenated blood into the muscle of the ventricles. The coronary veins return the deoxygenated blood to the right atrium.

corpus allatum An endocrine structure near the brain of immature arthropods.

corpus callosum A broad tract of nerve fibres within the brain of marsupial and placental mammals, connecting the two cerebral hemispheres.

corpus cardiacum One of a pair of separate or fused bodies of nervous tissue that function in the storage and secretion of brain hormone. It lies immediately behind the brain in insects.

corpus luteum (yellow body) Yellowish mass of tissue that forms from the granulosa cells in the cavity of a Graafian follicle in the ovary of mammal after the release of egg cell.
If ovulation does not result into fertilisation, then corpus luteum soon degenerates. If fertilisation occurs, corpus luteum persists and continue secreting progesterone during part or all of pregnancy.

corpus striatum (basal ganglia) A complex mass of nerve cell bodies and fibres lying deep within each cerebral hemisphere of the brain. It is highly developed in birds, in which it is the site of the highest brain function but less important in mammalian brain.

cortex

1. A primary tissue in roots and stems between the epidermis and vascular tissue.
2. The outermost layer of an organ or part. For example, the outer region of the kidney (renal cortex).

Corti cell Hair cells found in the organ of Corti in inner ear. Helps in detection of sound.

corticosteroid Any steroid hormone produced by the adrenal cortex. Its release is controlled by corticotrophin. These are of two type mineralo corticoid and glucocorticoids.

corticotrophin *See* ACTH

cortisol (hydrocortisone) Steroid hormone produced by adrenal cortex. It promotes the synthesis and storage of glucose maintains blood pressure thus is released in response to stress.

cortisone One of the hormones produced by cortex of adrenal gland; mainly a glucocorticoid. It is a corticosteroid it is used in treatment of autoimmune diseases and cancers.

corymb A raceme inflorescence with lower flowers having longer stems, resulting in a flat topped cluster of flowers, e.g. *Rowans.*

cosmids Hybrid vector, that is a combination of the plasmid vector and the *cos* site. It allows the target DNA to be inserted into λ head. It has high transformation efficiency and can carry up to 45 kb.

cosmoid scale Scales with outer layer of hard cosmin covered by ganoine (modified enamel) and inner layer of bones. It is characteristic of lungfish and coelacanths.

cotransmitter The substance released from a nerve ending along with primary neurotransmitter in order to modify its action.

cotransporter A type of transport protein that transports two or more substances across a cell membrane *via* concentration gradient.

cotyledon (seed leaf) An embryonic seed leaf. The cotyledons function in absorbing nutrients from the seed till the embryo is sufficiently developed to be capable of photosynthesis. Angiosperms are classified as either monocotyledons or dicotyledons, on the basis of the number of cotyledons present in the embryo.

counter shading The development of dark colours on the upper side and light colours on the under side. It is thought to be a form of camouflage.

courtship The special behaviour of animals in seeking mates. Varies from complex behaviour of birds and mammals to the random association of the sexes in many simpler animals.

Cowper's glands (bulbourethral glands) A pair of pea-sized glands, lying inferior to the prostate gland. These secrete alkaline fluids into the semen, that protects sperm.

coxa The basal segment of an arthropod leg with which the leg is joined on the ventral surface of the body.

coxal gland Excretory glands opening on the fifth segment of the body in arthropods such as king crab, scorpions, spiders and some insects.

coxopodite Joint of the protopodite nearest to the body in a crustacean.

C_3 pathway A metabolic pathway, characteristic of many plants of temperate region. In this, 3 phosphoglycerate is the first stable intermediate organic compound, (3 phosphoglycerate contains three carbon atoms, hence the name C_3).

C_4 pathway (Hatch and Slack pathway) A metabolic pathway, characteristic of tropical plants and plants that live in arid environment. The fixation of CO_2 with PEP form the 4-C compound, oxaloacetic acid.

CpG island A genomic region that contains a high frequency of dinulceotides containing cytosine and guaine.

C_3 plant A plant in which the first product of photosynthesis is a 3-carbon acid, phosphoglyceric acid, i.e. plants that follow C_3 pathway. Most temperate and many other plants are C_3 plants, e.g. mango.

C_4 plant A plant in which the first product of photosynthesis is a 4-carbon acid oxaloacetic acid. C_4 plants have evolved from C_3 plants by a modification in carbon dioxide fixation leading to more efficient photosynthesis, e.g. maize.

cranial Relating to the skull or brain.

cranial capacity The volume of an individual's braincase (usually measured in cubic centimeters). The average cramal capacity of humans is about 1300 cm^3.

cranial nerves Nerves that emerge directly from the brain. They supply the sense organs and muscles of head, neck and viscera. In humans, these are 12 pairs of nerves that can be seen on ventral surface of brain.

cranial reflex The reflexes mediated by pathways in the cranial nerves and brain, e.g. blinking.

craniata A clade of chordate animals with a cartilaginous or bony skull that protects the brain. Contains the hagfishes, lampreys and jawed vertebrates.

cranium (brain case) A portion of the skull enclosing the brain. It is formed by several fused bones.

Crassulacean Acid Metabolism (CAM) A photosynthetic carbon dioxide fixation pathway that takes CO_2 during the night. It is seen in plants that live in arid conditions, to reduce H_2O loss from evaporation.

C-Reactive Protein (CRP) An acute phase protein secreted by the liver into blood plasma. It is in response to inflammation and infection.

creatine (creatinine) A muscle compound synthesised from amino acid arginine, glycine and methionine.

creatinine phosphate An energy-rich compound especially prevalent in muscle tissue. It is converted to creatine during contraction and is excreted in wine.

C

cremocarp A dry fruit formed from two one-seeded carpels. The portions are termed as mericarps. Characteristic of some members of the Umbelliferae.

crenation Shriveling of cells due to the withdrawal of water, when the surrounding solution is hypertonic to cellular cytoplasm.

cretaceous period A geologic period that lasted from approximately 145.5 to 65.5 mya the third and most recent period of Mesozoic era. It was a period of warm climate and high sea levels. Its name was derived from the Latin word for chalk (*creta*), since some of the earliest Cretaceous formations studied were composed of this substance.

cretinism A condition in which a person has dwarfed stature and mental retardation as a result of a deficient thyroid gland during early childhood or fetal life.

Creutzfeldt-Jakob Disease (CJD) A rare degenerative fatal brain disorder. It is known to be caused by an abnormal prion protein, typically affects middle aged and elderly people. There is no treatment that can cure or control CJD.

crista The structure formed by folding of inner mitochondrial membrane containing oxidative enzymes.

critical group A large group of related organisms that cannot be divided into smaller groups, despite the presence of variations between them.

Cro-Magnon Man First modern humans (*Homo sapiens*) to reside in Europe; from the Cro-Magnon cave in South-Western France. In scholarly literature, this name has now fallen into disuse.

crop

1. A cultivated plant whose product is harvested by humans for use.
2. An enlarged portion of alimentary canal in animals, commonly used. For expandable pouch found in between the oesophagus and proventriculus of some birds.

crop-milk Secretion of crop epithelium in pigeons and doves on which nestling are fed. Like mammalian milk, its production is influenced by hormone prolactin.

crop rotation An agricultural practice in which different crops are cultivated in succession on the same area of land over a period of time, so as to maintain soil fertility and reduce the adverse effects of pests.

cross

1. A mating between two selected individuals to enhance the characteristics.
2. An organism resulting from such a mating.

cross fertilisation The union of gametes (sex cells) produced by different individuals, either in plants or animals.

Cross Over Value (COV) The percentage of linked genes that are exchanged during crossing over. It can be calculated by the percentage of offspring that show recombination.

cross-pollination The transfer of pollen grains to the stigma of the flowers on different plants of the same species. There is mixing of dissimilar characteristics.

crossing-over The exchange of genetic material between homologs during prophase-I of meiosis.

crustacean (Crustacea) An animal belonging to the major arthropod group–Crustacea, which includes crabs, shrimps and lobsters. Body consists of head, thorax and abdomen covered by a carapace and protected by hard exoskeleton.

cryobiology Study of the effects of low temperature on living organisms, tissues and cells.

cryophile A cold-loving organism.

cryophytes Plants that grow on ice and snow. Largely consisting of algae and also some mosses, fungi and bacteria. Algal forms may be so abundant as to give colour to substratum e.g. red snow, due to the presence of *Chlamydomonas.*

cryptic colouration The colouration of animals which helps to conceals it because of resemblance to the surrounding. It enable the animal to blend with the surrounding, e.g. stripes of zebras.

cryptic species *See* sibling species

cryptobiosis (anhydrobiosis) The state in the life of an organism when the metabolic rate is reduced to an extremely low level.

cryptochrome Proteins found in plants and animals that are sensitive to blue and UV light and change conformation in its presence. They play an important role in circadian rhythms.

cryptomonads An algae belonging to class cryptophyceae, which contain chlorophyll and phycobili protein photosynthetic pigments.

cryptonephridial system An excretory system arrangement in Coleoptera and Lepidoptera, where the distal end of Malpighian tubules are embedded in fat tissue surrounding the rectum. This arrangement serve to increase efficiency of solute processing.

cryptozoic Animals inhabiting crevices, (e.g. under stones, leaves) concealed or secluded places, e.g. woodlice, earthworms, etc.

crypts of liberkiihn (intestinal glands) The tubular glands lying in the villi of small intestine. The paneth cells in these glands secrete intestinal juice.

ctenidia The respiratory gills with comb-like structure present in molluscs.

Ctenophora An invertebrate phylum containing the combjellies or sea walnuts. Exclusively marine, they are biradially symmetrical and swimming by means of eight longitudinal rows of paddles. The property of bioluminescence is well marked.

CT scanner A device that uses completed tomography to produce detailed pictures of internal body structures from X-ray images.

cuboidal epithelium An epithelium composed of a single layer of cube-like cells. The free surfaces of the cells may be smooth or bear microvilli commonly found in ducts of glands, tubular parts of nephrons in kidneys, ovaries, etc.

cucumber A trailing or climbing hairy-stemmed plant, originally from Southern Asia. It is cultivated for its long, pulpy green fruits.

cultivar Any agricultural or horticultural variety of plant.

cultivation The planting and breeding of crop plants in agriculture.

culture A group of cells from microorganisms, tissue or other living matter in especially prepared nutrient medium.

culture medium The mixture of nutrients, which may be used in liquid form or solidified with agar. It is used to cultivate microorganisms, such as bacteria or fungi or support tissue cultures.

C

cumulus cells Cells from the Graafian follicle that surround and nourish the mammalian egg. They disperse quickly if sperms are present (30 min) but more slowly (2 h or more) if they are not.

cumulus oophorus The cell aggregation which immediately surrounds the mammalian egg within its Graafian follicle and after ovulation.

cupula A small inverted cup or dome-shaped cap over a structure, e.g. cupula optica in embryological development of the eye.

cupule A small cup-like structure, e.g. base of an acorn, a sucker on the feet of some flies.

curare Any of the arrow poisons originating from the bark of South American trees of genus *Strychnos* and *Chondrodendron*. It causes paralysis of voluntary muscles.

cusp

1. The projection of biting surface on mammalian molar.
2. A flap forming part of a valve.

cutaneous Of or pertaining to skin.

cuticle

1. An outermost layer of an organism, particularly the exoskeleton of an arthropod; usually made up of fibrous materials, such as collagen or chitin and in some cases strengthened by the addition of minerals, (e.g. calcium carbonate), (in zoology).
2. A waxy external layer, that prevents drying; found on the outer surface of many plants (in botany).

cutin A group of substances chemically related to fatty acids forming a continuous layer called the cuticle on the epidermis of plants, interrupted only by stomata or lenticles. It reduces water loss by plants and helps prevent the entry of pathogens.

cutinisation An impregnation of a plant cell wall with cutin.

cutis A term used for skin.

cutting A method of vegetative propargation. A part of a plant (stem, root, leaf, etc) is detached and inserted in soil to give rise to new daughter plants.

c-value An amount of DNA expressed in picograms present in a haploid nucleus.

cyanobacteria The photosynthetic prokaryote, belonging to eubacteria. Occur in all aquatic habitats. These fix atmospheric nitrogen and thus contribute to soil fertility, e.g. *Anabaena*.

Cyanophyta A phylum of blue-green algae.

cyanosis The bluish skin colouration due to the high levels of reduced haemoglobin in the blood.

cyclic AMP (cAMP) A form of adenosine monophosphate formed from ATP in a reaction catalysed by the enzyme adenylate cyclase. It acts as a secondary messenger in many biochemical reactions induced by hormones.

cyclic GMP (cGMP) Cyclic guanosine monophosphate. It is a derivative of the nucleotide GTP that acts as second messenger. It is formed from GTP by enzyme guanylate cyclase.

cyclic phosphorylation *See* photophosphorylation.

cyclin Any protein present in the dividing cells of many organisms. It acts as a control mechanism during cell division.

cyclitis An inflammation of the ciliary body.

cyclomorphosis (seasonal polyphenism) Phenotypic changes occurring seasonally in an organism through successive generations. Occurs in small aquatic invertebrates.

cyclosis *See* cytoplasmic streaming

cyclostomata Any of fishes of agnatha lacking jaws and true teeth and possessing a circular mouth comprising hagfishes and lamprey.

cyme Refers to an inflorescence in which the terminal bud is a first flower of the main stem and subsequent flowers develop as terminal buds of lateral stems, e.g. onion.

cymose inflorescence (cyme, definite inflorescence) A type of inflorescence in which the first flower develops from the growing region at top of flower stalk. It can be mono, di or polychasial.

cypsela Dry one seeded fruit of Compositae family similar to achene but is derived from an inferior ovary.

cyst

1. A membranous sac containing gaseous/liquid or semisolid substance.
2. A sac surrounding a larva in a resting or dormant state.
3. A sac or vesicle in body.

cysteine A non-essential amino acid, readily oxidised to cystine.

cysticercus A larval form of tapeworm, (*Taenia* sp.) consisting of a single scolex enclosed in a cyst.

cystine A compound consisting of two cysteine residues joined by a disulphide (S-S) linkage. It is a naturally occurring amino acid.

cystitis An inflammation of the urinary bladder.

cystocarp *See* carpogonium.

cystolith A stalked body, comprising ingrowth of cell wall having deposit of calcium carbonate. Found in epidermal cells of certain plants, e.g. *Ficus elastica.*

cytidine A nucleoside, comprises of cytosine attached to a ribose sugar molecule. The derived nucleotides cytidine mono di and triphosphate takes part in synthesis of variety of lipid derivatives.

cytochrome The ferrous protein containing haem group that play a role in electron transport in mitochondria and chloroplasts.

cytochrome oxidase An enzyme complex containing copper, that comprises the terminal two cytochromes (a and a_3) of the respiratory chain.

cytogenesis The origin and development of a cell.

cytogenetics Study of the physical appearance of chromosomes. It is a fusion science joining cytology with genetics.

cytokine A small protein released by cells that has a specific effect on the interaction between cells, or on behaviour of cells, e.g. interleukins, lymphokines, etc.

cytokinesis Division of the cytoplasm into two daughter cells that occurs after nuclear division in mitosis or meiosis. In animal cells it occurs by furrowing and in plants by cell plate formation.

cytokinins (kinin) The group of active growth substances (hormones) in plants which stimulate cell division in association with auxins. Chemically, cytokinins are purines. These act by increasing nucleic acid metabolism and protein synthesis in plants.

cytology The study of living cells, in particular, their physiological properties, structures, organelles and method of division.

cytolysis The lysis of cell usually by destruction of cell membrane.

cytomegalovirus Any of the group of highly host specific herpes viruses. It infects humans, rodents or monkeys producing unique large cells.

cytometer An instrument for counting the number of cells in a sample, especially blood cells.

cytoplasm All of the contents of a cell other than the nucleus (or in the case of a prokaryote, other than the nucleoid). It contains various cell organelles. All cellular activities like gaseous exchange, elimination of wastes, etc., occur in it.

cytoplasmic inheritance An inheritance of gene from the self-replicating organelles within the cytoplasm, rather than the nucleus. These are chloroplasts (in plants) and mitochondria. The female agg has a large amount of cytoplasm with many such organelles. These cytoplasmic organelles are thus not inherited from the male parent. This type of inheritance is exception to Mendel's law.

cytoplasmic streaming (cyclosis) Movement of the cytoplasm within certain cells and unicellular organisms. It allows movement of substances through the cell.

cytoplasmic trait A trait governed by genes found outside the nucleus in mitochondria or chloroplasts. Such traits are inherited only from one parent.

cytosine (C) A pyrimidine base found in DNA and RNA. In duplex DNA, it pairs with guanine.

cytoskeleton A fibrous network of microtubules and microfibres in eukaryotic cells that gives shape to the cell, anchors organelles, serves a variety of transport functions within the cell. In some cases, allows the cell to move or change shape.

cytosol (hyaloplasm) The semi-fluid soluble fraction of cytoplasm remaining after all organelles and insoluble components have been removed by centrifugation.

cytostome Cell mouth of certain Protozoa.

cytotaxonomy The use of chromosome number, size and shape in the classification of organisms.

cytotoxic Poisonous to the cells, e.g. drugs that inhibit cell division.

cytotoxic-T cell An essential part of the adaptive immune response. These are type of T-lymphocytes that recognises and eliminates body cells, infected with virus or other pathogens.

cyturia The presence of cells (of any type) in the urine of the host cell in unusual numbers.

D

2, 4D (2, 4 Dichlorophenoxyacetic acid) A synthetic auxin. It is used as a weedicide of broad leaved weeds.

dabble Bird behaviour pertaining to foraging for food on shallow water vegetation with their beaks or bills.

dacryagogue A substance causing lacrimation, causing lacrimal glands to secrete tears.

dacryelcosis Ulceration of the lacrimal sac.

dacryocystitis Inflammation of lacrimal sac, a condition that can result in obstruction of the affected nasolacrimal duct.

dactyl A digit (finger or toe) in human beings and corresponding body part in other vertebrates.

dactyloid Finger-shaped.

dance of the bees Movements performed by returning forager (worker) honeybees in order to indicate the direction and distance of a food source to other workers in the hive, e.g. waggle dance.

Daphnia A genus of Crustaceans belonging to branchiopoda and Cladocera. Contains waterfleas, with rounded body enclosed in a transparent shell having long, branched antennae.

dark adaptation The adjustment of the eye, physical as well as chemical that makes vision clear in darkness. It includes dilation of the pupil and increase in number of rod cells.

dark period The day length period changes considered to be critical in the responses of plant, e.g. onset of flowering.

dark reactions In photosynthesis, steps that can proceed in darkness by means of chemical energy produced by other steps (light reactions) that depends on the presence of light. During these reactions, CO_2 is reduced to carbohydrate in Calvin cycle.

Darwinian finches A group of finches (14 species) unique to the Galapagos islands, that were observed and studied by Darwin during his journey on HMS Beagle. Darwin believed all these finches were descendants of few that strayed from mainland, thus providing an evidence of evolution.

Darwinism The theory of evolution based on the idea of natural selection. First put forward by Charles Darwin.

daughter cells Cells produced by division of a single parent cell.

day neutral plants Plants in which flowering is not affected by length of photoperiod, e.g. tomato.

day vision Vision in bright light, in contrast to night or twilight vision.

DDT (Dichlorodiphenyl Trichloroethane) It is a colourless compound used to kill pests in agriculture. It is a stable compound which accumulates in the soil.

dead space Amount of space in the respiratory tract in which none or very little respiration takes place (the nasal passage, trachea, bronchi and bronchioles).

deaminase Enzyme which catalyses the removal of an amino-group from a compound.

deamination Chemical decomposition in which a compound such as an amino acid or alkaloid loses its amino group and forms ammonia.

death Termination of vital processes in the organism. In humans, it is cessation of heartbeat.

death phase The final phase in the bacterial growth curve that occurs when the rate at which bacteria die exceeds the rate of their production. It is due to decrease in level of nutrients and increase in toxin levels.

death rate (mortality) Number of deaths per 1000 population per year.

Decapoda An order of marine crustaceans including the shrimps, prawns, crabs, lobsters and crayfish. The members have five pairs of walking legs, head and thorax fused to form cephalothorax with the carapace.

decarboxylase An enzyme that catalyses the decarboxylation of carboxylic acids, including the conversion of amino acids to amines.

decarboxylation The removal of carbon dioxide from a molecule. It is an important reaction in several biochemical processes e.g. Krebs' cycle.

decidua Thickened and modified endometrium that lines the uterus during pregnancy in most mammals. Expelled soon after birth of baby.

deciduous Falling off leaves in certain plants at end of each growing season. It helps in retention of water by plants, e.g., rose.

deciduous teeth (milk teeth) First set of teeth of a mammal. They are temporary and soon fall out to be replaced by the permanent set, that are larger and more in number. In humans, they are 20 in number and there are no deciduous premolars.

decomposer An organism that helps in the decay of organic matter. They include bacteria, fungi, insects, protists and various types of worms and larvae. *See* saprotroph.

decomposition (decay) The breakdown of organic matter into its constituents by chemicals through the decomposers.

decompound A compound, i.e. more than thrice pinnate, giving rise to small axis on which leaflets are born.

decumbent Lying flat or creeping on ground but with their tip growing upwards.

decussate The arrangement of leaves on a stem in which the leaves are in opposite pairs and each pair is at right angles to the pairs above and below.

dedifferentiation Reversion of a specialised cell or tissue to an unspecialised form. It may occur in cancers, etc.

defecation The expulsion of faeces from the rectum due to the contraction in its wall. A sphincter muscle located at the end of rectum allows defecation and is under voluntary control.

defective ribosomal products *See* DRIPS

defective virus A virus that can reproduce only in the presence of another, so called helper virus that carries the genes necessary for the replication of the defective virus.

defensin Antimicrobial peptides involved in innate immunity, made within the body by neutrophils and macrophages.

deficiency diseases Diseases caused by inadequate intake or absence of an essential food substances, e.g. vitamins, minerals or essential amino acids.

definite (determinate) **inflorescence** A type of inflorescence in which flowers grow from terminal buds, e.g. cyme, glomerule.

definitive plumage Plumage of a bird attained after the shedding of all previous feathers that do not change significantly in colour or pattern, as long as the bird lives.

defloration

1. Rupture of the hymen.
2. The act of deflowering.

defoliant Chemical which induces premature leaf fall.

deforestation Process of cutting down of forests to extract timber and fuel wood or to clear the land for agriculture purpose.

degeneration

1. Reduction in size or complete loss of organs during evolution.
2. Changes in cells, tissues, etc due to disease or any other factor resulting in impairment or loss of function.

deglutition Act of swallowing in which bolus is conveyed into the pharynx and then into the oesophagus.

dehiscent Opening spontaneously to release spores or seeds, etc (in botany).

dehydrate To reduce or entirely remove the water content of a substance or body.

dehydrogenase An enzyme that transfers one or more hydrides from a substrate to an acceptor, usually NAD^+/ $NADP^+$ or a flavin cofactor such as FMN or FAD^+.

dehydrogenation Removal of hydrogen in a chemical reaction. Affected by dehydrogenases in biological systems.

delamination Separation of cell layers by splitting in blastoderm.

delayed plumage maturation A common phenomenon seen in male birds where the definitive plumage is delayed due to the number of factors.

deletion The loss or absence of one or more nucleotides from a chromosome.

deletion mapping Description of a chromosome that uses deletions as markers for ascertain the specific areas of mutation within that chromosome.

deme A small, geographically localised population, belonging to the same taxon, e.g. ecodeme.

demersal Fish or other animals living near bottom of the sea or lake.

demography Statistical study of human population.

denaturation Structural change in nucleic acid or protein. Denaturation leads to the loss or reduction of functionality. Common factors that can cause denaturation are changes in pH, salinity or temperature.

dendrite A short, branched process that conveys nerve impulses to the body of a neuron. (*See* neuron).

dendritic cell A type of antigen presenting cell's of immune system in mammal that activates cells. These cells are present in skin and inner lining of nose, lungs, stomach and intestines. Also found in blood in immature state.

Dendrobatidae A family comprised of the poison dart frogs. These are diurnal with brightly coloured bodies. Different species have varying level of toxicity.

dendrochronology An absolute dating method based on the pattern of tree rings or growth rings. It is widely used in the fields of paleoecology, archaeology and radiocarbon dating.

dendrogram A tree diagram, used to indicate some form of similarities between different organisms, e.g. cladogram.

dendrograph An instrument used to measure the periodic shrinking and swelling of tree trunks. (i.e. changes in tree diameter).

dendrology Study of wooden plants.

dendron The cytoplasmic process arising from the cell body of a motor neuron that branches into dendrites.

dengue (Breakbone fever) An acute fever due to a filterable virus transmitted by female *Aedes agepti* mosquito. It is characterised by sudden onset of acute fever with chill, severe headache, marked pain in the muscles and joints and profuse sweating. No specific treatment.

denitrification A chemical process in which nitrates in the soil are reduced to molecular nitrogen, to be used by organisms in atmosphere. The process is mediated by denitrifying bacteria (e.g. *Pseudomonas denitrificans*) which use nitrates as a source of energy. *Compare* Nitrification *See* Nitrogen cycle

de novo pathway The metabolic pathway in which a biomolecule is synthesised from simple precursor molecules, e.g. nucleotide synthesis.

dense body The structure found in the fibres of smooth muscle that anchors with the thin filaments within the muscle cell. Analogous in function to the Z lines of voluntary muscle.

density-dependent factor Size limiting factor of a population which depends on the number of individuals in the population, e.g. disease will have a greater effect in limiting the growth of a large population, since overcrowding facilitates its spread.

d

density-independent factor The limiting factor of a population whose effect is not dependent on the number of individuals in the population, e.g. earthquake will kill all members of the population regardless of whether the population is small or large.

dental caries Tooth decay, that involves the destruction of the enamel layer of the tooth by acids produced by bacterial action on sugar.

dental formula A representation for the number and kind of mammal teeth. Number in each jaw is written like a fraction with figures above the horizontal line showing number in the upper jaw and those below are the number in lower jaw, e.g. $\frac{2\,1\,2\,3}{2\,1\,2\,3}$ (human, 32 teeth).

dentary A bone present in the lower jaw of most vertebrates. In mammals, it is the sole bone of the lower jaw.

denticle *See* placoid scales.

dentine Main constituent of tooth. It lies beneath enamel. It is like bone in structure, but contains no cells and is made up of 70% inorganic material, 20% organic materials and 10% water. Formed from mesoderm of neural crest of embryo. Ivory that forms elephant tusk is dentine.

dentition The number, type, arrangement and development of teeth in any given species.

dentulous Toothed, having teeth.

deoxycorticosterone A steroid hormone, produced by the zona fasciculata and zona glomerula of adrenal cortex, having mineralo-corticoid activity. It is a precursor for the synthesis of cortisol and aldosterone.

deoxyribonuclease *See* DNAse.

Deoxyribonucleic Acid (DNA) A long thread-like nucleic acid. Structurally, its a double helix of polynucleotides held together by hydrogen bonds. It has deoxyribose sugar and the bases, adenine, guanine, cytosine and thymine. DNA is found mainly in chromosomes and is a carrier of hereditary information in majority of living organisms. The two strands of the helix are easily separated and can serve as a template for the synthesis of additional, complementary strands. It controls protein synthesis in cell and is also found in chloroplasts and mitochondria.

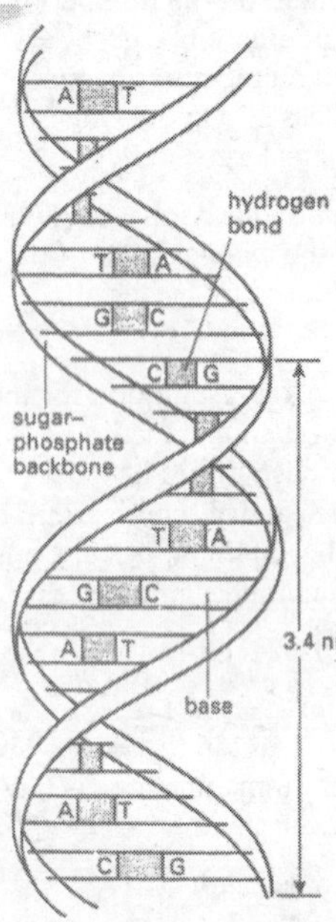

Double helical structure of DNA

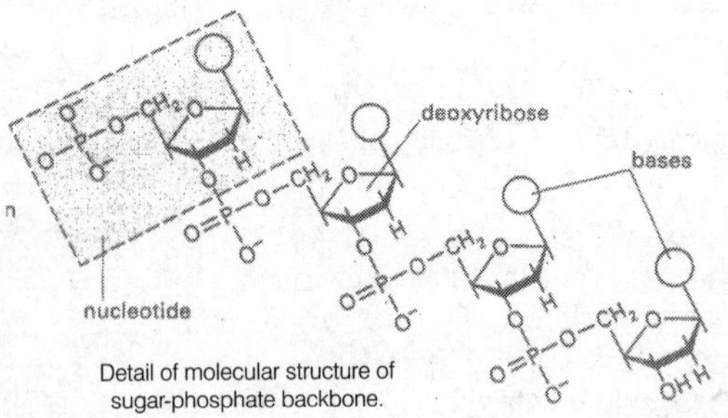

Detail of molecular structure of sugar-phosphate backbone.

Molecular structure of DNA

deoxyribose (2-deoxyribose) (Five-carbon) pentose sugar, component of DNA.

derived trait *See* apomorphic character.

dermal bone *See* membrane bone.

Dermaptera An insect order that consists earwigs. The members have long cylindrical body with cerci at abdomen tip.

dermatitis Inflammation of skin.

dermatogen External layer of a stem or root apex, one cell thick and giving rise to epidermis.

dermatology Study of the skin, nails and hair and the treatment of their diseases.

dermatome

1. Part of the mesodermal somites of vertebrate embryos that underlies the epidermis and develops into dermis.
2. An instrument used to cut thin slices of skin, e.g. for grafts
3. An area of skin mainly supplied by a single spinal nerve.

dermatophytosis A type of skin infection of widespread occurrence, caused by certain fungi (dermatophytes) feeding on keratin.

dermis (corium cutis) Innermost of the two layers of the skin of vertebrates, the outer being the epidermis. Much thicker than epidermis. It consists of connective tissue with abundant collagen fibres mainly parallel to surface, scattered cells, blood and lymph vessels and sensory nerves. Sweat glands and hair follicles project down from epidermis into dermis. Responsible for tensile strength of skin. May contain scales or bone.

dermotropic Attracted to or entering through skin.

derotremata A group of salamanders in which gill slits are present, but no external gills are present in the adult. It includes the giant salamanders family of Cryptobranchiata and the amphiumas.

descending tracts (spinal cord) Longitudinal nerve fibres of white matter. It consists of motor neurons, transmitting impulses from the brain.

desert biome An ecosystem characterised by low level of rainfall, it receives each year (less than 25 cm). It has sparse vegetation, xerophytic plants (e.g. cactus) and nocturnal animals.

desertification Persistent degradation of dryland ecosystems by variation in climate and human activities (e.g. deforestation, over intensive crop cultivation, overgrazing, etc).

desiccation

1. Process of drying.
2. To preserve food by removing moisture.

desmids Unicellular green freshwater algae of the order desmidiales, with elaborate chloroplast. These usually live in colonies in a branching net.

desmin A filament protein characteristic of striated, smooth and cardiac muscles.

desmosome (macula adherens) An animal cell junction structure that serves to attach adjacent cells in epithelia and cardiac muscles.

desmotubule A cylindrical membrane lined channel through a plasmodesmata. It links the cisternae of endoplasmic reticulum in the two cells.

desorption

1. To release through pores or interstices.
2. To remove an absorbed substance from a surface.

determinate growth Growth of a plant stem that is terminated early by the formation of a bud.

detoxification (detoxication) The process by which harmful compounds (such as drugs and poisons) are converted to less toxic compounds in the body. It is an important function of the liver.

detritivore An animal that feeds on detritus, e.g. earthworms, blowflies, maggots and woodlice. These species play an important role in the breakdown of organic matter from decomposing animals and plants.

detritus The raw material for decomposition, e.g. dead plant remains such as leaves, flowers, bark and dead remains of animals including faecal matter.

Detritus Food Chain (DFC) Food chain which begins with dead organic matter, being consumed by decomposers. These in turn are eaten by other animals. It is a large amount of energy flow through this chain. It helps in fixing inorganic nutrients.
Detritus → Detritivore → Carnivore

d

detrusor muscle Smooth muscle of the bladder wall. It is innervated with sympathetic fibres and contracts as a result of a reflex action, when the bladder wall tension has reached a certain level.

deuteranopia A type of colourblindness in which cones sensitive to green light are absent. (See colourblindness).

deuteromycota A large and heterogenous form of fungi in which sexual reproduction is absent. These are described as imperfect fungi.

deuterostome An animal of group deuterostomia in which the opening (blastopore) of embryo becomes the anus and mouth develops later. It includes the echinodermata, hemichordata and chordates.

deutoplasm These are nutritive substances or the yolk-laden cytoplasm at the vegetal pole of many eggs.

development The complex process of growth and maturation that occurs in living organisms. Cell division and differentiation are its important processes. In vertebrate animals, there are three developmental stages:

1. cleavage,
2. gastrulation
3. organogenesis

development threshold Body temperature of an organism below which no development takes place.

developmental biology Study of the processes by which an organism changes from a zygote into a mature, multicellular individual.

Devonian The fourth period of Paleozoic era, from about 416 to 359 million years ago. It is divided into 7 stages based on invertebrate fossil remains. Characterised by the development of lobe-finned fishes, the insects and amphibians appearance and the first forests.

dextrin A soluble polysaccharide obtained as an intermediate production during the conversion of starch to maltose. Used as gum and thickening agent in foods.

dextrocardia Presence of heart on the right side of the body.

dextrose *See* glucose.

diabetes insipidus A metabolic disorder marked by thirst and the passage of large amount of urine and no excretion of sugar. It is due to the deficiency of ADH.

diabetes mellitus (hyperglycemia) Disease caused by lack of insulin in which the body is unable to utilise glucose. Large amount of glucose accumulate in the blood and excreted in the urine. Can be treated by insulin injections.

diagenesis Changes in fossils or in sedimentary minerals, occurring after their deposition. It can result from any physical process (erosion, chemical change, distortion under pressure, bacterial infiltration, etc).

diageotropism The orientation of plant part by growth curvature, in response to stimulus of gravity so that its axis is at right-angles to the direction of gravitational force, i.e. horizontal.

diagnosis It is recognition of the nature of a disease from symptoms.

diakinesis Final stage of meiotic prophase-I. The chromosomes are fully condensed and the meiotic spindle is assembled to prepare the homologous chromosomes for separation. By the end of this stage, nucleolus disappears and the nuclear envelope also breaks down.

dialysis Technique for separating compounds with small molecules from compounds with large molecules by selective diffusion through a semipermeable membrane. Dialysis is performed naturally by the kidneys to extract wastes from the blood or in the case of lost or damaged kidneys, artificially by machine (dialyser).

diapause A period of suspended growth and delay of development in insects, etc., often connected with seasonal changes of atmospheric conditions.

diapedesis The passage of a blood corpuscle through vessel walls without rupture. (Squeezing of the leucocytes out of the capillary walls).

diaphragm

1. Any thin membrane.
2. A muscular wall separating the abdomen from the thoracic cavity; it contracts with each breath (inspiration) and relaxes with each exhalation (expiration).

diaphysis Shaft of a long bone in mammals. It contains the primary center from which the majority of a bone is ossified; the bone ends (epiphysis) are formed separately.

diapsid Reptile having a pair of openings in the skull behind each eye.

diarrhoea Increased frequency of bowled movements, the stool consisting of abnormally liquid faeces.

diastase An enzyme complex (about 5) which breaks down starch to glucose. It is found in milk.

diastasis

1. Final rest stage of diastole phase of heartbeat.
2. an injury involving separation, particularly of an epiphysis from its parent bone.

diastema The space created on the jaw by the absence of canines as in the orders of Lagomorpha (rabbit and hare), Rodentia (rat and squirrel) and many herbivorous mammals. The lip can be folded in through diastema, not allowing the food to fall out.

diastole Portion of the cardiac cycle during which the heart chambers dilate after contraction and fill the ventricles with blood.

diastolic pressure Lowest arterial pressure during relaxation and dilation of the ventricles of the heart when the ventricles fill with blood.

diathermy The method of medical treatment of heating the body tissues by the passage of high frequency electro-magnetic currents.

diatom Member of the class–Bacillariophyceae. They are unique in being enclosed in a silicaceous cell wall (frustule).

diatropism *See* plagiotropism.

dibiontic (haplodiplontic; diplohaplontic) Life cycle in which there is alteration of generation. Multicellular diploid and haploid stages occur.

dicer (endoribonuclease dicer, helicase with RNAse motif) Protein originally identified as the product of the dicer gene in *Drosophila*, that trims double- stranded RNA molecules to form short interfering RNAs (*si*RNAs) or mircoRNAs (*mi*RNAs).

dichasial cyme (dichasium) Cyme in which each branch gives rise to two other branches.

dichogamy A condition in which anthers and stigmas of a flower mature at different times, preventing self-pollination.

d

dichopatric speciation A form of allopatric speciation in which a formely contiguous closely connected population separates because of some geographic barrier, e.g. a mountain range.

dichotomous (dichotomous venetion, dichotomy) Dividing or branching into two equal parts, e.g. dichotomous branching in ferns and mosses.

dicots Flowering plants characterised by the presence of two cotyledons, floral organs are arranged in cycles of four or five and the leaves have reticulate veins, e.g. roses.

Dicotyledoneae An angiosperm class with two cotyledons within seed.

dicoumarol A toxic substance made by fungi from conversion of coumarin. If ingested, it causes haemorrhaging and is used as rodenticides.

Dictyoptera An insect order comprising cockroaches and mantids.

dictyosome Stack of membrane-bounded sacs (cisternae) that together with associated vesicles forms the Golgi apparatus in animal cells and rarely in plant cells. These are membranous vesicles in plant cells that modify proteins from ER and package materials for delivery to cell location.

diencephalon The part of forebrain, divided into thalamus, hypothalamus, epithalamus and subthalamus. It relays sensory information on between brain regions and control many autonomic functions.

diet The requirement of food in an organism. The food must contain mineral salts, vitamins, carbohydrates, water, etc.

dietary fibre *See* fibre.

differentiation Changes from simple to more complex forms undergone by developing tissues and organs so that they become specialised for particular functions. Differentiation occurs during embryonic development, regeneration and meristematic activity (in plants).

diffuse competition Weak interactions between species that are ecologically and distantly related.

diffuse porous wood Secondary xylem in which the vessels are distributed uniformly throughout a growth layer, when viewed in cross-section.

diffusion (passive transport) The net movement of a substance in the direction of its less concentration from a region of high concentration.

digestion Process by which large insoluble food is broken down into a form that can be absorbed by the body.

digestive system The system of organs which are involved in the process of digestion. In mammals, it consists of alimentary canal and accessory structure such as teeth, liver, etc.

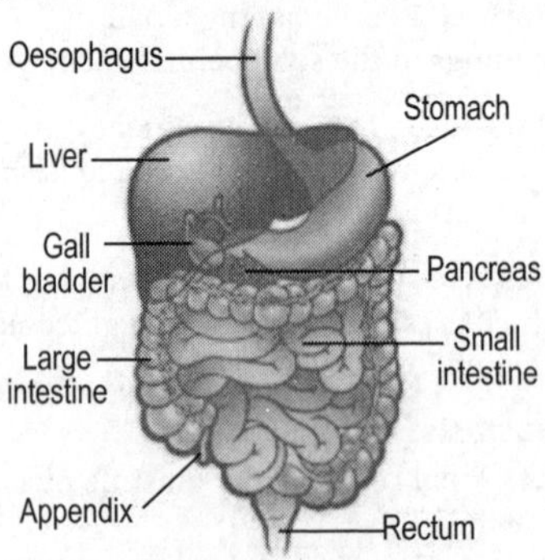

Digestive system

digitoxin Cardiac glycoside obtained from digitalis. Used in the treatment of certain heart problems.

digoxin Cardiac glycoside from digitalis. Used to treat heart failure and arrlalfthmia.

dihybrid cross A type of genetic cross in which the parents differ with respect to two genetically determined traits, e.g. a cross between tall pea plant bearing purple flower with dwarf pea plant bearing white flower.

dihybrid ratio A 9 : 3 : 3 : 1 phenotypic ratio among the offspring of a single dihybrid cross. Both involved factors are assorted randomly and independently and both parents are heterozygous for both the factors in consideration.

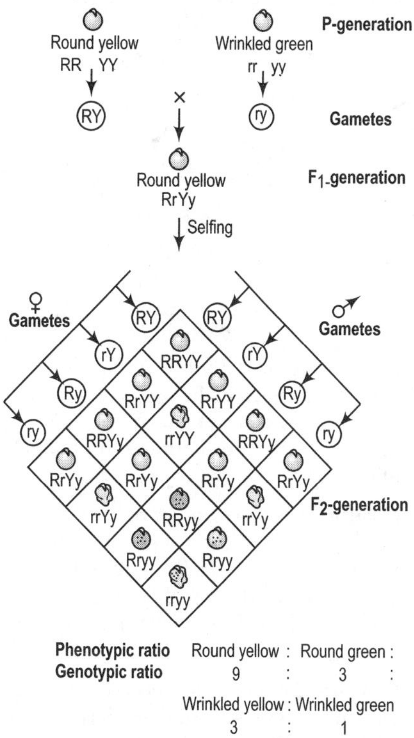

dihydroxyacetone Phosphate (DHAP) A compound involved in a wide variety of biochemical reactions. Its main role is in glycolysis.

dikaryon (dicaryon) A cell containing two different nuclei, arising from the fusion of two compatible cells, each with one nucleus, e.g. fungal mycelia.

dikaryosis The common form of heterokaryosis.

dikaryotic Having two nuclei per cell, e.g. Dikaryomycota of phylum-Fungi.

dilation Increase in the diameter of blood vessels which is mediated by the action of nerves on the muscle fibres.

dimegly These are possessing sperms of two sizes.

dimorphism Occurrence or existence of two forms within the same species, bearing distinct structure, features, coloration, etc. e.g. sexual dimorphism in which two sexes differ in colour, size, etc.

Dinomastigota (Dinoflagellata, Dinophyta) A Protista phylum; consisting of mainly photoautotrophic planktons. They contain xanthophyll as well as chlorophyll. Some of the species are bioluminescent.

dinosaur An extinct terrestrial reptile belonging to a group that constituted the dominant land animals of the Jurassic and Cretaceous periods, 190-65 million years ago. They belong to two order– Ornithischia, e.g. *Iguanodon* and Saurischia, e.g. *Tyrannosaurs.*

dinucleotide A compound consisting of two nucleotides.

dioecious A plant species having male (staminate) and female (carpellate) flowers on separate plants.

diotic Affecting or pertaining to both ears.

dipeptide A compound consisting of two amino acid units joined at the amino (–NH_2) end of one and the carboxyl (–COOH) end of the other. This peptide bond is formed by a condensation reaction that involves the removal of one molecule of water.

diphtheria An acute contagious disease caused by the bacterium *Corynebacterium diphtheriae.* It gains entry through respiratory passage.

diphygenic Having two modes of embryonic development.

diphyodont Dentition type in mammals (such as humans) in which temporary milk or deciduous teeth are replaced by a set of permanent or adult teeth.

diploblastic An animal with body wall with only two germ layers (ectoderm and endoderm), e.g. coelenterates.

diploid

1. A single-celled eukaryotic organism, having two complete sets of chromosomes.
2. A multicellular eukaryotic organism, having two complete sets of chromosomes in each somatic cell.

diplonema A stage of prophase-I of meiosis-I in which chromatids appear to repel each other. It is also known as diplotene stage.

diplont An organism with diploid somatic cell and haploid gamete cell.

diplophase Phase in the life cycle of an organism with diploid set of chromosomes. It begins with the formation of the zygote and ends with start of meiosis.

diplopia Double vision; seeing two images of one object.

diplopoda An arthropod class of worm-like terrestrial animals belonging to the subphylum –Myriapoda and comprising the millipedes. The members have distinct head, bearing a single pair of short antennae and 20 to over 60 body segments each bearing two pairs of legs.

diplosome A pair of centrioles in all eukaryotic cells, except for higher plants. Located near the nucleus, these are arranged perpendicularly.

diplotene Stage in prophase of first division of meiosis. Following pachytene, in which pairs of chromatids derived from homologous chromosomes begin to separate from each other except at certain points of connection (chiasmata), where interchange occurs between chromatid segment (crossing-over).

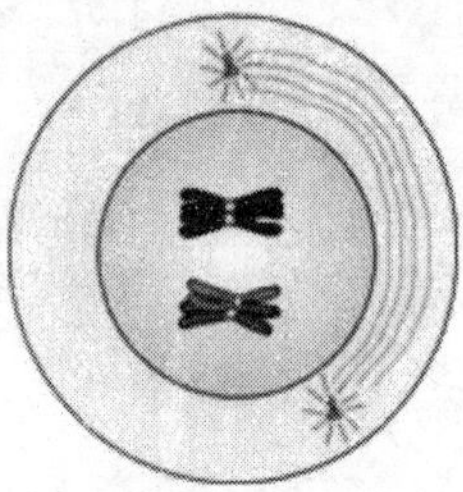

Diplotene

Diplura An insect order with members having elongated bodies, paired tail-like cerci and paired antennae. Body segment fixed and moulting continues throughout life.

Dipnoi Subclass or order of bony fishes that contains the lungfishes, which have lungs and breathe air. Dates from the Devonian era (416-360 million years ago) and share many features with the modern Amphibia.

Diptera An order of insects comprising the true or two-winged flies. Flies possess only one pair of wings-the forewings; the hindwings are modified to form small club-like halteres that function as balancing organs, e.g. housefly.

directed mutagenesis Hypothesis that proposes that organisms can respond to stress in environment by directing mutations to certain genes or area of genome.

directional selection Natural selection that favours the establishment of one particular mutation (that provides an advantage) within a population result in a change in phenotype in that direction, e.g. increase in darker forms of the peppered moth (*Biston betularia*) that occurred in industrial areas, against polluted tree trunks (see industrial melanism).

disaccharides Group name for complex sugars composed of two bonded monosaccharides. Examples of common disaccharides are sucrose, lactose and maltose.

disassortative mating Non-random mating in which individual selects mate with a dissimilar phenotype to themselves.

discicristates A group of eukaryotic protists. The members are characterised by disc-shaped cristate in their mitochondria and are shown to be related by molecular systematics. They include the euglenids, knietoplastids, and amoebomastigotes and the acrasid are cellular slime moulds.

Discomitochondria A phylum of protists consisting of motile single-celled organisms in few classifications. It is characterised by disc-shaped cristae in their mitochondria and the absence of sexual reproduction in their life cycles.

discontinuous replication The replication of lagging strand that results in the synthesis of Okazaki fragments (short DNA fragments) complementary to template strand. These fragments are bound covalently by DNA ligase, forming a chain of nucleotides (see continuous replication).

discontinuous variation (qualitative variation) Differences in a characteristic that are clearly defined to be observeable in a population. Characteristics that are determined by different alleles at a single locus show this type variation, e.g. garden peas are either wrinkled or smooth.

discoplacental Having a discoid placenta.

disease An interruption, cessation or disorder of a cell, tissue or organ. It is caused by various microorganisms and environmental agents.

disease resistance The genetic defence against any infection or its results.

dishabituation The recovery of a habituated response (see habituation) in an animal following the application of a different stimulus, e.g. siphon withdrawl in *Aplysia*.

disinfectant A substance that kills or inhibits the growth of disease-producing microorganisms. It is however toxic to human tissues, e.g. cresol, bleaching powder and phenol. Used to cleanse surgical apparatus, sick-rooms and household drains and as antiseptics if diluted.

disinhibition The tendency to exhibit displacement behaviour by an animal, at a point of equilibrium in situations where there are two or more conflicting drives.

dispersal Movement of offsprings of plants and animals. It provides better chance of survival to the organisms which are not mobile.

dispersive replication DNA replication in which the original DNA chain breaks and recombines in a random fashion before the double helix structure unwinds and separates to act as a template for messenger RNA synthesis. There is no evidence that it occurs in nature.

displacement activity An activity shown by an animal that appears to be irrelevant to its situation. Usually seen during a conflict between opposing drives.

display behaviour A conventional posture or movement that helps to influence the behaviour of another animal, e.g. male peacock spreads its tail in courtship.

disruptive colouration Colouration in animals which tends to break up their outlines and avoids visual predation.

disruptive selection Favouritism of extreme phenotypes in a population by natural selection.

distal Pertaining to an organ that is at the greatest distance from the point of its attachment to the body, e.g. hands are distal ends of arms.

distal convoluted tubule Section of the renal tubule where tubular secretion takes place (i.e. absorption of NaCl and other salts).

disulphide bridge (sulphur bridge) A covalent bond (S–S) formed between the thiol groups (–SH) of two cysteine residues, usually in the polypeptide chains of proteins, e.g. bonds in the tertiary structure of proteins.

diuresis Production of dilute (watery) urine in large quantities. It causes the osmotic pressure of the blood to rise and is counteracted by the antidiuretic hormone.

diuretic A compound which increases the activity of kidneys. (i.e. rate of urine formation).

diurnal Daily; denoting any event that occurs once every 24 hours.

diurnal rhythm Behaviour or physiological activity of animals or plants that happen in 24 hours periodicity.

divergence In terms of evolution, this is the process by which a single species evolves into two or more species because some of the population develops adaptations to one niche while others adapt to some other niche.

divergent evolution Separating from a common source.

diversity Wide range of plants and animals.

diverticulum Blind-ending, tubular or sac-like outgrowth from a cavity (tubular or hollow organ).

diving reflex A developed reflexive response to diving found in most aquatic mammals and birds that are characterised by complex physiological changes and adaptations, e.g. decreased heart rate.

division Category used traditionally in the classification of plants that consists of one or several similar classes, e.g. phyta (seed-bearing plants). Phylum has replaced the division in modern classification system.

dizygotic twins (fraternal twins) Twins (two individuals) developed as a result of simultaneous fertilisation of two separate ova. They differ in their genetic constitution and therefore, may or may not be of the same sex.

DNA *See* deoxyribonucleic acid.

DNAase *See* DNAse.

DNA-binding proteins Proteins that are able to bind to DNA and act as either activators or repressors of gene expression by controlling the binding of RNA polymerase to DNA during transcription (see transcription factor). They are also involved in DNA replication.

DNA blotting *See* Southern blotting.

DNA chip *See* DNA microarray.

DNA cloning *See* gene cloning.

DNA-dependent RNA polymerase RNA polymerase that catalyses the synthesis of RNA using existing DNA strand as template.

DNA filter assay A method used to determine the presence of recombinant DNA in cloned cells. The presence of recombinant DNA is revealed by the use of a gene probe and autoradiography.

DNA fingerprinting (DNA profiling; genetic fingerprinting) A molecular genetic technique to assess possible difference among DNA molecules of different samples. It involves identifying differences in some specific regions in DNA sequence called as repetitive DNA (a small stretch of DNA is repeated many time). The repetitive DNAs are separated from bulk genomic DNA as different peaks during density gradient centrifugation. The bulk DNA forms a major peak and the other small peaks are referred to as satellite DNA. These sequences show high degree of polymorphism and form the basis of DNA fingerprinting. Since, DNA from every tissue (as blood, hair, etc) from an individual differs, they become very useful tool in forensic applications. As the polymorphisms are inheritable, DNA fingerprinting is the basis of paternity testing.

The methodology includes:

1. DNA is isolated and digested by the **restriction endonucleases**.
2. DNA fragments are separated by **electrophoresis**.
3. Separated DNA fragments are transferred to synthetic membranes like nitrocellulase or nylon.
4. Hybridisation using labelled VNTR probe.
5. Hybridised DNA fragments are detected by **autoradiography**.

The sensitivity has been increased by use of Polymerase Chain Reaction (PCR).

DNA hybridisation A method to determine similarity of DNA from different sources. It is used to determine genetic relationships of species.

DNA gyrase An enzyme affecting the topology of DNA.

DNA library (gene library; gene bank) A collection of cloned DNA fragments representing the entire genetic material of an organism. These are the repositories of raw material for genetic engineering.

DNA ligase An enzyme that helps in joining together of two portions of DNA molecules.

DNA microarray (DNA chip) microarray containing numerous small DNA molecules, used, e.g. to analyse gene transcripts or to detect mutations of specific genes.

DNA photolyase An enzyme found in bacteria and certain lower eukaryotes (e.g. yeasts) that repairs damage to DNA induced by exposure to ultraviolet (UV) light. In the dark, it binds to thymine dimers formed by the UV light and on subsequent exposure to blue light, it absorbs the light energy and uses it to split the dimer and restore normal structure. This process is called **photoreactivation.**

DNA polymerase An enzyme that catalyses the synthesis of a complementary strand of DNA on a single-stranded DNA template during DNA replication.

DNA probe *See* gene probe.

DNA repair A variety of mechanisms that help to ensure that the genetic sequence, as expressed in the DNA is maintained and errors during DNA replication are not allowed to accumulate.

DNA repair genes Genes encoding proteins that can correct errors in DNA sequences.

d

DNA replication The process DNA makes exact copies of itself, which is controlled by the enzyme DNA polymerase. Replication in DNA strand occurs within a small opening of the DNA helix, known as replication fork. DNA-dependent DNA polymerases catalyse polymerisation only in one direction, i.e. 5′ → 3′. It creates additional complications at the replicating fork. Consequently on one strand (template 3′–5′), the replication is continuous, while on the other (template 5′–3′), it is discontinuous. The discontinuously synthesised fragments called Okazaki fragments are later joined by DNA ligase.

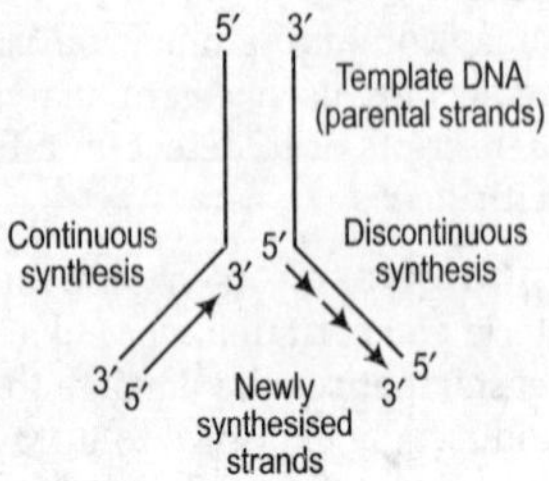

Replicating fork of DNA

DNA-RNA hybrid A duplex molecule composed of an RNA chain, hydrogen bonded to a complementary DNA chain.

DNAse (deoxyribonuclease; DNAase) An enzyme that catalyses the cleavage of DNA. DNase I is a digestive enzyme secreted by the pancreas, that degrades DNA into shorter nucleotide fragments. Many other endonucleases and exonucleases cleave DNA, including the restriction enzymes and enzymes involved in DNA repair and replication

DNA sequencing (gene sequencing) Process of elucidating the nucleotide sequence of a DNA fragment. The method now used almost universally is the Sanger method (named after Frederick Sanger), also called the dideoxy method.

DNA vector A virus, plasmid or other vehicle used to introduce DNA into a cell.

Dobson Unit (DU) Unit to measure thickness of the ozone in a column of air from the ground to the top of the atmosphere.

dodecandrous Flower having twelve stamens.

domain

1. Structurally compact unit of the tertiary structure of protein (in biochemistry).
2. The highest level of classification of organisms. Three domains are recognised Archaea, Eubacteria and Eukarya.

domestication Process of bringing about desirable genetic changes in wild plants and animals for yield, quality, hardiness, etc. through continuous breeding and selection over hundreds of years.

dominant

1. An allele that is expressed in the phenotype, when two different alleles of a gene are present in the cells of organism, e.g. 'T' allele in pea plant (in genetics).
2. The most conspicuously abundant and characteristic species in a community, e.g. pine trees in a pine forest (in ecology).
3. An animal that is allowed priority in access to food, mates, etc., by others of its species because of its success in previous aggressive encounters (in animal behaviour).

donor An individual whose organs or tissues are transferred to another person.

dopamine A catecholamine that is a precursor of epinephrine and nor-epinephrine. In mammals, it is found in highest concentration in the corpus striatum of the brain, where it functions as an inhibitory neurotransmitter.

High levels of dopamine are associated with Parkinson's disease in human.

dormancy Period of interrupted growth and metabolic inactivity, e.g. hibernation.

dorsal

1. Thoracic
2. Pertaining to the back or upper surface.

dorsal aorta Descending portion of the aorta in mammals. It carries oxygenated blood to the trunk and hind limbs. In fish, it is formed from the fusion of the six aortic arches. In tetrapods, it arises from the systemic arch.

dorsal root Part of a spinal nerve that enters the spinal cord on the dorsal side and contains only sensory fibres. The cell bodies of these fibres from the dorsal root ganglion.

dorsifixed Stamen in which the filament is fused to the back of the anther.

dot-blot Method for detecting a specific sequence of nucleotides in a DNA or RNA molecule. The nucleic acid sample is adsorbed onto a nitrocellulose filter. Then gene probe, which matches the specific sequence to be detected is added. After incubation, excess probe is washed off and the nucleotide sequence under investigation can be detected by autoradiography. This technique has been highly miniaturised.

double circulatory system Circulatory system that occurs in mammals, in which the blood passes through the heart twice before completing a full circuit of the body. Blood is pumped from the heart to the lungs and returns to the heart before being distributed to the other organs and tissues of the body. The heart is divided into two separate compartments to prevent oxygenated blood returning from the lungs from mixing with deoxygenated blood from the other parts of the body. See also pulmonary circulation; systemic circulation. compare single circulation.

double fertilisation An unique mode of fertilisation seen only in flowering plants where two male gamete nuclei fuses with two separate cells within the embryo sac. One cell contains the egg and form zygote. The other contains two polar nuclei to form triploid nucleus.

double helix A molecule having two similar polymeric chains coiled in the same direction about the same axis, e.g. in DNA or actin.

double recessive Individuals which is homozygous for a particular recessive gene. To test, whether an individual showing character of a dominant gene is homozygous or heterozygous, it is crossed to corresponding double recessive. If homozygous, all offsprings show dominant phenotype and if heterozygous, half show dominant and half show recessive phenotypes.

down feathers (plumules) Types of feathers in birds that cover the newly hatched bird; found beneath the contour feathers in adults. These are small and soft, covering the whole body and provide insulation.

Down's syndrome Congenital defect caused by trisomy of 21st chromosome pair (an extra chromosome). It is characterised by short broad face, slanted eyes, protruding tongue and mental retardation. It is also known as Mongolism.

draft sequence Preliminary assembly of data subsets to form a resemblance of a complete base sequence of an entire chromosome or genome (in bioinformatics).

drift Changes in the aggregate of genotypes in a small population resulting from the random extinction of allelomorphs of genes in regard to which the population was heterozygous.

DRIPS (Defective ribosomal Products) Peptides produced in cells by enzymatic digestion of the natural waste matter of protein synthesis, such as misfolded proteins and prematurely terminated polypeptides arising from errors in the synthetic machinery. DRiPs play a role in the immune system as an important source of self antigen for presentation at the cell surface in combination with MHC class I proteins.

dropsy A term used for swelling of soft tissues caused by accumulation of water.

drone A fertile male in a colony of social bees, especially the honeybee (*Apis mellifera*). The drones die after mating with the queen bee as the male reproductive organs explode within the female.

Drosophila Genus of fruit flies commonly used in genetics and cytology experiments; because of giant chromosomes in salivary glands and short life cycle.

drug Chemical substance that changes the physiology of an organism. Used in the prevention, diagnosis and treatment of disease, e.g. analgesics, antibiotics, etc. Some drugs are addictive and are often taken for their pleasurable effects only, e.g. LSD, narcotics.

drupe (pyrenocarp) Succulent fruit in which fruit wall (pericarp) consists of outer skin (epicarp), thick fleshy mesocarp and hard, stony endocarp, enclosing a single seed. Commonly called stone fruit, e.g. plum, cherry. In some plants like coconut, mesocarp is very fibrous. Pericarp consists of tough, leathery epicarp, very thick mesocarp and hard endocarp enclosing seed and forming with it the 'nut'.

dry mass Mass of a biological sample after eliminating the water content.

dryopithecine Any of various Miocene apes, formerly widespread in Eurasia.

Dryopithecus Genus of extinct apes, fossils of which have been found in Europe and Asia and dated to the mid-Miocene (about 9-13 million years ago). They split into several lines, some of which give rise to chimpanzees, gorillas, early hominids and organgutans.

duct Tube present in an organism which is used in the secretion and excretion of substances.

ductless gland A gland that secretes a 'hormone' or 'internal secretions' directly into the blood stream; (an endocrine gland).

ductus arteriosus Blood vessel present in mammalian foetus that links the pulmonary artery with the aorta. Derived from the outer portion of the sixth aortic arch, it closes at birth and remains as a solid strand in the adult.

duff An acidic humus, typically formed in free-draining acid soils of calcium concentration.

duodenum First part of small intestine of vertebrates. Leads out of stomach (junction being guarded by pyloric sphincter). Receives bile and pancreatic ducts. A region of active digestion.

duplex DNA Two single polynucleotide chains joined along their lengths side by side to form a double helix.

duplication Occurrence of extra genes or segments of a chromosome in the genome. It generally originates during crossing over in meiosis.

duramater (also dura) A tough, fibrous outer membranous sheath of the three membraned meninges that surround brain and spinal cord.

duramen *See* heartwood.

dwarfism Disorder occurring due to the underproduction of growth hormone from anterior pituitary gland. The affected person is short in stature.

dyad A pair of cells formed at meiosis, instead of tetrad; a chromosome with two chromatids.

dye Alternative name for stain.

dynactin (dynein) An eukaryotic cell protein with ATPase activity that binds to microtubules. It is essential to the motility of cilia and flagella.

dynorphin An opioid peptide produced as a neurotransmitter in brain and spinal cord. Their effect is similar to opiates such as opium.

dysentery An intestinal disorder characterised by frequent water stools; may have blood, pus or mucus caused by many factors, such as bacteria chemical irritants etc.

dysgenic Referring to any factor which tends to decrease the fitness of the race.

dyspepsia Indigestion.

dysphotic zone The zone of intermediate light intensity in a water body. The light is sufficient for behavioural response but insufficient for photosynthesis.

dysplasia Abnormal tissue development.

dyspnea Desire or hunger for air; manifested in strained, usually painful breathing.

dystrophic

1. Affected by or relating to dystrophy (in medicine).
2. (Of a lake) Containing high level of undecomposed organic matter, with brown acidic water, low in oxygen (in ecology).

dystrophy Disorder due to the defective or inadequate nutrition, e.g. muscular dystrophy.

E

ear A vertebrate sense organ that detects sound and maintains balance. Divided into outer ear, middle ear and inner ear. In mammals, the term ear is often used for pinna.

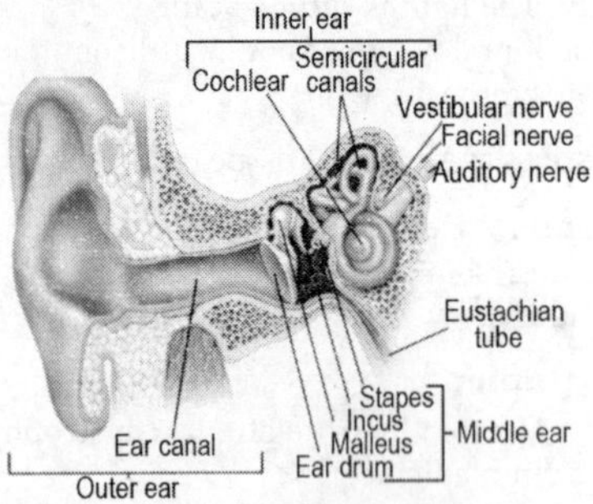

Structure of human ear

ear ossicles Bones in middle ear in mammals connecting ear drum to internal ear. Their function is to transmit ear drum vibrations caused by aerial sound waves to the fluid of inner ear. The bones are **incus** (anvil), **malleus** (hammer) and **stapes** (stirrup).

eardrum (tympanic membrane) *See* tympanum.

early wood (also springwood) Xylem lignified during the spring.

earthworm Common name for the largest member of Oligochaeta in the phylum–Annelida.

earwigs *See* Dermaptera.

ebractate Flower without bract. Bract is a leaf like structure at the base of flower.

eburnation Alteration of a bone to make it dense and ivory-like.

ecaudate Tailless; without a tail or tail-like appendage.

ecbolic Inducing labor or abortion, oxytocic.

eccrine glands Sweat glands which are linked to sympathetic nervous system and are widely distributed over the body surface of mammals.

eccrine (merocrine) **secretion** Secretion from the gland occurs without the secretory cells sustaining any permanent change.

ecdysis (moulting)

1. Periodic shedding of the hard outer cuticle or exoskeleton of arthropods (insects). This enables the insects to grow rapidly after each moult. Stages between each moult and next are known as instars. After a period of growth, a new cuticle is formed.
2. Shedding of outer epidermal layer periodically by reptiles (except crocodiles). It allows growth to occur.

ecdysis-triggering hormone Peptide hormone produced by neurosecretory cells in insects to initiate moulting of cuticle. It works in association with other hormones.

ecdysone Steroid hormone that causes molting in arthropods. It is secreted by prothoracic gland in thorax of insects. In crustaceans, it is formed by Y-organs and stimulate metamorphsis.

ecdysozoa A group of protostome animals, based chiefly on molecular systematics, whose members include the arthropods, nematodes onychophorans and tardigrades. All are characterised by their habit of periodic moulting.

e-cell A computer simulation of a living cell based on information obtained from genomics, proteomics, metabolomics, and other data drawn from real cells.

ecesis Establishment of colonising plants or animals in a new habitat.

ECG (Electrocardiogram) Recording of the electric current generated by action potentials of cardiac muscle cells of the heart.

echard Water present in the soil, which is unavailable to plants.

echinoderm Member of the phylum–Echinodermata. These are marine organisms with internal calcareous skeletons, five-part radial symmetry and a water vascular system. e.g. starfishes, sand dollars, crinoids and sea cucumbers.

Echinodermata Phylum of marine spiny-skinned invertebrates which are triploblastic, coelomate, unsegmented animals; adults pentamerically radial in symmetry but larva show bilateral symmetry, excretory system absent; high power of regeneration and a well developed water vascular system that provides power to tube feets, e.g. starfish.

echolalia Repetition of the vocalisation of others, e.g. an infant's repeating sounds uttered by others.

echolocation Process in which animals judge their surroundings in dark by producing high-pitched sounds, that bounce back off obstacles and are picked up by animal's receptor organ such as ear. From the direction of echo and the time period between reception and emission of the sounds, the object is located. Seen in animals such as bats, dolphins, etc.

echopraxia Mechanical, pathological involuntary repetition of the actions of others.

eclampsia Pregnancy associated toxemia. Symptoms include high blood pressure, albuminuria and edema. Extreme cases may involve convulsions or coma.

eclosion hormone A peptide hormone secreted by neurosecretory cells in the insect brain that leads to emergence of insect from pupa case or larva from egg. It is also involved in moulting along with other hormones.

ECM *See* extra-cellular matrix.

ecocline Series of biotypes within a species that shows a genetic gradient correlated with a gradual environment gradient.

ecodeme *See* deme

ecoid Skeleton of an erythrocyte.

ecological equivalents Unrelated organisms that occupy similar habitats and resemble each other, as a result from convergent evolution. e.g. sharks (fish) and dolphins (mammals).

ecological isolation Situation where closely related (sometimes virtually indistinguishable) species live in the same territory, but slight differences in their niche causes them to reproduce in isolation1 of the others.

ecological niche

1. Set of conditions (both with respect to the environment and association with other organisms) to which a particular type of organism is suited, to maintain a population in long term.
2. Role or status of an organism in its environment.

ecological pyramid Diagrammatic representation of the amount of biomass/energy/number of organisms found in a particular habitat at ascending trophic levels of a food chain. *See* pyramid of biomass, energy and number.

ecological release Progression in which a species expands its living habitat as well as the resources. It utilises into areas that have a lower density of species in terms of diversity.

ecological succession Process by which the composition and proportion of species within a community changes with time.

ecology Study of the interaction of organisms with each other and with their environment, both living and non-living.

ecomorphology Relation between an individual's ecological role and its form and structural adaptations.

ecophene (ecad) Range of phenotypes produced by one genotype within a particular habitat.

ecophysiology Study of tolerance limits of species which improves understanding with abiotic conditions.

ecospecies A group of ecotypes within a species, between which crossing is possible without loss of fertility in the offspring.

ecosystem Living organisms in a particular place, together with their physical environment, viewed as a functioning system. Organisms are classified in an ecosystem into various trophic levels.

ecosystem approach Method of resource management that acknowledges that the different components of an ecosystem (structure, function and species composition) are interlinked and this factor must be taken into consideration while restoring and protecting the ecosystem's natural balance.

ecosystem services Numerous benefits that are received from ecosystems by people. These can be classified into: (i) **provisioning services** (e.g. food, fuel, etc), (ii) **regulating services** (e.g. purification of air and water), (iii) **cultural services** (e.g. aesthetic experience) and (iv) **supporting services** (e.g. nutrient cycling, photosynthesis).

ecotone Boundary between two adjacent ecological communities of major rank.

ecotoxicology Study of the effects of chemicals on ecosystems.

ecotype Group of organisms within a species, adapted genetically to the combination of specific environmental factors in their habitat, but able to reproduce with other ecotypes belonging to the same species. Differences between ecotypes may be physiological or morphological.

ECT Electroconvulsive therapy.

ectoderm Outer cell layer of cells of an embryo (gastrula). During the course of development, it gives rise to the nervous system, the sensory organs and superficial structures such as the epidermis, hair, pigment cells and mammary glands.

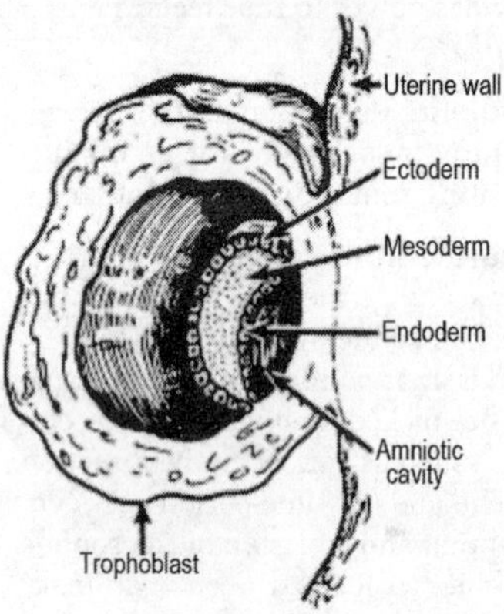

An embryo enclosed by extraembryonic membranes

ectogenesis The development of an organism in an artificial environment outside the body in which its growth occur naturally.

ectoparasite The organism such as a tick, that latches itself to the surface of its host, in order to survive. Ectoparasites have specially adapted mouth-parts for piercing the skin of the host in order to suck blood; their saliva contains an anticoagulant.

ectopia (ectopy) Abnormal placement of an organ or other bodily structure.

ectopic pregnancy (eccyesis) A life threatening condition in which the embryo implants outside the uterine cavity. Abdominal pain and vaginal bleeding are the signs.

ectoplasm Outer layer of the cytoplasm of a living cell. It is dense and is primarily concerned with cell movement.

ectopotomy Surgical removal of an ectopic foetus.

ectoprocta *See* Bryozoa.

ectosarc The outer semi-solid portion of the protoplasm in certain microorganisms (e.g. *Amoeba*).

ectotherm (poikilotherm) An animal that uses environmental temperature and behaviour to regulate its body temperature (commonly termed cold-blooded), e.g. all animals except mammals and birds.

ectothermy Refers to an organism's ability to maintain its body temperature by availing heat from the environment, either by absorbing radiation or through conduction.

ectotoxemia Toxemia resulting from a toxin that has been introduced into the body.

ectotrophic mycorrhiza Mycorrhiza form a network of hyphae around the root and grow into airspaces between the root cells (*See* mycorrhiza).

eczema Inflammatory skin condition characterised by erythema, itching and oozing, scaling lesions.

edaphic factors Environmental factors which are determined by physical, chemical and biological characterstics of the soil. e.g. alkaline soil may be edaphic factor limiting the plant growth.

edema (oedema) Swelling due to accumulation of excessive fluids in tissues or cavities of body. It is because of disturbance in physiological balance of body fluids (between intracellular and extracellular fluids and blood plasma). Fluid is drawn from the blood into the tissue where there is higher osmotic pressure.

edentata (xenarthra) An order of placental mammals, without teeth in the front part of the jaws and without enamel of the teeth; cervical vertebrae usually 6 or 8 unlike all other mammals, e.g. scaly ant eater, Pholidota. It includes the ant-eaters, sloths and armadillos.

edentia Lack of teeth.

edentulous Toothless.

EEG (Electroencephalogram) A graphic record of rhythmical waves of change in electrical potential occurring in vertebrate brain mainly cerebral cortex, which can be detected through intact skull.

effector Body structure (organ or cell) by which an organism generates a physiological response on stimulation by a nerve impulse. In humans, the main effectors are the muscles and glands.

effector neuron A nerve cell that transmits impulses from the central nervous system to an effector in order to bring about a physiological response to any environment change, e.g. motor neuron.

efferent Leading or carrying (impulses, blood, etc.) away from centre of body or organ towards peripheral regions.

egestion Removal of undigested material and associated microorganisms from the gut by the anus. The voided material is *egesta.*

EGF (Epidermal Growth Factor) A molecule important in the regulation of cell growth, proliferation and differentiation of mammalian cell. It acts by binding to the Epidermal Growth Factor Receptor (EGFR).

egg

1. Female sex cell or gamete (ovum).
2. Externally laid, fertilized ova in egg laying animals, e.g. birds and insects.

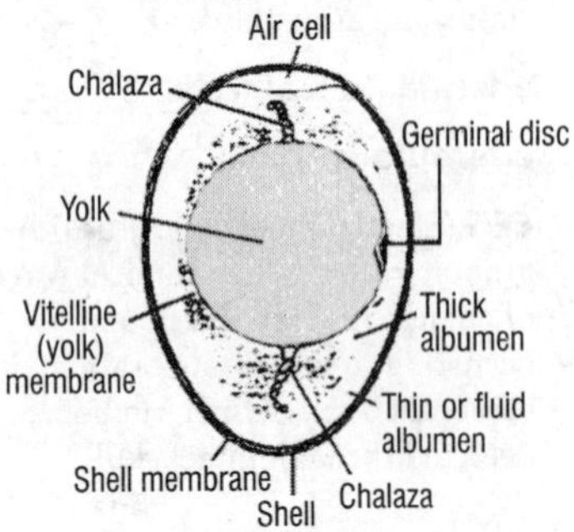

Internal structure of telolecithal egg

egg apparatus Functional female cell and the two synergids in the micropylar end of the embryo sac in angiosperms form egg apparatus.

egg dumping Laying of eggs by a bird in the nest of another, with the purpose of making the host bird to hatch and raise its young.

egg membrane Layer of material that covers an animal egg cell. These are primary membranes (develop in the ovary and cover the egg in addition to plasma membrane), called the vitelline membrane in insects, molluscs, birds and amphibians, the chorion in tunicates and fish and the zona pellucida in mammals. Insects have a second thicker membrane, also called the chorion. Secondary membranes consist of or formed by follicle cells around the egg. Tertiary membranes are secreted by accessory glands, oviducts etc. It includes albumen, shell membrane, egg jelly, etc.

eicosanoid Group of 20-carbon polyunsaturated fatty acids, particularly arachidonic acid and their derivatives. It consists of various biologically important molecules, e.g. prostaglandins, prostacyclins, thromboxanes and leukotrienes.

ejaculation Discharge of fluid from a duct, in particular ejection of semen from the male penis during orgasm.

elaioplast Colourless plastids (leucoplast) that stores oil common in liverworts and monocots.

elaiosome An oil rich seed or fruit body that attracts ants, which act as dispersal agents.

elasmobranchii Subclass of Chondrichthyes. Skin covered with placoid scales; endoskeleton cartilaginous, heterocercal tail, cloaca present, operculum absent and mouth ventral, e.g. sharks, skates and rays.

elastic cartilage A type of cartilage that has a yellow appearance due to numerous elastic fibres (elastin). This cartilage maintains the shape of certain organs, e.g. pinna.

elastin A protein that is the main component of yellow elastic fibres.

elastinase An enzyme that breaks down elastin.

elater

1. Elongated cell with one or more spiral bands of thickening internally. It occurs in numbers among spores in capsules of liverworts.
2. Appendage of spores of horsetails. Assist in spore dispersal.

electric organs Organs of certain fishes that produce electric currents by means of modified muscle cells (electroplate or electroplax cells), e.g. electric ray (*Torpedo*), electric eel (*Electrophorus electricus*).

Electromyogram (EMG) A record of the electrical activity of muscle fibres, obtained by means of electrodes inserted into the muscle, where they detect the action potentials of individual muscle units. The electrical signals are amplified by an apparatus (electromyograph) and displayed on an oscilloscope.

electron microscope A form of microscope that uses a beam of electrons instead of a beam of light to form a large image of a very small object, such as a cell organelle, a virus or a DNA molecule. These microscope uses wave properties of electron through the object held in a vacuum.

Electron Transport System (ETS) A series of coupled oxidation/ reduction reactions wherein electrons are passed quickly from one membrane-bound protein to another. It occurs in the inner mitochondrial membrane and is the final stage of aerobic respiration.

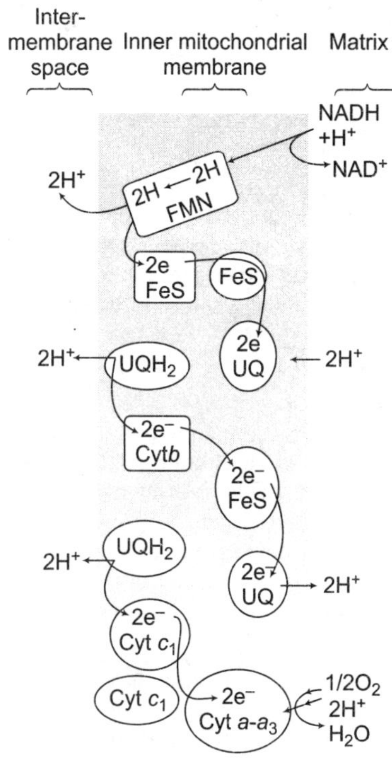

Electro-Olfacto Gram (EOG) Recording of the electrical activity of smell (olfactory) receptors in the nose or other olfactory organ. Electrodes positioned in the olfactory epithelium detect the electrical activity of sensory olfactory cells as they respond to smells (odarants) in the environment. The electrical signals are relayed to an apparatus (electro-olfactograph) that amplifies them and displays them on an oscilloscope.

electroplax A small group of modified muscle cells on the electric organs of certain fishes, e.g. electric eel.

electroreceptor An organ specialised for detecting electric currents, common in marine fishes. It helps them in detecting prey or potential attackers (i.e. electrolocation), e.g. ampulla of Lorenzini, groups of which are embedded in the head of sharks and rays.

elicitor A substance or any stimulus that causes a hypersensitive response in plants, e.g. small proteins or lipids, associated with cell wall of bacteria.

ELISA (Enzyme-Linked Immunosorbent Assay) Used for determining the concentration of proteins or other antigenic substances in biological samples by an enzyme catalysed colour change. Widely used in medical and veterinary diagnostics and in research.

elongation

1. Construction of a protein by the successive addition of individual amino acids (in protein synthesis). *See* translation.
2. Lengthening of any chain-like molecule by means of the successive addition of individual units.

elytra Thickened horny forewings of beetles (Coleoptera), covering and protecting membranous hindwings, when the insect is at rest.

emasculation Removal of stamens from hermaphrodite flowers before they have liberated pollen. Usually, it is as a preliminary step to artificial hybridisation.

Embden Meyerhof Pathway *See* glycolysis.

embedding Sealing of tissue prepared for permanent microscope slides in a solid block of paraffin wax prior to sectioning. Tissue prepared for electron microscopy may be embedded in Araldite and cut with a diamond knife.

embolism Obstruction of a blood vessel.

embryo

1. Earliest stage of development of animal formed after cleavage of the fertilised ovum. In humans, 'embryo' word is used up to the third month of pregnancy. Thereafter, once the basic body shape has formed, the term 'foetus' is employed (in zoology).
2. The organism that develops from the zygote in plants before germination (in botany).

embryo mother cell *See* megaspore mother cell

embryo sac A large oval cell in the nucleus of flowering plants. It corresponds to the female gametophyte of lower plants, although it is very much reduced. Typically, it contains eight nuclei formed by division of the megaspore mother cell. The egg cell which is associated with two synergid cells to form the egg apparatus, is fertilised by a male nucleus and becomes the embryo. The two polar nuclei fuse with a second male nucleus to form a triploid nucleus that gives rise to the endosperm. The three remaining nuclei form the **antipodal cells.**

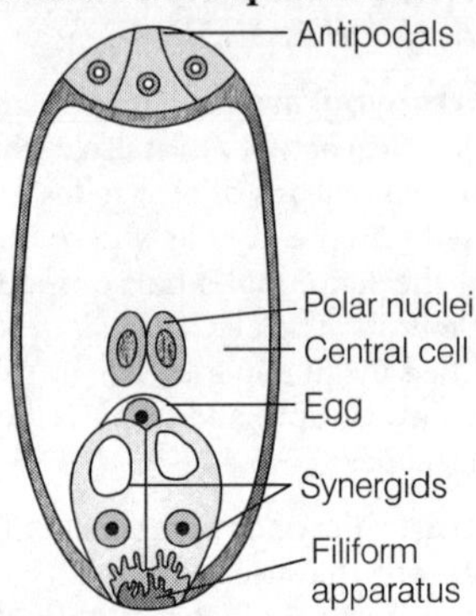

Micropylar end mature embryo sac

embryoblast Inner cell mass within a blastocyst.

embryogenesis Process in which zygote divides and redivides to form a mature embryo, i.e. formation and development of an embryo.

embryology Study of embryo development. Usually restricted to the period from fertilisation to hatching or birth.

embryonic disc A part of mammalian embryo where ectoderm and endoderm are in close contact with each other.

embryonic stem (ES) **cell** Embryonic cell that is capable of differentiating into many or various tissue cell. These are pluripotent and cultures of such cells provide replacement tissues and organs for medical use.

embryonic tissue Undifferentiated tissue with the potential to develop into any of the various specialised tissues.

embryophyte A plant that develops from an embryo and therefore is necessarily multicellular (i.e. a true plant). The term underlined the distinction between plants and algae, which lack embryos.

emigration Movement of individuals out of a population.

emphysema A condition of the lungs characterised by reduction in the surface area available for gas exchange due to deterioration of alveoli. It is due to cigarette smoking and inhalation of other toxic substances (e.g. smoke).

emulsification A process in which fat is broken down into smaller droplets in the duodenum. Bile juice emulsifies fats.

enaima Group which include animals with red blood cells. All chordates belong to this group. Enaima can be of two types—viviparous and oviparous.

enamel Hard white covering of exposed part (crown) of teeth. Formed by epithelium of mouth by ameloblasts, unlike dentine. Consists mainly of crystals of calcium phosphate–carbonate salt, with very organic substances.

enamel organ A soft pulpy tissue developing on the surface of a dental papilla which consists of fibres, fluid on the outer surface of the dentine.

enation Outgrowth from the underside of the veins of a leaf caused by virus infection.

encephalin *See* enkephalin.

Encephalisation Quotient (EQ) Measure of the relative size of the brain of a particular species compared with the expected value for members of the group to which it belongs. It is used to estimate the comparative intelligence of animals, particularly in anthropology.

encephalitis Inflammation of the brain.

encystment Formation of a walled, non-motile body from a swimming spore.

end organ Structure at peripheral end of a nerve fibre, usually either a receptor or a motor end-plate.

end plate A flattened nerve ending that occurs at the junction of a motor axon and a muscle cell. It transmits nerve impulses in a way similar to that of other synapses.

End Plate Potential (EPP) The electric potential measured in millivolts (mV), produced by depolarisation of the end plate at a neuromuscular junction in response to stimulation by the motor neuron. The size of the EPP depends on the amount of neurotransmitter released.

endangered species Entire population of organisms (plant or animal) that faces the immediate risk of extinction due to a steady reduction of their numbers. This may be the outcome of environmental changes, loss of habitat or predation.

endarch xylem Xylem in which oldest elements (protoxylem) are closer to the centre of the axis, characteristic of most stems.

endemic Plant on animal species confined to a particular region.

e

endergonic reaction A chemical reaction requiring energy, e.g. synthesis of organic compounds from water and carbon dioxide by green plants by means of energy absorbed by sunlight.

endocardium The inner surface of the wall of the heart that is continuous with endothelium lining the blood vessels.

endocarp Innermost layer of the pericarp of fruits. It may be hard and stony (e.g. drupe), membranous (e.g. citrus fruit) or indistinguishable from mesocarp (e.g. berries). *See* pericarp.

endochondral ossification Formation of bones by replacement of cartilage.

endocrine glands (ductless glands) These are glands secreting its product (hormone) into the bloodstream or lymphatic system (and not onto an epithelial surface) e.g. ovary.

endocrine system An interconnected system of glands that works with the nervous system to control the activity of various organs. The more diffuse hormonal outputs can take more time to reach effective concentrations and exerts effects over longer time scale.

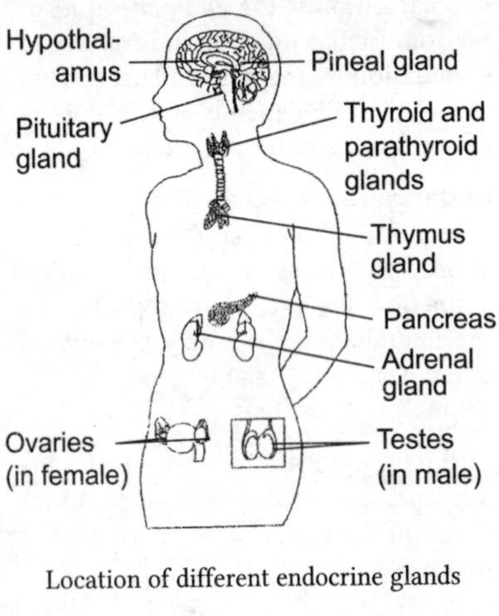

Location of different endocrine glands

endocrinology Study of the structure and function of endocrine glands.

endocytosis The transport of materials in bulk into cells across the plasma membrane. Both phagocytosis and pinocytosis are its forms.

endoderm Inner layer of cells of an embryo. During the course of development, it gives rise to the digestive system and its associated glands (including the liver), the respiratory organs, the urethra, urinary bladder and the vagina. Also called entoderm.

endodermis The innermost layer of the root cortex of a plant, lying immediately outside the vascular tissue. It consists of a single layer of cells. Various modifications its cell walls enable them to regulate the passage of materials both into and out of the vascular system. An endodermis may also be seen in the stems of some plants.

endogamy

1. Inbreeding.
2. Pollination between two flowers of the same plant.
3. Union of two sister gametes, both female.

endogenous Describing a substance, stimulus, organ, etc. that originates from within an organism. e.g. endogenous virus. It is a virus inherited from a parent (because copies of it were present in the parent's genome).

endolithic Microorganisms which live in rock.

endolymph Fluid that is present in the membranous sacs of the vestibular apparatus and cochlear duct of inner ear.

endomembrane system A series of interacting organelles (ER, Golgi bodies, vesicles) within the cytoplasmic matrix. It's main function is to make lipids, enzymes and protein for insertion into cell's membranes or secretion to external environment. The system also destroys toxins, recycle wastes, etc.

endometrium The glandular mucous membrane lining the uterus. It develops to receive and nourish any developing embryo present. If fertilisation does not occur, it either returns to it natural state or in case of humans and some primates, breaks down and is discharged at menstruation.

endomitosis (endoreduplication) Doubling of the chromosomes without the division of nucleus, resulting in polyploidy.

endomixis Reorganisation of nuclear material in Protozoa.

endonuclease Enzyme that catalyses the hydrolysis of internal bonds of polynucleotides (such as DNA or RNA) producing short segment of oligonucleotides. See DNAse.

endoparasite An organism which lives inside the body of its host and benefits from it, e.g. tapeworm lives in human gut.

endopelon Group of algae living and moving within mud.

endopeptidase An enzyme that divides polypeptide chain between amino acids.

endophenotype Part of a complex phenotype.

endophyte Any species, particularly bacteria and fungi that grows inside a plant, symbiotically.

endophyton Group of algae which grows between the cells of other plants.

endoplasm The fluid (less dense) part of cytoplasm which contain most of the cell's structures.

Endoplasmic Reticulum (ER) An organelle present in eukaryotic cells. It is a system of membranes forming tubular channels and flattened sacs (cisternae) running through the cytoplasm and continuous with nuclear envelope. It is the site of protein synthesis and is also concerned with the transport of proteins and lipids within the cell. *See* rough ER, smooth ER.

endopterygote Any winged insect having holometabolous development. Includes beetles, ants, bees, wasps, flies, butterflies and moths.

endoreduplication Repeated duplication of the entire complement of genes within a single nucleus *See* endomitosis.

endorphin (encephalin, enkephalin) A group of peptides produced in the brain and other tissues that are released after injury and have pain-relieving effects, similar to those of opiate alkaloids, such as morphine. Other larger endorphins occur in pituitary, pancreas, adrenal gland and other tissues.

endosarc Inner portion of the protoplasm of certain unicellular organisms (e.g. *Amoeba*).

endoscopic Type of development of a plant embryo in which the inner cell formed by the first division of the zygote develops into the embryo, while the outer cell develops into the suspensor. It is seen in many Pteridophytes and in all seed plants.

endoscopy The study of the interior of hollow organs, such as the intestines and the reproductive organs, using a flexible probe (endoscope), which contains optical fibres for transmitting images onto a screen.

e

endoskeleton An internal skeleton, as opposed to an exoskeleton, such as the bony skeleton of vertebrates or the spicules of a sponge. The function of an endoskeleton is to support the body and in vertebrates it also protects the organs and provides a system of levers on which the muscles can act to produce movement.

endosperm A tissue that surrounds and nourish the embryos of seed plants. It develops from nuclei in the embryo sac and its cells are triploid. In endospermic seeds, it remains and increases in size; in non-endospermic seeds, it disappears as the food is absorbed by the embryo, particularly the cotyledons.

endosperm nucleus Forms when second male nucleus fuses with the primary endosperm nucleus present in the centre of embryo sac.

endospore Resting spore formed within the bacteria under unfavourable conditions. It is surrounded by a thick coat.

endosporic Development of gametophyte (alternation of generations) within spore wall.

endostosis Formation of a bone within a cartilage.

endostyle A ciliated, glandular groove in the ventral wall of pharynx for ciliary feeding in Urochordata, Acrania and Ammocoete larva of lamprey. Produces threads of mucus to which food particles attach and then pass around pharynx and backwards into gullet by ciliary action. Thyroid of adult lamprey develops from the duct of endostyle of Ammocoete larva. Thyroid of vertebrates is therefore considered to be homologous with endostyle.

endosymbiont theory A theory now largely accepted, that eukaryotic organisms evolved from symbiotic associations between bacteria.

endosymbiosis A special form of symbiotic association between cells of two or more different species in which one type of the cell lives inside another.

endothecium

1. Inner mass of cells in the early development of the sporangium of the Bryophyta.
2. Fibrous layer in the wall of an anther with secondary wall thickenings.

endothelin (ET) Any of several related peptides, one of which, **endothelin** 1 (ET-1), is released by the endothelium of blood vessels. It is a potent vasopressor, causing constriction of vessels and raising blood pressure. The other endothelins resemble ET-1 structurally but are synthesised in other tissues.

endothelium Single layer of smooth, thin flattened cells lining heart, blood vessels and lymph vessels in the vertebrates. In capillaries, it is the only layer. It is derived from the mesoderm.

endotherm (homoiotherm) An animal that can generate and maintain heat within its body to regulate its temperature (commonly termed as warm-blooded), e.g. birds and mammals.

endothermic

1. Relating to endotherms.
2. Denoting a chemical reaction that takes heat from its surroundings.

endotoxin Toxic substances formed inside the cells of Gram negative bacteria, e.g. *Salmonella* and released on disintegration of the cell. They are heat-stable polysaccharide-protein complexes causing non-specific effects in their hosts, e.g. fever.

endotrophic mycorrhiza A mycorrhiza, where fungal hyphae grow both between and within the cells of the host root cortex.

endozoon Community of algae that lives within the animals.

energy flow The flow of energy that occurs across a food chain. Trapping of solar energy by photosynthetic organisms and its transformation into chemical energy and passage through various trophic levels of food chain in an ecosystem. It progresses from autotrophic to heterotrophic organisms.

enhancer A site on eukaryotic DNA at which a protein may bind and turn on transcription of a gene which may be either close by or at relative distance on same chromosome.

enkephalin *See* endorphin.

enolase Enzyme catalysing the formation of phosphoenol pyruvic acid from 2-phosphoglyceric acid. It is the ninth and the penultimate step of glycolysis.

enology Scientific study of wine and wine-making.

Ensembl A joint project between EMBL (European Molecular Biology Laboratory), EBI (European Bio-informatics Institute) and the Welcome Trust's Sanger Institute to develop a software system which produces and maintains automatic annotation on selected eukaryotic genomes.

enteritis Inflammation of the intestines, especially of the mucous membranes of the small intestine.

enterocoely Method of coelom formation within pouches of mesoderm budded off from embryonic gut wall. Seen in Echinoderm, hemichordates and brachiopods, etc.

Enteroendocrine Cells (EECs) These are the endocrine cells of the gut mucosa, e.g. humans have G cells in the stomach.

enterogastric reflex A nervous reflex whereby stretching of the wall of the duodenum results in inhibition of gastric motility and reduced rate of emptying of the stomach. It is a feedback mechanism to regulate the rate at which partially digested food (chyme) leaves the stomach and enters the small intestine.

enterokinase (enteropeptidase) Enzyme (peptidase) secreted by vertebrate small intestine which activates trypsinogen into trypsin. It is a component of succus entericus.

enterology Study of the intestinal tract.

enteron (coelenteron) *See* gastrovascular cavity.

enteropathy Any disease of the intestines.

enterorrhaphy Stitching of an intestinal wound or of the intestines to some other structure.

enterorrhexis Rupture of the intestines.

enterotoxins Toxins acting on small intestine and cause loss of fluid which results in diarrhoea.

entoderm *See* endoderm.

entomogenous That grows in or on the body of an insect

entomology Study of insects.

entomophagus Animals that feed chiefly on insects; insectivorous.

entomophily Pollination by insects. Various structures and mechanisms have evolved to attract insects, e.g. colourful petals and nectar.

e

entoprocta An invertebrate phylum of pseudocoelomate and mostly marine animals. The members have a cup-shaped body bearing tentacles and attached to the substrate by means of a stalk. The entire animal is usually less than 10 mm long.

entrez Information retrieval service or browser provided by the National Centre for Biotechnology Information (NCBI), a division of the US National Library of Medicine. It gives access to a collection of NCBI databases, both bibliographic and biomolecular, including PubMed, MEDLINE and OMIM.

envelope A term used for the membranes (two) that surrounds the nucleus.

environment The sum total of conditions, under which an organism lives. It includes the physical, chemical and biological factors.

environmental resistance The sum total of the factors that prevent populations from continually growing and therefore tend to keep populations at constant levels, e.g. predators, disease and a shortage of any requirements such as food, water, etc.

environmental selection Natural selection within a population resulting from the influence exerted by the environment. It leads to a change in the composition of genes within a population.

enzyme A protein catalyst which accelerates a specific reaction within a cell or an organism. They enhance the rate of a chemical reaction without themselves being changed or used and are efficient in very small amounts. Enzymes are highly specific as each of them catalyses only a specific reaction, e.g. maltase acts only on maltose. These can be denatured by heat (thermolabile or heat sensitive). Human enzymes are active at 35-40°C, denature at 50-55°C. The enzymes of bacteria living in hot springs have optimal temperature of 70°C. For a chemical reaction to proceed, the substrate (S) must bind to enzyme at the active site within a cleft or pocket. The substrate is a metabolite upon which an enzyme acts. During conversion of substrate into a product, formation of an enzyme- substrate complex takes place.

$$\text{E (enzyme)} + \text{S (substrate)} \longrightarrow \text{ES complex}$$

This complex formation is a **transient phenomenon**. Thus, during the state, when substrate is bound to the enzyme active site, a new structure of the substrate, called **transition state** structure is formed. After the expected reaction of breaking and making of bond is completed, the product is released from the active site and the substrate gets transformed into the structure of product.

enzyme inhibition *See* inhibition.

enzyme substrate complex Formation of intermediate after the reaction between substrate molecule and active site of the enzyme.

eocene Second oldest epoch of the tertiary period. It lasted from 55.8 ± 0.2 to 33.9 ± 0.1 million years ago. It is characterised by predominance of early hoofed mammals, the ungulates. Many other mammals and birds are also present.

eosin An acid stain that acts as a fluorescent red dye, much used in microscopy; stains collagen, cytoplasm and muscle fibers; most often used as a counterstain to haematoxylin. Haematoxylin and Eosin (H&E) staining is a primary technique in histology.

eosinophil A leukocyte with lobed nucleus and cytoplasmic granules that stains red with eosin. Eosinophils target parasites and play a role in allergic reactions.

ependyma Epithelial lining of the spinal canal and brain cavity.

ependymal cell A type of neuroglial cell that forms the ependema.

ependymitis Inflammation of the ependyma.

ephedrine Alkaloid used in medicine to treat cold, asthma, etc.

ephemeral A plant which completes its life-cycle in a very short time, having several generations in a year. e.g. groundsel.

ephemeroptera An the order of exopterygote insects that comprises the mayflies, in which the adult stage lasts for only a few hours since they have rudimentary mouthparts.
The aquatic nymphs are long lived, which may moult upto 23 times.

ephiphytotic An epidemic among plants.

ephyra A small medusa-like stage in the life cycle of scyphozoans (jellyfish). After breaking off from the scyphistoma, it swims freely and grows into an adult jellyfish.

epibenthos Organisms living below the low-tidemark and above the 100 fathom line.

epiblast Outer layer of cells on an embryo, that gives rise to ectoderm at gastrulation.

epiboly A process seen in the eggs of some amphibians and most teleost during embryonic development. The blastomere or later cleavage product spread out from the animal pole of a blastula to engulf the vegetal part of the egg, including its yolky cells.

epicalyx A ring of bracts or bracteoles below a flower that resemble calyx e.g. in the sweet William flower.

epicarp Outer layer of the fruit wall, that may be tough and hard. *See* pericarp and endocarp.

epicotyl The part of the plumule above the cotyledons.

epidemic Large scale increase in prevalence of a disease.

epidemiology Study of diseases affecting large numbers within a population. These include both epidemics of infectious diseases and also diseases associated with environmental factors and dietary habits (e.g. lung cancer, some forms of heart disease, etc).

epidermis

1. The outermost layer of cells covering a plant. It is overlaid by a cuticle and its functions are principally to protect the plant from injury and to reduce water loss. Some epidermal cells are modified to form guard cells or hairs of various types (in botany).
2. The outermost layer of the cells of the body. In invertebrates the epidermis is normally only one cell thick and is covered by an impermeable cuticle. In vertebrates, the epidermis is the thinner of the two layers of skin may bear hairs, feathers, etc. (In zoology).

epididymis A long narrow coiled tube attached to the surface of the testis in reptiles, birds and mammals. It acts as a temporary storage organ for spermatozoa received from the seminiferous tubules, until their release to the vas deferens and the exterior during mating.

epigamic Any feature tending to attract the opposite sex, e.g. colours of feathers and birds song.

epigeal

1. Seed germination in which cotyledons appear above ground and function as true leaves, e.g. lettuce, tomato.
2. Of animals, inhabiting exposed surface of land as distinct from underground.

epigenesis Process by which a developing organism increases in complexity, which is brought about by interaction between parts of the nuclear genetic program, the organised cytoplasm of the developing egg and the egg's environment.

epigenetic Describing heritable changes that are not the result of changes in DNA sequence. e.g. pramutation, gene imprinting.

epiglottis A cartilaginous flap on ventral wall of pharynx in mammals. It prevents food etc. from entering the trachea.

epigynous Flower in which ovary is inferior and other floral parts are above the ovary, e.g. apple.

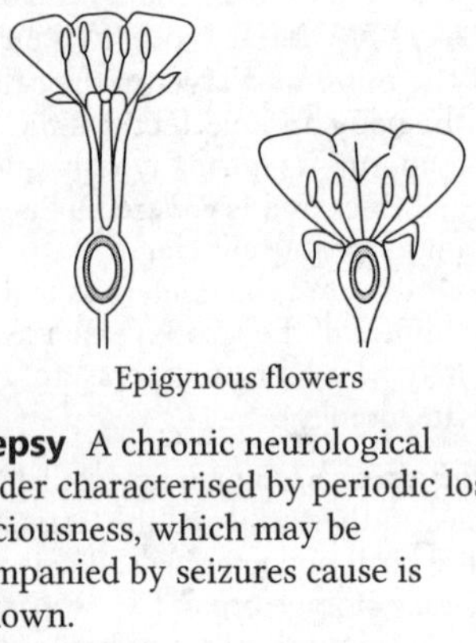

Epigynous flowers

epilepsy A chronic neurological disorder characterised by periodic loss of consciousness, which may be accompanied by seizures cause is unknown.

epilimnion Upper layer of water in a lake.

epilithic Growing on the surface of rocks.

epinasty Rapid growth of the upper surface of plant organ, compared with the lower causing the organ to bend downwards.

epinephrine (adrenaline) A hormone secreted by adrenal medulla, during stress. It causes rise in heart beat, constriction of arterioles, dilation of pupil and a marked increase in metabolic rate. Thus, preparing the body for emergency.

epipelagic zone *See* euphotic zone.

epipelon Community of algae which occurs in water where sediments accumulate on which light penetrates.

epipetalous Stamens borne on the petals with stalks (filaments) more or less fused with petals and appearing to originate from them, e.g. *Petunia.*

epiphysis

1. Ends of a long bone in mammals. It is separated from the bone shaft (diaphysis) by cartilage. New bone is produced on the side of the cartilage facing the diaphysis, while new cartilage is produced on the other side of the cartilage disc. When the bone reaches adult length, the epiphysis merges with the diaphysis.
2. Synonym for pineal gland.

epiphyte A plant that grows upon another plant that is not parasitic to it, but merely using it for physical support, e.g. lichens, mosses, etc.

epiphyton Community of organisms which lives by attaching itself to other plants.

epipsammon Community of algae which lives by attaching itself with sand grains either in fresh water or marine water.

episome Genetic element inside a cell, mainly in bacteria (plasmid) that integrate with chromosome and replicate.

epistasis Masking of action of one gene by another non-allelic (hypostatic) gene, e.g. a gene causing baldness would mask the effect of a hair colour gene.

epithelium A tissue consisting of a sheet (sheets) of cells that covers a surface or lines a cavity. The cells are close together with very little cement substance between them.

Epithelia may be cubical, columnar, ciliated or squamous depending on the shape of cells.

epitope Region of an antigen molecule that is unique to the antigen and therefore, responsible for its specificity in an antigen-antibody reaction. The epitope combines with the complementary region on the antibody molecule.

epitreptic behaviour Behaviour of one individual which causes the approach of a member of the same species.

epizoic (epizoite, epizoon) Growing or living on the external surface of a living animal.

epoch A portion of a geological period.

eponychia (sing. eponychium) In the embryo, the horny structures from which the nails develop.

EPSP *See* excitatory postsynaptic potential.

Epstein-Barr virus DNA containing virus, which is the causal agent of glandular fever and also implicated in the cancer Burkitt's lymphoma and cytomegalovirus.

equatorial plate Area across the centre of the nuclear spindle where the chromosomes become attached at metaphase.

equifinality The ability of certain structures or behaviour patterns to form by more than one route in a developing embryo or young animal.

Equus Genus that includes horses, zebras and asses.

ergocalciferol Chemical name of vitamin-D_2, a form of vitamin-D, (See vitamin-D).

ergosterol A sterol in plants, which upon exposure to ultraviolet light is converted to vitamin-D_2 in animals.

ergot A fungus *Claviceps purpurea* that infects various cereal plants forming compact black masses of branching filaments that replaces many grains of plants. It is the source of medicinally important alkaloids and of lysergic acid.

erythema Abnormal redness of the skin, as a result of dilation of blood vessels (as in sunburn).

erythroblast Cell of myeloid tissue in red bone marrow; precursor of red blood corpuscles. It is nucleated, undergoes mitosis and gradually develops haemoglobin.

erythrocytes (red blood corpuscles) Cells that function in transporting oxygen to the body's tissues due to the red pigment haemoglobin. Mammalian erythrocytes are disc-shaped and lack a nucleus, while vertebrates have an oval and nucleated RBC. In humans, the number of erythrocytes in the blood varies between 4.5 and 5.5 million per cubic millimetre. They survive for about four months and are then destroyed in the spleen and liver.

erythropoiesis The formation of red blood cells (erythrocytes), which occurs in the red bone marrow. The earliest precursor that can be distinguished microscopically is the proerythroblast (from haemopoietic stem cell) → Early erythroblast → Intermediate erythroblast → Late erythroblast (synthesise haemoglobin). In mammals, the nucleus is then forced from the cell, which assumes a biconcave shape and is known as a **reticulocyte**. Reticulocytes are released into the blood and develop into mature erythrocytes within two days.

erythropoietin The hormone that stimulates red blood cell production in response to reduced oxygen availability in the tissues. It is released from the kidney and speeds up the process of erythropoiesis.

e

Escherichia coli **(E. coli)**
A rod-shaped bacterium found in the lower digestive tract. The most common bacterium used in genetic engineering due to lack of pathogenicity, small genome size and easy propagation.

Esophagus A tube through which food and other liquids, etc. pass from the pharynx to the stomach; commonly known as the gullet. It has three coats :

1. An internal or mucous coat.
2. It is a middle or areolar coat.
3. It is the external or muscular coat. The muscular coat is made up of two thick layers of muscle fibres, an internal circular and external longitudinal layer. The areolar coat provides a loose connection between the internal and external coats. The mucous coat is thick and has longitudinal folds that disappear when the passage is distended with food.

essential amino acid Amino acid required in the diet of organisms as it cannot be synthesised. In humans, these amino acids are histidine, lysine, etc.

essential fatty acids Fatty acids that must normally be present in the diet of certain animals, including humans. Essential fatty acids belong to the *n*-3 and *n*-6 classes of polyunsaturated fatty acids and include arachidonic, linoleic and linolenic acids.

essential oil Natural oil with a distinctive scent secreted by the glands of certain aromatic plants. Terpenes are the main constituents. These are extracted from plants by steam distillation, extraction with cold neutral fats or solvents (e.g. alcohol) or pressing and used in perfumes, flavourings and medicine, e.g. citrus oils, flower oils (e.g. rose, jasmine) and oil of cloves.

esthesiology Scientific study of sensations and sense organs.

estivation

1. Way in which sepals and petals are folded in a flower bud before expansion.
2. Period of inactivity in animals during dry hot season.

estradiol Most active estrogen produced by the body. It promotes the proliferation of the endometrium of the womb during the first half of the estrous cycle to prepare the womb for ovulation and is also important in female development. It is metabolised to estrone and then estriol.

estriol A metabolite of estradiol and estrone produced mainly by the placenta during pregnancy, when it occurs in the urine in high concentrations. It has weak estrogenic activity.

estrogen A hormone produced by the ovaries, that prepares the uterus for embryo implantation and is responsible for typical female sexual characteristics.

estrone Naturally occuring weak estrogen hormone produced by the ovary and by peripheral tissues, with action similar to estradiol.

estrus Period of heat in estrous cycle. Ovulation occurs, stimulated by a rise in the secretion of pituitary lutenising hormone and copulation takes place. In some mammals, such as the cat and rabbit, ovulation is delayed until copulation has occurred.

estrus cycle Rhythmic cycle, varying from 5 to 60 days, occurring in sexually mature females of most mammal species in the absence of pregnancy. It can be divided into follicular phase, estrus phase, luteal phase and regression phase. Estrous cycle can occur either throughout the year or only during breeding season. See also menstrual cycle.

etaerio (of fruits) An aggregation of fruits formed from the unfused carpels of a single flower, e.g. an etaerio of drupes in blackberry.

ethmoid bone A sieve-like bone of cranium filled with air cells, forming the roof of the nasal cavity and floor of the anterior portion of the braincase.

ethology Study of animal behaviour.

etiolation An abnormal condition developing in vascular plants in suboptimal light or in darkness. The leaves become small and yellowed and the internodes are abnormally long.

etiology Study of cause of a disease.

etioplast Modified chloroplast formed from proplastids in leaves grown in total darkness. Instead of the normal chloroplast membrane, etioplast contains a highly organised semi-crystalline array of tubular membranes. Radiating from the prolamellar body are single thylakoids. A normal membrane system develops from this body, once the plant is exposed to light.

euapogamy Development of a saprophyte from the tissue of a gametophyte and not from the zygote resulting from sexual fusion.

euapospory Complete failure to form spores.

eubacteria A large and diverse group of bacteria, principally distinguishable from the other major group of prokaryotes (the arachaebacteria) by the presence of a peptidoglycan layer in their cell walls and their membrane lipids contain fatty acids in ester linkage to glycerol.

eucarpic A fungus in which the thallus is differentiated into vegetative and reproductive regions.

eucaryote *See* eukaryote.

euchromatin Chromatin that is uncoiled (not condensed) during interphase. Euchromatin reaches maximal condensation during mitotic metaphase.

eudicot Any dicotyledonous flowering plant whose pollen has three apertures (i.e. triaperturate pollen), through one of which the pollen tube emerges during pollination.

eugenics The study of methods of improving the genetic quality of human populations by the application of selective breeding (positive eugenics). Negative eugenics aims to eliminate harmful genes (e.g. those causing haemophilia and colour blindness) by counselling any prospective parents who are likely to be carriers.

eugenol Colourless compound obtained from essential oils such as cloves. Used in perfumes and as an antiseptic.

euglendia A class of mostly unicellular protists (including *Euglena*) that move by means of undulipodia (flagella). Most euglenids are photosynthetic and inhabit freshwater. They are covered with a proteinaceous pellicle (no cell wall) and some forms are colourless and thus ingest food, since they cannot photosynthesise.

euglenophyta Group of 800 species of algae present in freshwater and marine.

eukarya A domain containing all eukaryotic organisms, including protists, fungi, plants and animals. In three-domain classification system, the other two (prokaryotic) domains are Archaea (comprising the archaebacteria) and Eubacteria.

eukaryote (eucaryote) Organisms both unicellular and multicellular, in which there is a membrane-bounded cell nucleus, that contains the genetic material as well as other well-developed organelles. All organisms other than viruses and prokaryotes (bacteria) are eukaryotes.

eumetazoa *See* metazoa.

eumycota In some earlier classifications, a division or phylum, containing the 'true fungi' as distinct from slime moulds.

euphotic zone (epipelagic zone) Zone near surface of sea (roughly upper 100 m) into which sufficient light penetrates for photosynthesis.

euploidy Normal state in which an organism's chromosome number is an exact multiple of the haploid number, characteristic of the species, e.g. if the haploid number is 7, the euploid number would be 7, 14. 21, 28, etc., and there would be equal numbers of each different chromosome.

eurybathic Bottom-dwelling, in either deep or shallow waters.

euryhaline Able to subsist in a broad range of salinities.

eurypterid A sea scorpion. Eurypterids now extinct, were perhaps the most characteristic marine animals of the Silurian Period, a time when they were abundant and often gigantic (meters long) in size.

eurythermal (eurythermous) Tolerating a broad range of temperatures.

eusocial Organisms that live in complex highly organised groups characterised by division of labour and castes either temporal or morphological. Examples include wolves, humans, all ants, many bees and wasps as well as termites.

Eustachian tube Narrow passage connecting the middle ear with the upper pharynx. It is normally closed, but during swallowing it opens to allow air into the middle ear, which equalises the pressure on each side of the tympanum (ear drum).

eustele The stele arrangement found in most gynosperms and dicots in which the vascular tissue is arranged into a ring of discrete bundles, all contained within a stele ring of endodermis and pericycle.

eutherian (placentalia) A member of the major division–Eutheria of Class–Mammalia, the other two major divisions being marsupials and monotremes. Eutherian mammals nourish their young ones within the uterus by placenta, to more advanced stage of development. The young ones are thus fully protected during their embryonic development and kept at a constant temperature. Placental mammals evolved during the Cretaceous period (about 125 million years ago).

eutrophication Superabundant growth of plants in lakes and rivers resulting from the release of nitrogen and phosphorous-containing fertilisers. This stimulates excessive growth of algae (algal bloom). The accelerated decomposition of these by bacteria, use up oxygen causing fish and other aquatic organisms to die of suffocation, due to oxygen depletion.

evapotranspiration Loss of water to the atmosphere through soil and vegetation.

evocation The ability of experimental stimuli (e.g. chemicals or tissue implants) to cause unspecialised embryonic tissue to develop into specialised tissue.

evolution Production of new forms of life over time. It is believed to have been continuing for atleast past 3000 million years.

evolutionary biology Branch of biology concerned with the study and identification of processes producing new forms of life.

evolutionary transformation series A pair of homologous characters from which one is derived from the other.

evolutionary tree (phylogenetic tree) A diagram depicting the evolutionary relationships between species or groups of organisms. It is usually in the form of dichotomously branching lines to form a tree-like pattern. In some trees, changes in the relative abundance of a lineage are indicated by the width of the branch.

ewe Female sheep.

exact compensation Increase in initial density is balanced by increase in death rate or decrease in birth rate.

exaptation Morphological or physiological character that predisposes an organism to adapt to a changed environment or lifestyle.

exarch A stele in which the metaxylem develops to the inside of the protoxylem.

excavates An assemblage of eukaryotic protists characterised by one or more feeding grooves equipped with undulipodia (flagella) that propel food to the mouth, e.g. Giardia.

excision repair A form of DNA repair found in living cells in which damaged or mismatched bases in one of the two strands of DNA are cut out and replaced with the correct bases using the other strand as a template. It is performed by various enzymes, e.g. DNA glycosylase.

excitation

1. The act or process of stimulating a cell, especially a nerve cell.
2. The response of a cell that has been stimulated.

Excitatory Postsynaptic Potential (EPSP) The electric potential that is generated in a postsynaptic neuron during the transmission of a nerve impulse (see synapse). It is caused by depolarisation of the postsynaptic membrane when a neurotransmitter (such as acetylcholine), released from the presynaptic membrane binds to the postsynaptic membrane. This will induce an action potential in the receiving neuron if the EPSP is large enough.

excretion Process of expelling waste materials (such as H_2O, CO_2 and nitrogenous compounds) from the body or from an individual cell. It plays an important role in maintaining the homeostasis of organism.

excretory system One of the major body systems in animals. It regulates the volume in addition to the molecular and ionic constitution of internal body fluids and eliminates metabolic waste products from the body.

exergonic reaction A chemical reaction in which energy is released, the products have less energy than the starting materials.

exine Outer tough covering of a pollen grain.

exocarp *See* epicarp.

exocrine gland A gland that secretes its contents onto an epithelial surface or into body cavity such as gut (and not into the bloodstream lymphatic system), e.g. mammary gland.

exocytosis The expulsion of macromolecules from a cell to the cell surface within membrane bound vesicles.

exodermis The outermost layer of thickened or suberised cortical cells that sometimes replace the epidermal layer in the older parts of roots, if the epidermal cells have died.

exoenzyme Enzyme that functions outside the cell which produces it, e.g. pepsin.

exogamy Fusion of gametes of organisms that are not closely related, i.e. outbreeding.

exogenesis One of the two major hypothesis concerning the origin of life on Earth. It proposes that life on this planet arrived from elsewhere in the universe, presumably in the form of frozen particles containing prokaryotes and perhap, other simple organisms. This hypothesis is sometimes called panspermia, but this latter term, strictly speaking refers to the claim that the "seeds of life" are spread throughout the universe and begin to develop wherever suitable environment exist. The other major hypothesis opposed to exogenesis is primordial- abiogenesis.

exogenote DNA that a bacterial cell has taken up through one of its sexual processes.

exogenous Substance which originate outside an organism.

exon A protein-coding portion of a gene. Most genes in eukaryotes consist of exon interrupted by introns (non-coding sequences).

exon shuffling Recombination or rearrangement of exons encoding functional domains of proteins by introns. It permits rapid evolution of proteins with novel permutations of functional group.

exonuclease Enzyme that catalyses the hydrolysis of terminal linkages of DNA and RNA.

exopeptidase Protein-digesting enzyme that cleaves amino acids from the ends of a polypeptide chain. e.g. carboxypeptidase in the small intestine.

exopodite *See* biramous appendage.

exopterygote Any winged insect showing hemimetabolous development, i.e. the eggs hatch into young (called nymphs) that resemble the adults but lack wings; these develop gradually passing through a series of stages called instars, e.g. mayflies, grasshoppers, etc.

exoscopic Type of development of a plant embryo in which the apex of the sporophyte develops from the outer cell formed by the zygote's first division, e.g. whisk ferns and bryophytes.

exoskeleton A hard case on the external surface of many invertebrates, such as the shell of a mollusc or the cuticle of an insect.

exosporic Formation of spores outside the spore producing organ, e.g. basidiospores.

exosporium *See* exine.

exotoxins A group of heat-labile proteinaceous toxic substances produced by bacteria inside their hosts. These cause diseases such as tetanus.

experiment A manipulation done on the natural world in order to test a hypothesis. Experiments usually involve comparing a control group with an experimental group where the two groups are as similar as possible except that the experimental group receives a treatment not given to the control group.

experimental taxonomy (biosystematics) The experimental study of diversity, especially at the species level.

expiration (exhalation) Process by which gas is expelled from the lungs. In mammals, the volume of the thoracic cavity is reduced by contraction of the internal intercostal muscles and relaxation of the muscles of the diaphragm, assisted by upward pressure of the abdominal organs.

As a result, pressure in the lungs exceeds atmospheric pressure and gas flows out of the lungs, allowing the pressure to equalize. *Compare* inspiration.

expiratory centres Subcentres within the ventilation centre that control expiration. See ventilation centre.

explantation Culture of the isolated tissues of adults or embryos in an artificial environment, usually *in vitro*, for the maintenance, growth and for differentiation.

Expressed Sequence Gag (EST) Unique DNA sequence (200-400 bp), derived from *c*DNA library. It can be mapped by a combination of genetic mapping procedures to a unique locus in the genome and helps in identifying that gene locus.

expression system Host cell which contains a gene, so that it can be expressed.

expression vector Vector used in genetic engineering that enables a particular gene to be not only cloned but also expressed in a host cell.

extensor Muscle that causes a limb to extend.

exteroceptor Any receptor that detects external stimuli, e.g. thermoreceptors in the skin, which monitor the temperature of the external environment.

extracellular Located or occurring outside the cell.

Extra cellular Matrix (ECM) The viscous watery fluid that surrounds cells in animal tissues. Secreted by the cells themselves, it is the medium through which they receive materials (e.g. nutrients, hormones) from elsewhere in the body and *via* which they communicate with other cells.

extrachromosomal DNA DNA in eukaryotic cells that occur within self-perpetuating organelles in cytoplasm (mitochondria, chloroplasts) and is responsible for cytoplasmic inheritance.

extrachromosomal inheritance Inheritance of genetic factors that does not form part of a chromosome.

extraembryonic membrane (embryonic membranes) The structures and membranes developed by embryos for the purpose of nutrition or protection and not directly involved in development of embryonic structures. The membranes of human embryos are known as foetal membranes.

extrafusal muscle fibres Standard skeletal muscle fibres that make up large mass of skeletal (striated) muscle and allow for skeletal movement. They attach to bones by tendons.

extranuclear genes Genes included in the DNA present in organelles other than the nucleus, such as the mitochondria and chloroplasts, some of which code for the synthesis of proteins. The DNA of these organelles is inherited by the offspring *via* the cytoplasm of the gametes (see cytoplasmic inheritance).

extremophile An organism that thrives under extreme conditions, such as extreme heat or salinity, e.g. certain archaebacteria.

extrorse Anthers in which dehiscence lines are to the outside of the flower, promoting cross-pollination.

eye Receptor organ for light perception or the organ of sight. In invertebrates, it is either a simple scattering of light-sensitive pigment spots in the general epithelium but often forming an optic cup of receptor cells with screening pigment cells (ocellus).

Higher molluscs and arthropods have compound eyes that is more advanced. In vertebrates, it consists of a complex spherical structure filled with fluid. Light is refracted by the cornea through the pupil in the iris and onto the lens, which focuses images onto the retina. These images are received by light-sensitive cells in the retina which transmit impulses to the brain *via* the optic nerve.

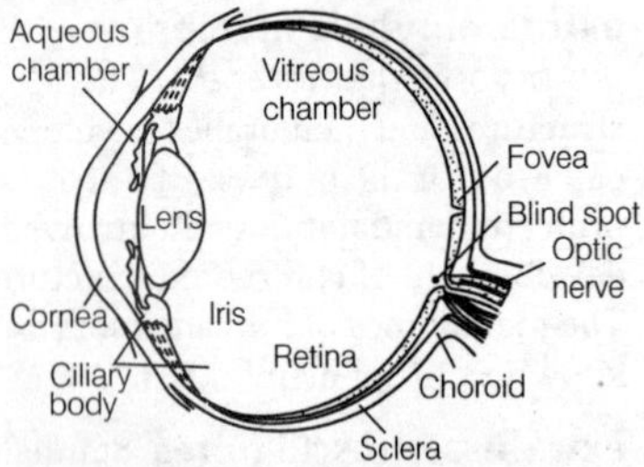

Diagram showing different parts of an eye

eye muscles

1. Extrinsic (outside the eyeball). The set of six muscles which move each eyeball, characteristic of all vertebrates with functional eyes. Consist of anterior (i.e. near to nose) pair of oblique muscles and four more posterior (i.e. nearer to ears), rectus muscles. Supplied by third, fourth and sixth cranial nerves.
2. Intrinsic (inside the eyeball).

eye spot (stigma)

1. Light-sensing organelles present in certain protoctists and invertebrates animals. It contain globules of orange or red carotenoid pigments in unicellular and colonial algae and their gametes and zoospores. It controls locomotion, ensuring optimum light conditions for photosynthesis.
2. Spot of pigment found in some lower animals, e.g. jellyfish.
3. Eye-like marking on the wings of certain insects which is revealed in startle displays.

eye tooth A canine tooth in the upper jaw.

eyepiece (ocular) The lens or system of lenses in an optical instrument that is nearest to the eye. It usually produces a magnified image of the previous image formed by the instrument.

F

F_1 (first filial generation) The generation produced by an initial cross between two distinct types of parents.

F_2 (second filial generation) The generation produced by interbreeding of individuals belonging to the F_1-generation. In general, the F_n-generation is produced by interbreeding of individuals belonging to the F_{n-1} generation.

fab fragment One of the fragments produced by partial digestion of immunoglobulin with the help of proteolytic enzyme papain. It corresponds to an arm of the Y-shaped molecule and consists of an entire light chain paired with part of a heavy chain, complete with the antigen-binding site.

F(ab')$_2$ fragment One of the fragments produced by partial digestion of immunoglobulin with the help of proteolytic enzyme pepsin. It consists of both arms of the Y-shaped molecule complete with antigen-binding sites and the disulphide bond of the hinge region that joins them together.

fabry's disease Lysosomal storage disease inherited in an X-linked manner. It is caused by the deficiency of alpha-galactosidase and result in an accumulation of globotriaosylceramide in the renal and cardiovascular system. The disease is characterised by skin lesions, renal failure and disturbances of the cardiovascular, gastrointestinal and central nervous system.

facial nerve Seventh cranial nerve in vertebrates. It is partly sensory but mainly motor in mammals going to the superficial muscles of face, salivary glands and to taste buds of front of tongue.

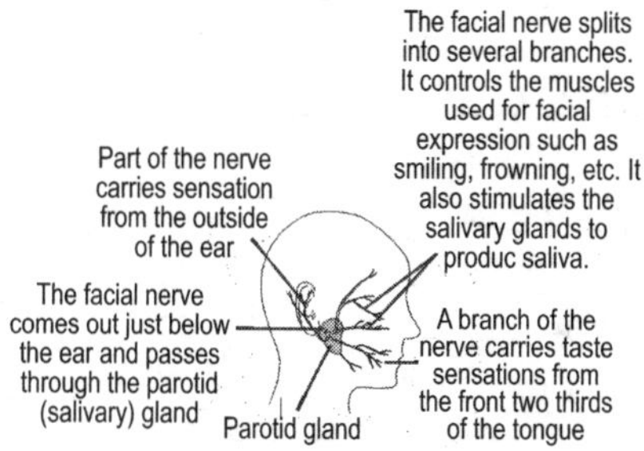

The facial nerve

facial shield (frontal shield) Some birds such as the Eurasian coot, have a hard plate on their forehead called a facial shield, which serves as a display ornament.

facilitated diffusion Transport of molecules across a plasma membrane with the assistance of carrier proteins located within the plasma membrane. No energy is expended in this process.

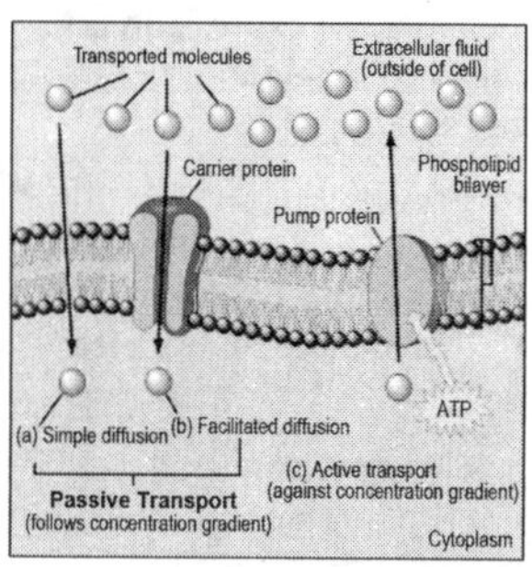

Figure (a) Simple diffusion
(b) Facilitated diffusion
(c) Active transport

facilitation

1. Increase in response of a post synaptic membrane to successive stimuli, each one making the membrane more responsive to the next.
2. The phenomenon in which the presence of one species increases the likelihood or speed of colonisation by a second species. The first species brings about some change in the environment that make it more favourable for the second.

faciocervical Pertaining to the face and neck.

faciolingual Pertaining to the face and tongue.

facioplasty Facial plastic surgery.

facultative Indicating the ability to live under altered conditions or to behave adaptively under altered circumstances.

facultative anaerobe An organism that engages in aerobic respiration when oxygen is present, but relies on fermentation in the absence of oxygen.

facultative gene

1. A gene transcribed only when needed or when a cell receives a signal from its surroundings or,
2. A gene whose expression is influenced by environmental factors.

FAD (Flavin Adenine Dinucleotide) A derivative of riboflavin that is a coenzyme in various biochemical reactions, e.g. in electron transfer reactions.

faeces (only used in plural) Excrement (body waste ejected from the intestines). It consists of undigested residue to food, epithelium, intestinal mucus, bacteria and waste material from blood.

falciparum malaria The most dangerous type of malaria, caused by one of the genus *Plasmodium* (*P. falciparum*) and carried by infected mosquitoes of the genus *Anopheles.* This parasite uses red blood cells to complete its reproductive cycle. Common symptoms of an attack include high fever, chills, sweat and bodyache.

fallopian tubes (oviduct) Ducts by which ova pass from the ovaries to the uterus in female mammals.

fallout The phenomenon in which birds need to land while migrating in areas they would not normally inhabit, due to harsh weather.

false berry Refers to fleshy fruits which looks like a berry. These are formed from an inferior ovary plus the hypanthium of the floral tube that surrounds the ovary e.g. cranberry.

false fruit A fruit formed from other parts of the flower as well as the gynoecium, e.g. apple formed from thalamus.

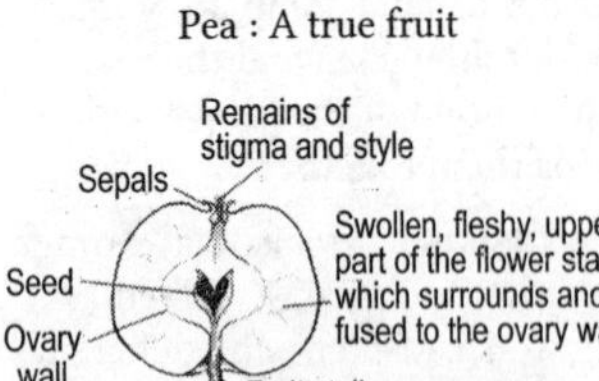

Pea : A true fruit

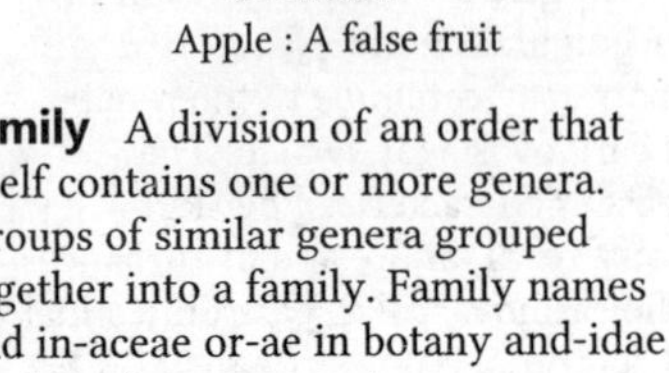

Apple : A false fruit

family A division of an order that itself contains one or more genera. Groups of similar genera grouped together into a family. Family names end in-aceae or-ae in botany and-idae in zoology (in taxonomy).

fangs Long, pointed tooth located in front of the mouth. In mammals, fangs are the canine teeth and are used for tearing flesh. In snakes, they are used to inject venom into the victim.

fas signal pathway An intracellular pathway that initiates the process of programmed cell death (apoptosis). Fas is a cell surface receptor, that binds to fas ligand changing shape of the fas protein.

fascia A sheet of fibrous connective tissue, e.g. layer of adipose tissue under human dermis and sheet of rough connective tissue around muscles.

fasciation Coalescing of branches, stems, etc. to form abnormal thick growths.

fascicle

1. A small bundle or cluster, especially of nerve, tendon or muscle fibres.
2. *See* vascular bundle.

fascicular cambium A strand of cambium arising between the xylem and phloem of a vascular bundle.

fasciculation

1. An arrangement of fasciculi.
2. Small local involuntary muscular contraction. It represents spontaneous discharge of a number of fibres innervated by a single motor nerve filament.

fast green A green dye frequently used to stain plant tissue (cellulose, cytoplasm and collagen) in optical microscopy.

fat A triacylglycerol ester of three fatty acids and glycerol that are solid below 20°C. During hydrolysis, fats are broken down into glycerol and fatty acids, a process that provides energy to cellular tissues. Fats occur widely in plants and animals as a means of storing food energy, having twice the calorific value of carbohydrates.

fat body

1. Abdominal organ in many amphibians and lizards containing adipose tissue, used during hibernation.
2. Diffuse tissue between organs, storing fat, proteins and glycogen and uric acid occassionally (in insects).

fat cell (adipocyte) Any of the cells of adipose tissue, in which fats (triglycerides) are stored. Fat cells contain enzymes (lipases) that can breakdown fat into glycerol and fatty acids, which can be transported in the blood to the liver, where they are used in fatty acid oxidation.

fate map A map of the probable fates of the various regions of the blastula surface of animals with indeterminate development, usually shown as colours. The three major germ layers occupy large discrete areas of the fate maps of vertebrates.

fatigue A decline in the level of response of tissue (such as muscle), cells, etc. to nervous stimulation, which occurs after prolonged and continued stimulation of these structures.

fatty acid Any molecule produced by replacing one of the hydrogen atoms of a hydrocarbon with a carboxyl group, e.g. palmitic acid (saturated fatty acid), oleic acid (unsaturated fatty acids).

fatty acid oxidation (β-oxidation) The metabolic pathway in which fats are metabolised to release energy. Fatty acid oxidation acts as a major source of energy after the animal's carbohydrate resources are exhausted, e.g. during starvation. Fatty acid oxidation occurs chiefly in mitochondria in animal cells and in peroxisomes in plant cells.

fauces Passage from mouth to pharynx.

fauna The animal life as opposed to plant life (flora) of an era, geographic region or biome.

faunal region A region of the earth with a distinct and characteristic assemblage of animal taxa. There are six commonly recognised faunal regions used in zoogeography; the **Ethiopian region**, Oriental region, Palaearctic region, Nearctic region, Neotropical region and Australian region.

Fc fragment See fab fragment

feather Epidermal outgrowth found in birds. It is used for flight, to provide heat insulation and to give the body its streamline shape.

febrifuge A fever-reducing agent.

fecundity The organism's ability to reproduce. In higher animals, it refers to a female's potential capacity to reproduce, based on the number of gametes (eggs), seed set or asexual propagules.

feedback *See* homeostasis.

feedback inhibition A condition in which a process producing a particular product tends to occur at a slower rate when that product is present at high concentration.

feeding *See* ingestion.

feeding brushes The group of bristles located on either side of the mouth in mosquito larvae which help in swallowing tiny aquatic vegetation.

female

1. The gamete (sex cell), that during sexual reproduction fuses with a male gamete in the process of fertilisation. Female gametes are generally larger than the male gametes and are usually immotile (see oosphere; ovum).
2. An individual or organism whose reproductive organs produce only female gametes.

female haploidy Conditions in which females arise from fertilised eggs and have two parents.

female reproductive system The system for reproduction in females. It consists of pair of ovaries, uterine tube, uterus, vagina and external genitalia lia and mammary glands.

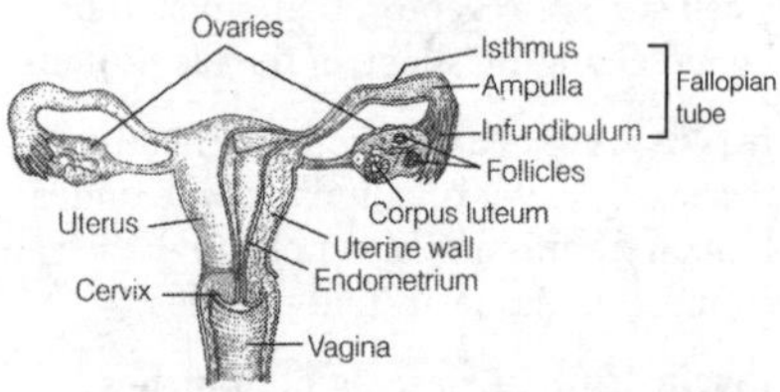

Female reproductive system (sectional view)

femoral Of or relating to the thigh or the femur (thigh bone). For example, the **femoral artery** runs down the front of the thigh.

femorotibial Pertaining to the femur and the tibia bones.

femur

1. Thigh-bone of terrestrial vertebrates. It articulates with the pelvic girdle at the hip joint and with tibia at the other. In humans, it is the largest bone of the body.
2. In insects, the third segment from the base of the leg.

fenestra Either of the two delicate membranes between the middle ear and the inner ear. The upper membrane is the **fenestra ovalis** (*See* oval window); the lower membrane is the **fenestra rotunda** (*See* round window).

fenestration The condition of tissues or other structures within an organism of being perforated by slits (like windows).

feral animal Any domesticated animal living in a wild state.

fermentation Enzymatic and anaerobic catabolic process by microorganisms to yield simpler organic products.

The term is often used synonymously with anaerobic respiration, e.g. alcohol production by yeasts (alcoholic fermentation) and lactic acid production by animal cells (e.g. muscles in humans).

fern A pteridophyte. There are about 10,000 species from moss-like plants to trees. They mostly have large leaves called fronds, on which spores are formed. Ferns reproduce from these spores.

Ferredoxin (Fd) (Iron-sulphur proteins) A carrier molecule mediating electron transfer. There are many different types of ferredoxins that participate in a wide variety of metabolic reactions.

fertilisation (syngamy) Union of a male and a female gamete to form a zygote. It is an essential process of sexual reproduction. It can be external (typically in aquatic forms) or internal (typically as an adaptation to terrestrial life).

fertiliser Chemicals added to soil to increase yield of crop plants, more commonly nitrogen, phosphorus and potassium.

fertility

1. Capability of an organism to reproduce. In plants and animals, it is the number of fertilised eggs produced in a given time.
2. The relative ability of a soil to support the growth of plants.

ferretin Iron storing protein (e.g. in spleen, liver and bone marrow). The iron (Fe^{3+}) is made available when required for haemoglobin synthesis, being transferred by transferrin.

fertilizin A chemical substance of egg cortex, necessary for normal fertilization.

fetal Of or pertaining to the fetus.

fetal membranes
See extraembryonic membrane.

fetus (foetus) An embryo of a mammal or other viviparous vertebrate, after the embryonic stage and before birth. In humans, the embryo from eight weeks to birth is called a fetus.

Feulgen's test Divised by German chemist R Feulgen (1884-1955). A histochemical test to observe the distribution of DNA in chromosome of dividing cells.

F-factor (short for fertility factor also sex factor or F-plasmid) A plasmid allowing bacteria to form sex pili and therefore, to carry out DNA transfer during conjugation.

fibre

1. Elongated, tapering, thick-walled sclerenchyma cell of vascular plants, whose walls may or may not be lignified or possess a living protoplast (in botany).
2. Thin elongated cell (e.g. nerve fibre, muscle fibre) or characteristic structure adopted by molecules of collagen, etc., (in zoology).
3. The part of food that cannot be digested and absorbed to produce energy. Dietary fibre falls into four groups: cellulose, hemicelluloses, lignins and pectins (dietary fibre).

fibril Any small fibre or thread-like structure.

fibrin An insoluble protein network formed from fibrinogen into fibrin when the blood clots.

fibrinogen (factor I) The precursor of fibrin and structural element of blood clots present in blood.

fibrinolysis Destruction of blood clots as a result of dissolution of fibrin by the enzyme plasmin (fibrinolysin).

fibroblast A cell present in dermis that secretes the necessary constituents of extracellular fibres.

Fibroblast Growth Factor (FGF) Any of a family of proteins, characterised by a 120 amino acid core region in the molecule, that influence cell division, migration or differentiation in a wide range of tissues. They are crucial in embryonic development, the formation of blood vessels (angiogenesis), the differentiation of nerve cells and wound healing.

fibrocartilage A cartilage in which bundles of collagen fibres predominate. Found in areas as the intervertebral discs. It is stronger and less elastic.

fibronectin A cell adhesion molecule that guide cell migration in embryos.

fibrous protein (fibroin) Proteins consisting of long coiled strands or flat sheets, which confer strength and elasticity. Generally insoluble in water, e.g. keratin, collagen.

fibrous root system It is a root system made up of many thread-like roots of more or less equal length as in grasses.

fibula A long, slender bone located on the lateral side on the lower hindlimb of tetrapods. In humans-the calf bone.

field capacity Point at which the soil contains all the water it can hold by capillary and chemical attraction. Any more water added to soil at field capacity would drain away by gravity.

filament

1. Stalk of the stamen in flower (in botany).
2. A long, slender hair-like structure such as barbs (in zoology).
3. The vegetative body to the filamentous algae.

filariasis A parasitic disease of the tropics caused by filaria worm *Wuchereria bancrofti* (a nematode) transmitted by *Culex* female mosquito which serves as intermediate host. High fever, lymphangitis and swelling of lymph nodes are the symptoms.

filicinophyta (Pterophyta; filicales) A phylum of seedless vascular plants comprising the ferm.

filiferous Thread-bearing.

filiform Thread or filament-shaped.

filoplumes Hair-like feathers consisting of a bare shaft (rachis) tipped with a few barbs that are not held together by barbules.

filopodium Slender extension of the cell membrane (50 pm long and 0.1 μm wide), protruding from the surface of migrating cells or growing nerve axons.

filter feeder Organisms that feed by sieving water for food particles, with the help of special filtering structures in their mouths. Clams, sponges, krill and baleen whales use this method.

fimbria (pl. fimbriae)

1. Finger-like projections which surround the ovarian end of uterine tube.
2. Small bristle-like surface structures of the bacteria that do not play role in motility. In some cases, it help to attach the bacteria to rocks in streams and also to the host tissues.

fin The locomotory organs of aquatic vertebrates. In fish, there are typically one or more dorsal and ventral fins (sometimes continuous), whose function is balance; a **caudal fin** around the tail, which is the main propulsive organ and two paired fins; the **pectoral fins** attached to the pectoral (shoulder) girdle and the **pelvic fins** attached to the pelvic (hip) girdle, which are used in steering. These paired fins are homologous with the limbs of tetrapods. Fins are strengthened by a number of flexible fin rays, which may be cartilaginous, bony and jointed, horny or fibrous and jointed.

finger domain A finger-shaped structure produced in a protein, when a series of the constituent amino acids combines with a metal atom. Finger domains are often found repeated in transcription factors.

fingerprinting *See* DNA fingerprinting

fir An evergreen conifer found in the Northern Hemisphere, they have flat needles and upright cones, which fall to pieces when their seeds are ripe.

first convoluted tubule
See proximal convoluted tubule.

fish *See* Chondrichthyes (cartilaginous fish); Osteichthyes (bony fish); pisces.

fission Asexual reproduction by sub-division of the parent body into two or more parts which are approximately equal and become independent individuals. Common form of reproduction in protozoans. It can be binary (forming two) or multiple (more than two) fission.

fission track dating Method of estimating age of glass and other mineral objects by observing the tracks made in them by the fission fragments of radioactive uranium nuclei that they contain.

fissipedia Suborder of Carnivora, that includes all land carnivorous mammals having pointed and large canines.

fixation

1. First step in making permanent preparations of organisms, tissues, etc. for study. Solution of formaldehyde and osmium tetraoxide is often used (in microscropy).
2. Spread of an allele of a gene through a population until it comes to occupy 100% of available sites (of genes).
3. The conversion of an inorganic source of element to organic source, e.g. C-fixation occurs in photosynthesis; N-fixation occurs in soil by bacteria, algae, etc.

Fixed Action Pattern (FAP) A sequence of stereotyped behaviour in response to a stimulus in the environment. These are movements that once started, are performed in the same way, unaffected by external stimuli.

flaccid A soft or relaxed condition of a cell, usually due to lack of water within the cell.

flagellar apparatus Whole complex of flagellar basal bodies and microtubular roots present in a flagellate cell.

flagellar canal A narrow infolding of the plasmalemma from the floor of which flagellum arises and emerges to the exterior.

flagellar roots Root-like structure consisting of microtubules that are connected to the flagella present inside the cell.

flagellum A movable slender process arising from a cell. An organ of locomotion of certain one-celled organisms occurring singly or in groups of two or four.

flame cells (solenocyte) Hollow cup-shaped cells with a bunch of cilia which work in the lumen. Flame cells are usually connected together by canals which ultimately open to the exterior. Occurs in several groups– Platyhelminthes, Rotifera, Annelida, larvae of Molluscs, etc. These are also called excretory cells.

flapping flight Type of flight in birds, in which wings are moved up and down, i.e. upstroke and downstroke movements. It provides thrust that counteracts drag and increases speed.

flatworms *See* Platyhelminthes.

flavin A derivative of riboflavin occurring in the flavoproteins, e.g. FAD and FMN.

flavin adenine dinucleotide *See* FAD.

flavonoid One of a group of naturally occurring phenolic compounds many of which are plant pigments. They include the anthocyanins, **flavonols** and **flavones**. Plant- derived flavonoids in foods are strong antioxidants in their natural state, but are poorly absorbed from the intestine and rapidly metabolised and excreted.

flavoproteins Dehydrogenases containing flavin as prosthetic group. They play a key role in the cellular respiration of plants and animals.

fleas *See* Siphonaptera.

fledge Stage in a young bird's life, when the feathers and wing muscles are developed and the bird is capable of flight. Alternately, it is also used to describe the act of raising chicks to a fully grown state by the parents.

fledgling A young bird that has recently fledged, but is still being fed and cared for, by the parents.

flexor A muscle that causes limb to bend by bringing the two parts of limb together, e.g. biceps.

flies *See* diptera.

flocculation Aggregation of soil particles into crumbs. Flocculation of clay particles in soil can be induced by the addition of calcium salts. It is also often observed in cultures of bacterial and yeast cells.

flora

1. Plant life, as opposed to animal life (fauna) of an era, geographic region or biome.
2. A gut flora: microorganisms living in the digestive tract.

floral apex Apical meristem that develops into a flower or inflorescence.

floral diagram Diagram illustrating relative positions and number of parts in each of the sets of organs comprising a flower.

A flower and its floral diagram

floral formula A summary of the flower characteristics K–calyx, C–corolla, P-perianth, A–androecium (stamens), G-gynoecium (carpels). The number following them represents the number of parts in each whorl. Parentheses means that the parts are fused, e.g. C(7) means 7 fused petals. If they are in separate groups or whorls, the number is split, e.g. C2 + 2 indicates a corolla of two whorls each of two petals). The symbol G has a line above it to indicate an inferior ovary and a line beneath it to indicate a superior ovary.
e.g. the buttercup flower has the following floral formula:

$$\oplus\ K5\ C5\ A\infty\ \underline{G}\infty$$

i.e. it has a calyx of five sepals, a corolla of five petals, an androecium of many stamens and a superior ovary consisting of many carpels.

floral tube Tube formed by fusion of basal parts of sepals, petals and stamens.

florigen Hypothetical plant hormone suggested to explain transmission of flowering stimulus (which is in response to photoperiod) from leaf, where it is perceived to growing point. Attempts to isolate florigen have have so far been unsuccessful.

floristic equilibrium Composition of flora, when there is no change either internally or externally.

floristics Study of composition of vegetation in terms of the species present in it over geographic areas.

flow cytometry Analysis of cells or chromosomes by evaluating their light absorption or fluorescent properties as they stream through a laser beam. A machine that carries out this process is known as a Fluorescence-Activated Cell Sorter (FACS).

flow karyotyping Use of flow cytometry to analyse and separate chromosomes.

flower A reproductive part of an angiosperm. A perfect flower consists of an axis or receptacle bearing sepals, petals and stamens. The gynoecium is borne either above receptable or enclosed within it. Flower structures are extremely variable and show a number of adaptations to promote flowering and seed dispersal.

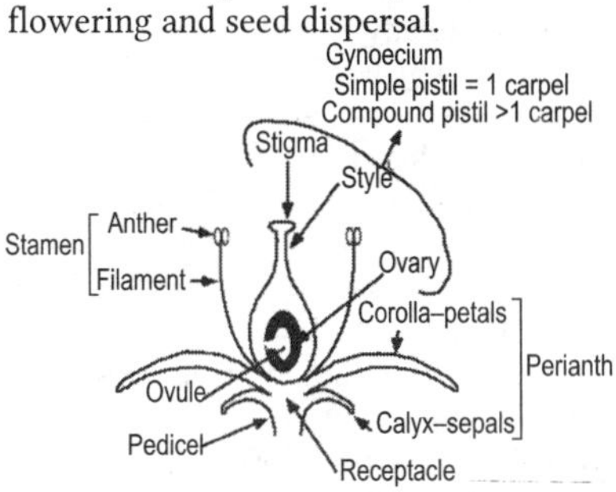

Structure of a flower

flowering Production of flowers by flowering.

fluid mosaic model A widely accepted model proposed by SJ Singer and GL Nicholson in the 1970s. It treats the cell membrane as a phospholipid bilayer within which is embedded proteins randomly with various functions. According to this, the quasi-fluid nature of lipid enables lateral movement of proteins within the overall bilayer of cell membrane. This is measured as its fluidity.

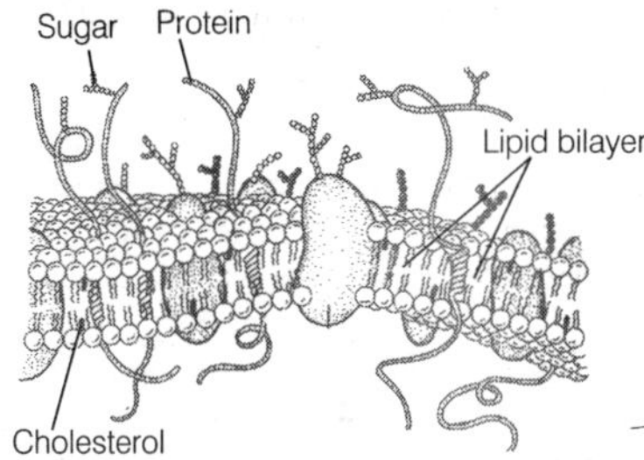

Fluid mosaic model of plasma membrane

flukes *See* trematoda.

Fluorescence In Situ Hybridisation (FISH) A technique in which a DNA probe labelled with a fluorescent dye, base-pairs (hybridises) with the complementary base sequence of a target nucleotide. It is used in genetic mapping for locating specific genes within a chromosome set and also for locating particular messenger RNAs (*m*RNAs) within cells. If just a partial sequence of a gene is known, a DNA probe for the gene can be constructed.

fluorescent antibody technique In this, cells or tissues are treated with antibody that indicates the presence and location of the antigen with which it combines.

fluoridation The process of adding very small amounts of fluorine salts (e.g. sodium fluoride, NaF) to drinking water to prevent tooth decay.

FMN (Flavin Mononucleotide) A flavin; derivative of riboflavin, that is a coenzyme in electron-transfer reactions.

foetal ejection reflex Signals for parturition which originate from the fully developed foetus and the placenta which induce mild uterine contractions.

This initiates the release of oxytocin from posterior pituitary, thus causing stronger contractions.

foetus *See* fetus.

foliar Of, pertaining to or *via* leaf.

folic acid (also folacin pteroylglutamic acid) A B-vitamin (B_9) present in food such as liver, green vegetables and yeast. A lack of folic acid in the diet causes anaemia because it is essential in the production of erythrocytes. In the liver, it is converted to dihydrofolic acid and then into tetrahydrofolate.

follicle

1. Any small, bulbous group of cells forming a cavity, e.g. follicles within the ovary contain developing ova (in zoology).
2. A dry dehiscent fruit formed from one carpel that splits along one edge to release its seed, e.g. columbine.

Follicle-Stimulating Hormone (FSH) A gonadotropin hormone secreted by the gonadotropin cells of the anterior pituitary gland in vertebrates. Its function is much broader than its name suggests. Although FSH does stimulate the production of ovarian follicles, it also plays an important role in growth and development of the sexual organs and spermatogenesis (in males).

follicular phase Phase of menstrual cycle in which the follicle grows, produces certain hormones which reactivate uterine wall to become thickened and supplies blood. Prepares uterus to receive fertilised egg.

follistatins Family of glycosylated proteins that modulates pituitary FSH.

fomentation Application of a hot, moist substance for the relief of pain or discomfort.

fomes

1. Any substance that transmits infection.
2. A genus of bracket fungi forming corky or woody shelf-like sporophores of large size (cause destructive heart rot in trees).

fontanel (fontanelle) A gap in the cranium where the bone has not yet formed, being closed only by a membrane, e.g. the anterior fontanle is found in newborn babies situated on top of the head between the frontal and two parietal bones. It is completely closed after about 18 months.

food Material that contains nutrients like proteins, fats in order to get energy for development of an organism.

food additive Substance which is added to the food to improve its quality, texture or stability.

food chain A chain of organisms, existing in any natural community through which food energy is transferred. Each link in the food chain obtains energy by eating the one preceding it and is in turn eaten by the organism following it in the link. At each transfer, a large energy (about 90%) is lost as heat, therefore the number of links are limited to 4 or 5. Food chains are of two basic types, i.e. grazing food chain and detritus food chain (*See* trophic level, food web).

food poisoning An acute illness caused due to consumption of poisonous or contaminated food containing a large number of disease-causing organisms.

food preservation Prevention of food from getting spoiled using various techniques.

food production *See* agriculture.

food reserves Reserves of fat, carbohydrate or (rarely) protein in cells and tissues that function as an important store of energy that can be released and used in ATP production when required by the organism, e.g. in animals fat is stored in adipose tissue and carbohydrate in the form of the storage compound **glycogen** is stored in liver and muscle cells. In plants, starch is a major storage compound.

food supply Production of food for the consumption of humans.

food vacuole An intracellular digestive vacuole in most protozoans (e.g. *Amoeba*).

food web A set of interconnected food chains which exist within an ecosystem. In a food web, a particular organism may feed at more than one trophic level.

footprinting A technique for detecting regions where a protein is bound to DNA, e.g. where RNA polymerase is bound to the DNA of a gene. The protein protects the nucleic acid (i.e. the DNA) from digestion by a nuclease enzyme, so that after such treatment the undergraded DNA, the 'footprint' can be isolated and characterised.

foramen (pl. foramina) A hole, opening or orifice in a bone (or that connects two cavities of an organ).

foramen magnum Large opening in the base of the skull (piercing the occipital bone) through which the spinal cord passes.

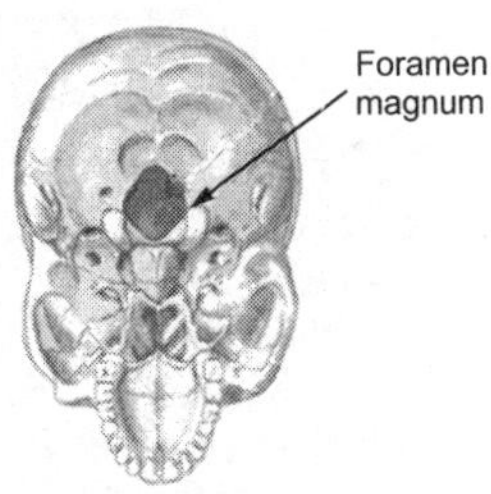

Location of foramen magnum in human skull

foramen ovale Opening between left and right atria of hearts of foetal mammals, normally closing at birth.

foraminifera An order of mainly marine protozoans whose shells form important component of chalk and deep sea oozes.

forb A non-woody plant that is not a grass.

forebrain (prosencephalon) Anterior of the three divisions in brain of vertebrates. It consists of cerebral hemispheres, hypothalamus and thalamus.

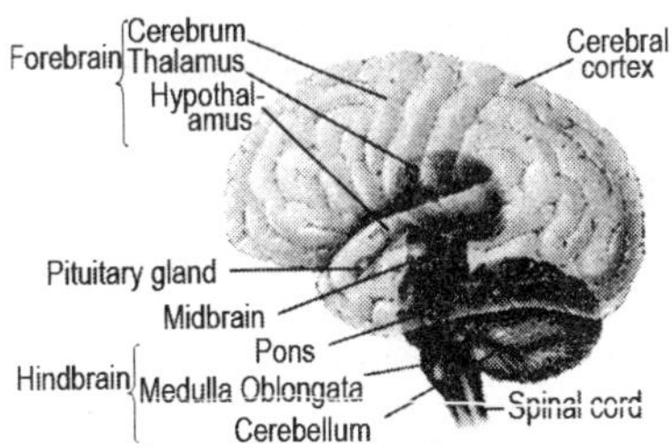

Sagital section of human brain

foregut First part of the alimentary canal of arthropods and vertebrate animals.

foreskin Loose fold of skin over the enlarged end of penis called glans penis.

forest Area in which plants are dominant.

form

1. Lowest taxonomic group, ranking below the variety level. Subforms may also be recognised (in botany).
2. A vague term used when the appropriate taxonomic rank is not clear (in zoology).

formamide A molecule that causes denaturation of double-stranded DNA by combining with the free NH_2 groups of adenine.

forward genetics The traditional approach to genetic investigation, in which the aim is to identify the gene that governs a particular phenotype of an organism.

forward mutation A mutation in a gene that changes it from wild type to an abnormal allele (non-functional or mutant).

fossil Any remain of or impressions of an organism buried in the earth or rocks by natural causes in past geologic periods. Remains of organisms are direct evidence of its presence in rocks. Generally, hard parts are preserved usually partly or wholly replaced by minerals deposited from circulating water, as impressions or casts.

fossil DNA *See* DNA.

fossil fuels Remains of organisms embedded in the surface of earth. These are used by man as fuels (e.g. coal, oil, natural gas, etc.).

fossil hominids Ancestors of any member of primate family-Hominidae.

fossiliferous Bearing or containing fossils.

fossorial Suited for digging; living in burrows.

founder effect In terms of describing the genetic outcome of a new population it is a population being established by a very small number of individuals from a larger population. The founder effect thus increases the likelihood of evolutionary divergence from the main species and eventually the formation of a new species.

fovea (fovea centralis) Shallow pit in retina of some vertebrates, which is a place of greatest acuity of vision. Contains no rods but a large concentration of cones. There are no blood vessels and no thick layer of nerve fibres interposed between cones and incoming light as in rest of retina. In binocular vision, the two eyeballs are oriented so that image of one object falls on both their foveae. Also known as yellow spot.

F-pili These are hair-like projections on bacterium involved in anchorage during conjugation and through which DNA passes. Also called sex pili.

fragmentation

1. Form of asexual reproduction in which an organism breaks or is broken into pieces and the separate pieces go on to live as distinct individuals.
2. When large habitats are broken up into small fragments due to various human activities and other reasons leading to population decline.

frameshift mutation An alternation in the sequence of DNA bases read as base triplets during transcription, caused by the deletion (or addition) of a single nucleotide in the DNA sequence. The missing (or additional) base results in an abnormal triplet and also causes each subsequent triplet group to be altered, e.g. the deletion of uracil (U) in the second codon of the following *m*RNA sequence:
AUU CAU CGG UAG ACC
results in the frameshifted sequence:
AUU CAC GGU AGA CC

fraternal twins (dizygotic twins) (non-identical twins) These are two individuals born at the same time from a single pregnancy but developing from separate fertilised egg.

freemartin A female member of unlike sexed twins that develops male traits due to the influence in the uterus of male hormones, from its male sibling when the foetal circulations are continuous (the term is usually used in connection with cattle).

free-nuclear division Stage of development in which unwalled nuclei result from repeated division of primary nucleus.

free-nuclear endosperm Endosperm in which there are many nuclear divisions without cell division before cell walls start to form.

freeze drying (lyophilisation) The removal of liquid from heat-sensitive materials. The material is frozen, placed under a high vacuum and maintained at a low temperature (−40 °C or below). It is used to preserve tissues (e.g. blood plasma), foods and to concentrate solutions.

freeze-etching It is the technique for examining the inner surface of membranes.

freeze fracture A method of preparing material for electron microscopy that allows the visualisation of the interior of plasma membranes and organelles.

frond A term applied to large well-divided leaves found in ferns and other cryptogams.

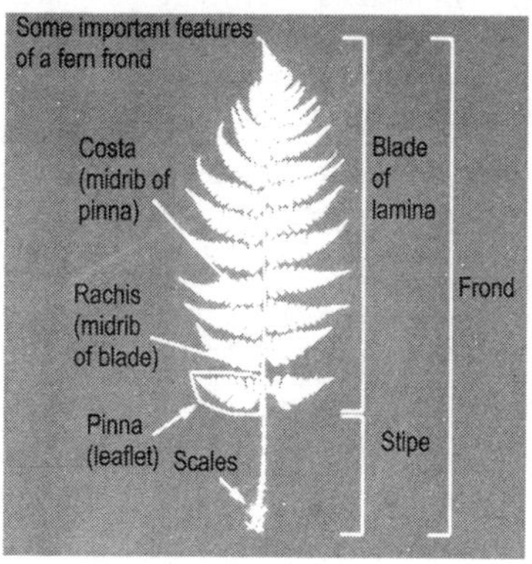

Frond of a fern

frontal bone Bone underlying the forehead.

frontal lobe Major part of the cerebral cortex of mammalian brain that is responsible for motor activities, speech and thought processes.

frontoparietal Pertaining to the frontal and parietal bones.

fructification Reproductive organ in context to fungi, bacteria and myxomycetes.

fructose (fruit sugar) A simple monosaccharide present in corn syrup, green plants, honey and many fruits. Fructose is an isomer of glucose.

fructose-1-phosphate (F1P) Product of phosphorylation of fructose catalysed by hepatic fructokinase.

fructose-6-phosphate Part of the glycolysis metabolic pathway. It is produced by isomerisation of glucose- 6-phosphate. The β-D-form of fructose-6-phosphate is abundant in cells.

frugivorous (fruit-eating) Any organism whose preferred food type is fruits.

fruit Ripened ovary of a flower, usually formed after fertilisation of the ovule. Other part may also contribute as in false fruit.

fruit fly *See Drosophila.*

frustule The cell wall of a diatom which is impregnated with silica. The wall consists of two halves that overlap one another.

FSH Follicle-stimulating hormone.

fucoxanthin Carotenoid pigment present along with chlorophyll in brown algae.

fugitive species A species that is able to coexist with a competitively superior species due to its better dispersal capabilities. The fugitive species is quicker to exploit any vacant patches in the environment that become available (e.g. due to a fire or storm), colonising and reproducing before the more competitive species occupies the patch and excludes it.

f

fumaric acid An unsaturated dicarboxylic acid which occurs in many plants. The fumarate ion participates in several important metabolic pathways, e.g. Krebs' cycle, purine pathway and the urea cycle.

functional response Relationship between prey and predator, or deviation in the rate of exploitation of prey by an individual predator due to change in prey density.

functional trait Any characteristic of an organism that influences its performance (e.g. growth rate, competitive ability) and so ultimately affects its ability to survive and reproduce (i.e. its fitness). e.g. include leaf photosynthetic rate, leaf size, stomatal density and herbivore resistance in plants and beak size, feeding rate and milk yield in animals.

fungi A kingdom of non-photosynthetic eukaryotic organisms. They are saprophytic or parasitic characterised by cell wall made of chitin and development occurs directly from spores. They may be unicellular or composed of filaments (hyphae). The cells of most fungi grow as filamentous structures (hyphae), which may contain multiple nuclei and grow from their tips. The most familiar fungi are mushrooms and toadstools.

fungi imperfecti *See* Deuteromycota.

fungicide A substance destructive to fungi.

funiculus (funicle) In animals, tracts of nerve fibres in the central nervous system of vertebrates. In plants, stalk which attaches the seed to the placenta of the fruit.

furanose A sugar that has a five-membered ring (four carbon atoms and one oxygen atom).

furrowing Cytokinesis characteristic of animal cells and a constriction of the cytoplasm into two portions.

fusiform initials Vertically elongated cells in vascular cambium that give rise to cells of the axial system in secondary xylem and phloem.

fusion Union or combining together of cells, nuclei or cytoplasm. (*See* cell fusion, fertilisation).

G

G$_1$-phase (Gap-1) The phase of the cell cycle between the mitosis and initiation of DNA replication. The cell becomes metabolically active.

G$_2$-phase (Gap-2) The phase of cell cycle just before prophase of mitosis starts during which growth occurs in the cell for a short duration. Repair of damaged DNA and duplication of mitochondria, centrioles and plastids take place during this phase.

GABA *See* Gamma-Aminobutyric Acid.

galactagogue A stimulant of milk flow (lactation).

galactose (Gal) A monosaccharide or simple hexose sugar; an isomer of glucose, and a component of lactose, as well as various glycoproteins.

Galapagos finches *See* Darwinian finches.

galectin A category of lectins, with specificity for galactose.

gall (cecidium)

1. An abnormal outgrowth of tissues; (in plants) caused by a variety of parasites, ranging from insects to fungi and bacteria.
2. A term used for bile.

gall bladder Small sac-like extension in many vertebrates arising from bile duct near liver, storing bile. Contractile walls empty bladder, when there is food, especially fat, in intestine. The release is controlled by hormone cholecystokinin.

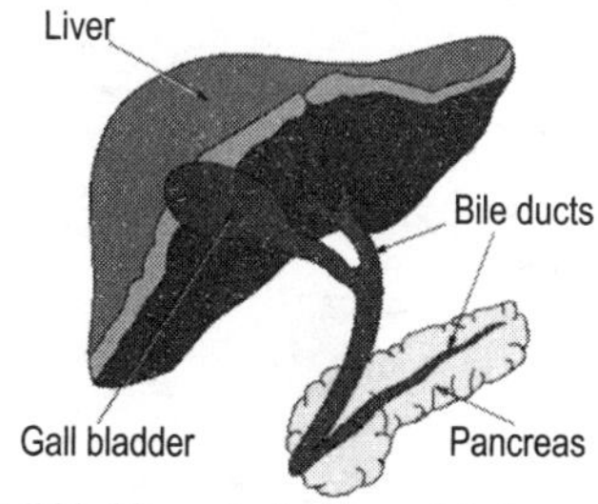

Gall bladder and other associated organs

gallstone A hard ball of material formed in the gall bladder from surplus cholesterol, which precipitates in the bile. Gallstones may lodge in and obstruct the bile duct.

gametangium (pl. gametangia) Gamete producing cells most commonly used to refer a compartment in which ferns, mosses and fungi develop their gametes.

gametes Haploid reproductive cells. Two gametes, one of each sex, fuse in fertilisation to produce a zygote in sexual reproduction. A female reproductive cell is an ovum (egg) and a male reproductive cell is a spermatozoa (See anisogamy and isogamy).

gametic number The haploid number of chromosomes.

gametocyte A cell from which gametes are formed by meiotic division (*especially*, spermatocyte or oocyte).

gametogenesis The process of formation of gametes. Gametes are normally formed by meiosis but sometimes by mitosis (as in the gametophyte generation of the ferns). In mammals, gametogenesis in the female is known as oogenesis and occurs in the ovaries; in the male, it is known as spermatogenesis and occurs in the testes.

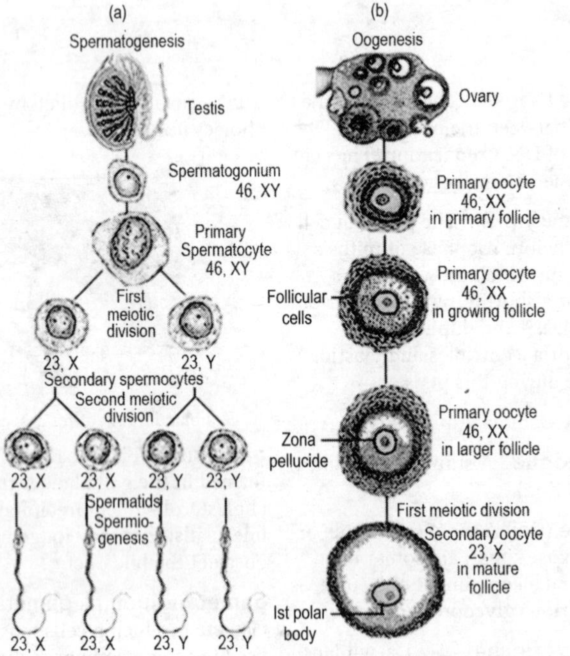

Gametogenesis in humans (a) spermatogenesis (b) oogenesis

gametophore A fertile stalk bearing sex organs in bryophytes.

gametophyte A multicellular haploid form, occurring in the life cycle of a plant or fungus undergoing alternation of generations that produces haploid gametes *via* mitosis. The prothallus of pteridophyte is the gametophyte.

Gamma Amino Butyric Acid (GABA) A neurotransmitter synthesised by decarboxylation of amino acid glutamate that inhibit central nervous system and is capable of increasing permeability of postsynaptic membranes.

gamma diversity (gamma richness) The measure of biodiversity, which means the total species richness within an area.

gamma globulin (Immune serum globulins) The class of globular serum proteins. These include the immunoglobulins responsible for immune responses.

gamone A chemical substance secreted by a gamete that bring about fusion of gamete in brown algae, e.g. *Ectocarpan.*

gamont A cell that form gametes in certain protozoans.

gamopetalous (sympetalous) Describes a flower in which the petals are fused to form a corolla tube, e.g. primrose.

gamophyllous Having leaves by united.

gamosepalous Describing a flower that possesses a calyx consisting of fused sepals.

ganglion A small mass of nervous tissue composed of numerous cell bodies, bound by a sheath or capsule. In vertebrates, the ganglia are mostly located outside the CNS in invertebrates, ganglia occur along the major nerve cords and are centres of nervous integration.

ganglion cells Cells present in retina, that along with bipolar cells transmit the nerve impulse generated by rods and cones to the optic nerve and finally to brain.

ganglioside A type of glycolipid common in nerve cell membranes.

gangrene Death of soft tissue due to interrupted blood supply, disease or injury.

ganoid scale The dermal scale typical of primitive Actinopterygii. Diamond-shaped and closely arranged in diagonal rows like tiles in a floor, e.g. *Polypterus* and *Lepisosteus.* The outer layer consists of ganoin, a calcareous substance secreted by the dermis and the inner layer consists of isopedin.

Ganong's potometer An instrument to measure rate of water taken up by plants.

gap junction (nexus) A type of junction between animal cells that allows current or cellular material to pass from one cell to another.

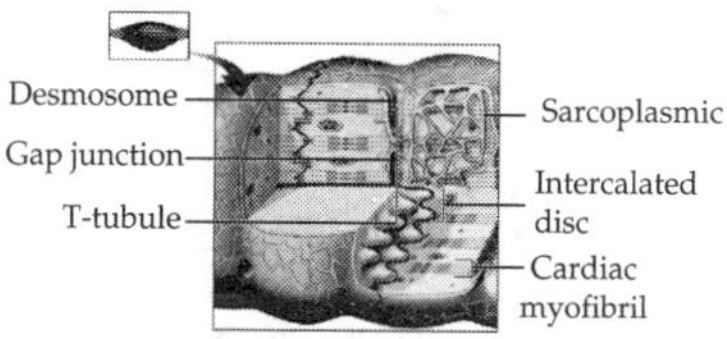

Gap junctions in a cell

gas bladder *See* swim bladder.

gas vacuole A large cavity found in cells of certain algae and protozoas which contain gas or viscous substance and help in increasing buoyancy.

gaseous exchanges The transfer of gases between an organism and the external environment in either direction.

gastric Relating to the stomach.

gastric gland *See* gastric juice.

gastric juice The digestive juice in the stomach secreted by gastric glands, situated in the thick stomach wall. Contains pepsin which breaks proteins into polypeptide chains and in milk-feeding young animals, chymosin (rennin) which coagulates caseinogen to form casein. It also contains mucus and HCl. It has an acidic pH. The chemical and mechanical stimulation of stomach lining by food causes secretion of gastric juice and gastrin.

gastric mill (proventriculus) A type of gizzard occurring in many crustaceans, situated in the anterior region of the stomach and consists of a set of bones (ossicles) and muscles that grind food particles. The food particles are then filtered by bristles in the posterior section of the stomach.

gastrin A hormone, produced by G cells in the mucosa of the stomach and first part of the duodenum, that controls the release of gastric juice. The secretion of gastrin is stimulated by the presence of food in the stomach. It is one of the hormone that integrates and controls digestive processes (*see* also secretin).

gastrocoel *See* archenteron.

gastrodermis An epithelial lining of digestive tract of certain invertebrates, (nematode worms and coelenterates), it serves for both digestion and circulation.

gastroenteritis The inflammation of the stomach and intestines.

gastroesophageal sphincter A ring of smooth muscle fibre at the junction of the oesophagus and the stomach which remains closed except during swallowing of food, etc., to prevent the stomach contents from entering the oesophagus.

gastrointestinal Pertaining to the stomach and intestine.

gastroliths

1. Stones found in the stomach of certain reptiles, fishes and birds that helps in digestion.
2. A stony mass formed in the stomach.

gastropoda A mollusc class that is comprised of the snails and slugs. They are characterised by a well developed head with tentacles and eyes, a single shell and a large flat foot.

gastrovascular cavity A body cavity of coelenterates, also called coelenteron. It has one opening serving both as mouth and anus.

gastrula An early stage in the development of an animal embryo during which its shape is restructured by cell migration, the primary germ layers are laid down as a result of morphogenetic processes. The exact process differs in different animals, but in the more complex organisms, it typically involves an infolding of the blastula to form a two or three-layered ball of cells.

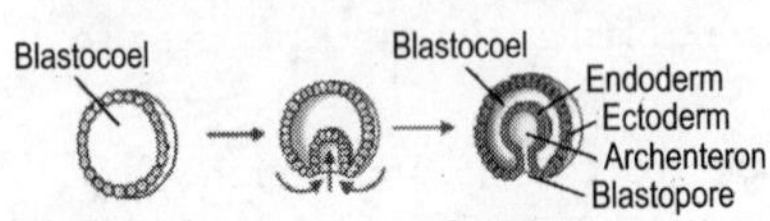

Formation of gastrula

gastrulation A process that forms a gastrula.

GC-rich region The region of a DNA sequence where the base pair G-C occurs at an elevated frequency; often indicates a gene-rich region.

geitonoga The transfer of pollen grains from the anther of one flower to the stigma of another flower of the same plant.

gemma (pl. gemmae) An organ of vegetative reproduction in mosses liverworts and certain pteridophytes.

gemmation A type of asexual reproduction in which small clumps of undifferentiated cells (gemmae) develops on the surface of a plant.

gemmule In sponges, a bud formed internally as a group of cells, which may become free by the decay of parent and subsequently from a new individual. It is an asexual reproductive structure.

GenBank The genetic sequence database of the US National Institutes of Health. Started in 1982, it now contains nearly 70 million annotated sequence entries and over 70 billion base pairs. It can be searched using the Entrez browser.

gene Unit of hereditary information located on a chromosome, composed of DNA, that determines a particular characteristics. It can exist in different forms called alleles, which determine which aspect of the characteristic is shown (e.g. tallness or shortness for the characteristics of height).

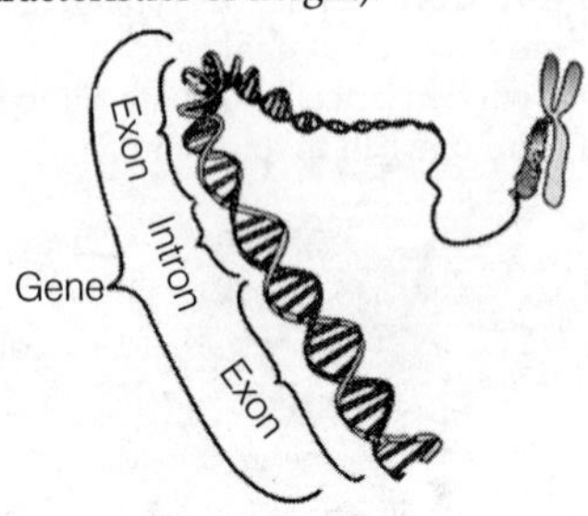

Structure of a gene

gene amplification The multiple replication of a section of the genome, which occurs during a single cell cycle and results in the production of many copies of a specific sequence of the DNA molecule, e.g. amplification of viral oncogenes in tumour cells.

gene bank *See* DNA library.

gene cloning (DNA cloning) The technique of genetic engineering whereby a gene sequence is replicated, giving many identical copies. The gene sequence is isolated by using restriction endonucleases or by making a complementary DNA from a messenger RNA template using reverse transcriptase.

gene conversion The non-reciprocal transfer of genetic material from one homologous chromosome to another, thereby converting a normal (wild-type) allele to a mutant allele or *vice versa.* It is believed to occur as a result of recombination between DNA strands of sister chromatids following a double-strand breakage in one chromatids and subsequent mismatch repair using the 'wrong' DNA strand as template.

gene dosage The number of copies of a gene in a cell or organisms.

gene expression The presence of a genetically determined phenotypic character, resulting from transcription and translation of the information in a gene. Expression of a single gene is regulated by other genes in genome and is influenced by the environment.

gene family (also multigene family) A set of genes, presumed to be descendants of a single ancestral gene that through duplication of loci and subsequent mutation has become a related set of genes occurring at different loci.

gene flow The movement of genes between populations of a species through interbreeding.

gene frequency The frequency of occurrence of a given kind of gene (allele) in a population in relation to frequency of all its alleles.

gene gun The method by which DNA is delivered on microscopic particles into target tissues and cells.

gene knockout The replacement of a normal gene by its mutated copy.

gene library A collection of cloned DNA fragments derived from the entire genome of an organism. Each fragment is cloned using a vector (e.g. plasmid or bacteriophage) inside a suitable host, such as a bacterium *E.coli* particular genes are identified by a suitable DNA probe.

gene manipulation *See* genetic engineering.

gene mapping The determination of the location of genes on a chromosome (or plasmid).

gene pool The total number and variety of genes existing within present sexually reproducing population or spcices at a given point of time.

gene prediction The computer prediction of identifying possible genes in an unannotated nucleotide sequence.

gene probe (DNA probe) The single-stranded DNA or RNA fragment used in genetic engineering to search a particular gene or other DNA sequence. The probe has a base sequence complementary to the target sequence and will thus attach to it by base pairing. By labelling the probe with a radioactive isotope or fluorescent label it can be identified on sub-sequent separation and purification.

gene product The RNA or protein, produced from a gene. The activity of a gene is measured in terms of the amount of gene product produced by that gene.

gene sequencing The determination of the order of bases of a DNA molecule making up a gene. Sequencing requires multiple cloned copies of the gene; long DNA sequences are cut into more manageable lengths using restriction enzymes.

gene splicing

1. Joining of exons after the intron sequences have been removed to produce functional messenger RNA.
2. In genetic engineering, the joining of DNA fragments by the action of enzyme DNA ligase.

gene therapy An application of genetic engineering techniques to alter or replace defective genes. A defective gene may result from an incorrect sequence of bases in the DNA molecule or an inability of the gene to code for the expression of a particular polypeptide. First gene therapy was done on a 4 years old girl with Adenosine Deaminase (ADA) deficiency by Blease and Andresco in 1990s.

gene tracking A method used for determining the inheritance of a particular gene in a family. It is used in diagnosis of genetic disease.

genecology The study of genetical composition of populations in relation to their habitat.

generation The group of organisms of nearly same age within a population.

generation time An average time between the cell division of parent and daughter cells within a population of cells.

generative nuclei Two gametic nuclei found in the pollen tube. One fuses with the egg cell to form a zygote while the other either degenerates or fuses with polar nucleus to give primary endosperm nucleus.

genetic

1. Having a hereditary basis, as opposed to an environmental cause.
2. Of, relating to, produced by, or being a gene or genes.

genetic balance The hypothesis that the characters of an organism are each determined by the interaction of a large, but unknown number of genes. Some affecting development in one direction and some in another, so that ultimate result is a balance struck between the total effect.

genetic code The correspondence between nucleotide triplets in DNA or RNA (codons) and the amino acids and protein. A gene's DNA sequence determines the *m*RNA sequence produced from it and the genetic code then determines the amino acid sequence in the protein produced from that *m*RNA. The genetic code is apparently the same in all organisms (with the exception of a few mitochondrial systems).

First position	Second position: U	C	A	G	Third position
U	UUU Phe UUC Phe UUA Leu UUG Leu	UCU Ser UCC Ser UCA Ser UCG Ser	UAU Tyr UAC Tyr UAA Stop UAG Stop	UGU Cys UGC Cys UGA Stop UGG Trp	U C A G
C	CUU Leu CUC Leu CUA Leu CUG Leu	CCU Pro CCC Pro CCA Pro CCG Pro	CAU His CAC His CAA Gin CAG Gin	CGU Arg CGC Arg CGA Arg CGG Arg	U C A G
A	AUU Ile AUC Ile AUA Ile AUG Met	ACU Thr ACC Thr ACA Thr ACG Thr	AAU Asn AAC Asn AAA Lys AAG Lys	AGU Ser AGC Ser AGA Arg AGG Arg	U C A G
G	GUU Val GUC Val GUA Val GUG Val	GCU Ala GCC Ala GCA Ala GCG Ala	GAU Asp GAC Asp GAA Glu GAG Glu	GGU Gly GGC Gly GGA Gly GGG Gly	U C A G

Codons for the various amino acids

genetic counselling Services provided by specialists related to human genetic disorder.

genetic disorder Disorders in human beings, associated with the inheritance of change or altered genes or chromosomes, e.g. haemophilia.

genetic drift (Sewall Wright effect) The random change in gene frequency from one generation to another in a population. It cause loss of genetic diversity if there are no counteracting factors.

genetic engineering Altering the genetic material by inserting genes from another organism into its DNA it allow organism to perform new functions or make new products of an organism to perform new functions or make new products. It is the technique has many applications, ranging from the commercial production of insulin and other hormones, vaccines, etc., to the creation of transgenic animals and crop plants in agriculture. Also known as recombinant DNA technology.

Genetic Engineering Approval Committee (GEAC) An organisation of Indian government which makes decisions regarding the validity of GM (Ggenetic Modifi- cation) research and the safety of GM organisms for public services.

genetic equilibrium The condition of a population in which successive generation consists of the same genotype with the same frequencies, with respect to particular genes or arrangement of genes.

genetic fingerprinting *See* DNA fingerprinting.

genetic load The concentration of deleterious genes carried within a population.

genetic marker

1. A probe used to mark a gene, chromosome or nucleus.
2. A phenotypic trait indicating the presence of a particular allele.

genetic mosaic An organism in which different cells differ at the genetic level; results either from fusion of embryos or mutation during development.

genetic polymorphism The existence of two or more distinctly different forms within a species as a result of inherited characteristics, e.g. sickle-cell disease.

genetic redundancy Two or more genes performing the same function, where inactivation of one will have no effect on phenotype.

genetic scalpels (Restriction enzymes) A segment of DNA is cut with the help of specific enzymes called restriction enzymes.

genetic screening The process by which the genome of a human or other organisms is analysed for genetic markers that indicate the presence of particular genes, especially ones that cause or predispose to certain diseases. Increased knowledge of the human genome and technological advances have simplified genetic screening in persons with a family history of certain inherited diseases, e.g. certain forms of breast cancer.

genetic variation The variation due to the differences in genetic constitution. It is inherited and the most important sources are mutation and recombination.

Genetically Modified Organisms (GMOs) Organisms whose genomes incorporate and express genes from another species. Genetically modified (or transgenic) individuals are created by genetic engineering, using suitable, vectors to insert the desired foreign gene into the fertilised egg or early embryo of the host.

genetics The study of heredity and variation.

genome An entire virus, plasmid, bacterial chromosome or haploid set of eukaryotic chromosomes, considered as a single informational unit.

The term is generally applied to all the chromosomes of an organism, to an entire virus, or to some class of chromosome or plasmid, such as a mitochondrial genome.

genomic library A set of clones composed of overlapping, randomly generated DNA fragments representing an organism's entire genome. Also known as clone bank.

genomics The study of entire genomes.

genotoxicocity The condition resulting from the interaction of toxic agents (X-rays, acridine virus) with DNA molecules in genes. The toxicity is heritable.

genotype The genetic constitution of an organism, as opposed to its physical appearance (phenotype).

genotype frequency The percentage of individuals in a population that possess a specific genotype. The genotype frequency can be calculated using the Hardy-Weinberg equation (see Hardy-Weinberg equilibrium).

genucubital Pertaining to the knees and elbows.

genus (pl. genera) In taxonomy, a division of a family that itself contains similar or closely related species.

geobiology The science that combines geology and biology to study the interactions of organisms with their environment (biosphere with lithosphere/atmosphere).

geocarpy The ripening of fruits underground. The young fruits are pushed into the soil by a post-fertilisation curvature of the stalk, as occurs in the peanut.

geographical isolation The separation of two populations of the same species or breeding group by a physical barrier, such as a mountain. It may ultimately lead to the populations becoming separate species by adaptive radiation.

geophyte A class of Raunkiaer's system of classification. It consists of plants possessing perennating buds below ground as corm, bulbs or rhizome.

geotaxis The taxis (responsive movement) in which stimulus is gravitational force.

geotropism (gravitropism) The response of a plant part to the stimulus of gravity. A main root is positively geotropic and a main stem negatively geotropic, growing downwards and upwards, respectively, irrespective of the positions in which they are placed.

geriatrics The medical treatment and study of the diseases of the elderly.

germ

1. A microorganism, esp. a disease causing.
2. A bud, offshoot or seed.

germ cell Any of the cells in animals that produces gametes. In mammals, oogonia and spermatogonia are the germ cells.

germ layer The three major body layers, i.e. ectoderm, mesoderm and endoderm. These develop in the embryos of most animals during gastrulation. (See ectoderm, endoderm and mesoderm).

germ line The term used to refer to cells that give rise to gametes and so, to descendant generations.

germ line gene therapy An experimental method of preventing genetic disease, in which genes are inserted into germ cells or zygotes to prevent the disease in offsprings.

germ line mutation A heritable mutation, transmitted through germ line from parent to progeny.

germ pore A pore, pit or the thin-walled area on outer wall of a pollen grain or spore through which a pollen tube or germ tube makes its exit on germination.

germicide The substance capable of destroying germs.

germinal epithelium An epithelium lining of seminiferous tubules of vertebrate testis, from which arise spermatogonia and the Sertoli cells arise during sexual maturity. Applied also to epithelium covering coelomic surface of vertebrate ovary, which probably forms follicle cells. Though, it is doubtful whether it forms oogonia during sexual maturity.

germination

1. Process by which of plant embryo in a seed ends dormancy and resumes normal metabolism and development. It involves the emergence of radicle or coleoptile through the testa.
2. First signs of growth of a reproductive body such as spore or pollen grain.

germplasm *See* Weismannism.

gerontology The study of old age, it is the process of ageing and the problems of old people.

gestation period The period in viviparous animals from the fertilisation of the egg to birth of the young (parturition). In humans gestation is known as pregnancy and takes about nine months (40 weeks).

ghrelin A peptide hormone secreted chiefly by cells in the stomach lining and increases hunger by stimulating the release of neuropeptide Y, a potent stimulant of appetite, from the arcuate nucleus in the hypothalamus. It also stimulates release of growth hormone by binding to receptros in the anterior pituitary.

giant chromosome *See* polytene.

giant fibre A nerve fibre of very large diameter. Occur in many invertebrates, e.g. annelids, crustacea, etc. It helps in rapid conduction of nerve impulses that is required in emergencies (e.g. for escape).

Gibberellic Acid (GA_3) A common gibberellin and one of the first to be discovered in 1954 from fungal culture (*Gibberella fujikuroi*).

gibberellin An important group of phytohormones. Common in higher plants but restricted to only a few species of bacteria and fungi. Occur in stem apex, young leaves and seeds, etc. They are capable of converting even a genetically dwarf plant into a tall plant. Gibberellins favour flowering in long day plants, break dormancy, promote seed germination, substitute cold treatment, induce parthenocarpy, etc.

gigabase A measure of nucleotide sequence length equal to one billion base pairs.

gigantism The abnormal growth (over 7-8 feet) due to hypersecretion of human growth hormone from pituitary during childhood and adolescence.

gill

1. Gas exchange organs found in many aquatic animals. Fishes have internal gills, formed as outgrowths from the pharynx wall and contained within gill slits. The gills of most aquatic invertebrates and amphibian larvae are external gills, which project from the body so that water passes over them as the animal moves (in zoology).
2. Lamella one of many thin plate-like spore producing structures radiating outwards from the stalk on the undersurface of the cap of mushrooms and toadstools (in botany).

gill arches Basic structural unit of gills consisting of rod-like gill bars made up of cartilage.

gill bar Skeletal structures providing support to tissues separating adjacent gill-slits in lower chordates, containing blood vessels, nerves and skeletal support.

gill book *See* lung book.

gill cleft (visceral cleft) An opening through which water leaves body of a cartilaginous fish after passing through the gills.

gill rakers Gill bars give rise to a series of elongated, stiff projections at the point where gill slit opens into the pharynx. These skeletal projections are the gill rakers. Act as screen and protect gills from any damage.

gill slit An opening leading from the pharynx to the exterior in aquatic vertebrates and lancelets. In lancelets they function in filter feeding. In fish they contain the gills and are usually in the form of a series of long slits. They are absent in adult tetrapod vertebrates (except for some amphibians). Their presence in the embryo vertebrates is a characteristic of phylum-Chordata.

gills pouch (visceral pouch) Of several pairs of compartment lying laterally to the respiratory tube in cyclostomes. They contain gill filaments.

gingiva Gum that covers the jaw bones and is continuous with the lining of the mouth. It consists of stratified squamous epithelium covering a dermal layer of connective tissue containing nerves etc, which gives pink colour to the gum.

gingivoglossitis An inflammation of the gums and tongue.

Ginkgophyta (Ginkgos) Gymnosperms possessing active cambial growth and fan-shaped leaves with open dichotomous venation.

girdling The ring-like round cut in the stem of a healthy potted plant or around a thick twig of a tree, deep enough to penetrate the phloem and cambium but not the xylem. It result in the death of the area above the girdle over time.

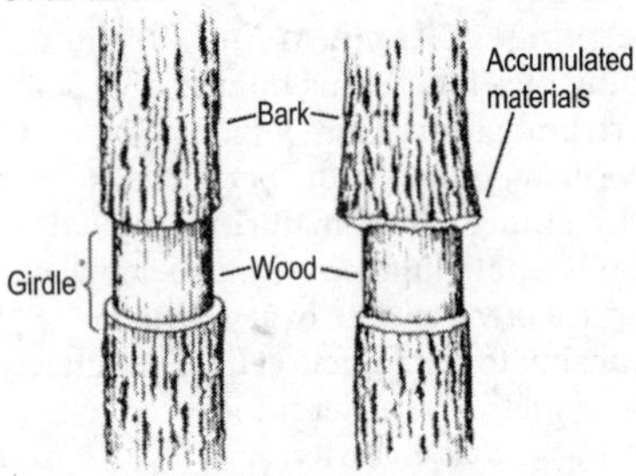

Girdling in a stem

gizzard A highly muscular region in the gut of many animals, where food is broken up prior to main digestion. It may often contain hard teeth (e.g. crustaceans) or contain grit, stones, etc (e.g. birds and reptiles) see also gastric mill.

gland A group of cells or a single cell in animals or plants, i.e. specialised to secrete a specific substance. In animals, the glands are either exocrine or endocrine. In plants these may be superficial discharging secretion externally (e.g. glandular hair) or embedded in tissue as isolated cells or as layer of cells, e.g. resin canal of pine.

Glaucophyta A small phylum of single-celled freshwater algae whose primitive chloroplasts (cyanelles) resemble cyanobacteria.

glenoid cavity The cup-like hollow on each side of the pectoral girdle into which head of humerus fits, forming shoulder-joint, in tetrapod vertebrates.

glial cells (glia, neuroglia) Non-conducting supporting cells of the nervous system. There are four classes of glial cells: astrocytes, oligodendrocytes, ependymal cells and microglia. Other functions of these cells include providing nutrients for neurons and controlling the biochemical composition of the fluid surrounding the neurons.

global stability The tendency of a community to return to its original state even after large disturbances.

global warming The gradual continuous increase in average temperature of surface of the earth as a result of increase in concentration of greenhouse gases (CO_2, CHs, etc) All these gases allow the heat waves to reach the earth but prevent their escape and the earth becomes warm. (See greenhouse effect)

globin *See* haemoglobin.

globular protein *See* protein.

globulin Simple globular proteins, which are insoluble in water, but dissolve in certain salt solutions. Present in blood plasma in vertebrate animals may function like antibodies and in seed in plants.

glomerular filtrate The fluid in the lumen of the Bowman's capsule of the nephron that has been filtered from the capillaries of the glomerulus (see ultra filtration). The glomerular filtrate has the same composition as the plasma except that it does not contain any of the larger components, such as plasma proteins or cells.

glomerular filtration The first step in urine formation. It is filtration of blood, which is carried out by the glomerulus of kidney.

Glomerular Filtration Rate (GFR) The volume of fluid that is filtered from the capillaries of the glomeruli into the kidney tubules per unit time. It is influenced chiefly by blood pressure and by changes in the diameter of arterioles in the glomeruli. In humans GFR is typically about 125 mL/min, although some 99% of this filtered volume is reabsorbed as it passes along the kidney tubules to form urine.

glomerulus (pl. glomerule)

1. A cluster of capillaries, particularly within the Malpighian body of the kidney, which is surrounded by a cup-shaped Bowman's capsule. It filters the compounds excreted in urine out of the blood (renal)
2. A globular cluster of axons and dendrites found in the olfactory bulb and concerned with relaying signals from the olfactory epithelium to brain.

glossalgia The pain in the tongue.

glossitis (also glottitus) Inflammation of the tongue.

glossodynia A burning or painful sensation in the tongue.

glossopharyngeal nerve Ninth cranial nerve of vertebrates. It is supplied to posterior part of mouth cavity, including tongue and controls swallowing reflex.

glottis The opening from the pharynx to the trachea (windpipe). In mammals it also serves as the space for the vocal cords. See also epiglottis; larynx.

glucagon The hormone antagonistic to insulin in action, converting glycogen into glucose in the liver to raise blood-sugar level. Secreted by alpha cells of the islets of Langerhans in the pancreas.

g

glucan Any polysaccharide composed of only glucose residues, e.g. starch (also known as glucosan).

glucocorticoid Hormones secreted by adrenal cortex of adrenal gland. Regulate carbohydrate, protein and fat metabolism (e.g. corticosterone). Accelerate the formation of glucose from protien and breakdown of glycogen.

gluconeogenesis The synthesis of glucose from non-carbohydrate source (such as fat and protein). It occurs in the liver, when glycogen supplies are exhausted. The reactions are essentially of glycolysis in reverse and the pathway is stimulated by the hormone glucagon.

glucosamine An amino sugar and an important precursor in the synthesis of glycosylated proteins and lipids. Glucosamine is an abundant monosaccharide present in the polysaccharide chitin, a major constituent of the cuticle of arthropods and cell wall of fungi.

glucose (dextrose; grape sugar) The most important carbohydrate in body metabolism; used by cells as a source of energy being transported in body in blood, lymph, etc. Energy is released by glycolysis. A monosaccharide ($C_6H_{12}O_6$) is formed during digestion, especially from starch and sucrose. Glucose is an isomer of fructose and is present in the sap of plants, in fruits and in honey and is also a constituent of many polysaccharides, most notably of starch and cellulose.

glucosuria (glycosuria) The excretion of glucose in the urine due to high blood glucose level.

glucuronic acid A compound, $OC_6H_9O_6$, derived from the oxidation of glucose. It is an important constituent of gums and mucilages. It can combine with hydroxyl (—OH) carboxyl (—COOH), or amino (—NH_2) groups to form a glucuronide.

glucuronide *See* glucuronic acid.

glume One of a pair of dry bracts at the base of or enclosing the spikelet of grasses.

glutamate The anion of the amino acid glutamic acid. It functions as a neurotransmitter at excitatory synapses in the vertebrate central nervous system and at excitatory neuromuscular junctions in insects and crustaceans. In plants, glutamate is the initial acceptor molecule in the assimilation of the ammonium ion (NH_4^+), combining with it to form glutamine in a reaction involving ATP.

glutamate receptor (GluR) Any receptor protein that binds the neurotransmitter glutamate as a ligand.

glutamic acid A non-essential amino acid.

glutamine *See* amino acid.

glutathione A tripeptide of cysteine, glutamic acid and glycine, important as an intracellular antioxidant. It exists in reduced (GSH) and oxidised (GSSG) states. GSH is regenerated from GSSG by glutathione reductase.

gluten A mixture of protein present in wheat flour. It is composed mainly of gliaden and glutelin.

glycaemic Index (GI) The indication of how eating a particular food will affect the concentration of glucose in the blood. GI is determined relative to the effects of a standard test meal of (usually) 50 g of glucose in a fasting individual, which is given the value 100.

glycan *See* polysaccharide.

Glyceraldehyde-3-Phosphate (GALP) Present in all organisms, it is an intermediate in various important metabolic pathways (e.g. glycolysis and Calvin's cycle). It is a phosphate ester of the 3-carbon sugar glyceraldehyde.

glycerate 3-phosphate A phosphorylated three carbon monosaccharide that is an intermediate in the Calvin cycle of photosynthesis and also in glycolysis. It was formerly known as 3-phosphoglycerate or phosphoglyceric acid (PGA).

glyceride (acylglycerol) A fatty acid ester of glycerol. Esterification can occur at one, two or all three hydroxyl groups of the glycerol molecule producing monoglycerides, diglycerides, and triglycerides respectively. Triglycerides are the major constituent of fats and oils found in living organisms. Alternatively, one of the hydroxyl groups may be esterified with a phosphate group forming a phospholglyceride) or a sugar forming a glycolipid.

glycerine *See* glycerol.

glycerol (Propane -1, 2, 3 triol) A precursor of triacylglycerols and phospholipids. When the body extracts energy from fat, glycerol and fatty acids are released into the blood. In the liver, glycerol is converted to glucose, which provides energy for cellular metabolism.

glycine (Gly or G) The smallest of the 20 amino acids found in proteins. It is non-essential in the human diet because it is produced in the body from serine. It is the main inhibitory neurotransmitter for fast synapses in the spinal cord of vertebrates.

glycobiology Study of carbohydrates and its complexes.

glycocalyx (cell coat) A rough coat of adhesive oligosaccharides on the external surface of most eukaryotic cells.

1. It plays a role in cell-cell adhesion and in regulating the exchange of materials between a cell and its environment.
2. The outermost layer of a bacterium, typically consisting of numerous polysaccharides plus various glycoproteins.

glycogen (animal starch) The polysaccharide that serves the same energy-storing function in animals that starch does in plants. It is the main carbohydrate store of animals, largely found in the liver and in muscle. Also known as animal starch.

glycogenesis The manufacture of glycogen from glucose.

glycogenolysis The breakdown of glycogen into glucose. It occurs in liver.

glycolipids Complex lipids which consist of compounds of fatty acids with carbohydrates. Found particularly in the outer half of phospholipid bilayers of plasma membranes.

glycolysis The production of pyruvic acid by the splitting of glucose. It is the single metabolic pathway occurring in all types of living cells. One molecule of glucose breaks down into two molecules of pyruvate. The pyruvate in the cytoplasm of cell. also known as Embden Meyerhof pathway.

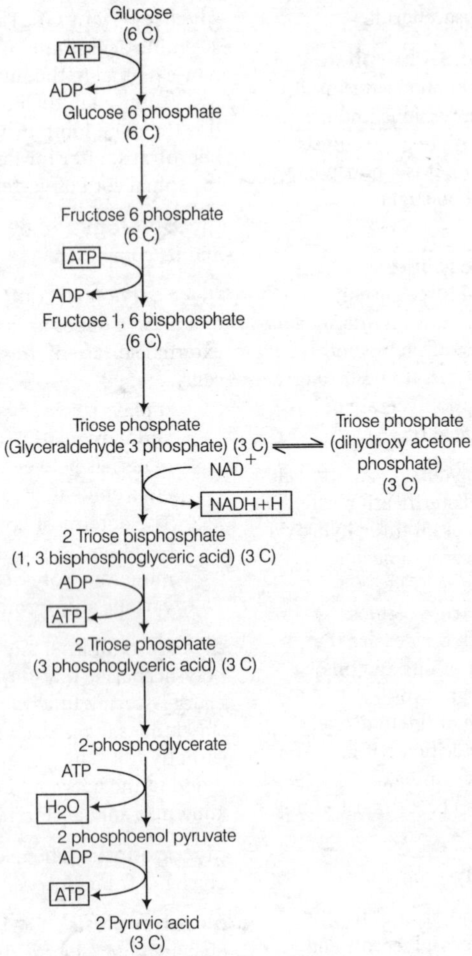

Steps of glycolysis

The overall reaction of glycolysis can be depicted as

- Glucose + 2Pi + 2ADP + $2NAD^+ \rightarrow$ 2 Pyruvate + 2ATP + 2NADH + $2H^+$
- Two molecules of NADH on oxidation produce 6 molecules of ATP. Therefore, a net gain of 8 ATP molecules occurs during glycolysis.
- The fate of glycolysis depends upon the availability of oxygen in the cell. In the presence of oxygen, pyruvic acid will enter the mitochondrion and undergo complete oxidation of glucose to CO_2 and H_2O in aerobic respiration (Krebs' cycle).
- One the other hand in the absence of oxygen, the pyruvic acid will undergo anaerobic respiration (lactic acid fermentation or alcoholic fermentation).

glycomics The study of the entire set of oligosaccharides (i.e. the glycome) synthesized by an organism.

glycoprotein A protein-carbohydrate containing oligosaccharide chains covalently bonded to polypeptide side-chains. Formed in the Golgi apparatus in glycosylation.

glycosaminoglycans (GAGs) Any one of a group of polysaccharides that contain amino sugars (such as glucosamine). Formerly known as mucopolysaccharides, they include hyaluronic acid and chondroitin which provide lubrication in joints and form part of the matrix of cartilage.

glycoside A derivative of a pyranose sugar (e.g. glucose) in which there is a group attached to C-atom that is joined to — CHO group. The C-OR replaces C-OH in glycoside, e.g. anthocyanin pigments.

glycosidic bond The bond formed by dehydration between two carbon atoms of two adjacent monosaccharides in di-oligo and polysaccharides.

glycosylation Bonding of a sugar residue to another organic comopund, e.g. glycoproteins formed in lumen of rough ER.

glyoxylate A modified form of Krebs' cycle in plants, fungi, protists and fungi. It utilises fats as a source of carbon and enables the synthesis of carbohydrate from fatty acids by avoiding the stages of the Krebs' cycle in which carbon dioxide is released.

glyoxysomes A small membrane bound sacs in plants which contain powerful oxidative enzymes such as peroxidase, catalase and urate oxidase. In animals, they are called peroxisomes.

glyphosate A herbicide that cells plants and has low toxicity to animals.

Gnathostomata The group of chordates under the sub phylum–Vertebrata with true jaws, paired fins or limbs, includes fishes, amphibians, reptiles, birds and mammals.

gnotobiosis The use of microbiologically monitored environments in experimental work. e.g. rearing of animals in laboratory in a germ-free environment.

goblet cell The pear-shaped cell, present in columnar epithelium of intestinal mucosa and in shin of animals such as earthworm. It secretes mucous from time to time on the surface.

goitre The enlargement of thyroid gland caused by deficiency of iodine in the diet. It results in swelling of the neck.

golgi apparatus (Golgi complex, Golgi body) A major complex protoplasmic structure being made up of many flat, disc-shaped sacs or cisternae (0.5-1.0 μm) in diameter.
It occurs in all cells except prokaryotes (i.e. PPLO, bacteria and cyanobacteria) and some eukaryotes such as human RBCs, sieve tubes of plants, etc.
The Golgi apparatus is involved in the formation of lysosomes, vesicles that contain proteins and remains within the cell, performs the function of packaging material, acts as an important site for the formation of glycoproteins and glycolipids helps in the production of complex carbohydrates other than glycogen and starch and in the formation of cell wall.

Golgi tendon organ A receptor organ located at the junction between a skeletal muscle and its tendon that detects contraction of the muscle.

gonad The reproductive organ of animals. It produces sex cells (gametes) and even hormones. Female gonad is ovary and male gonad, the testis. Some invertebrates have both male and female gonads.

Gonadotropin Releasing Hormone (GnRH) The hormone produced by the hypothalamus. GnRH in turn controls the secretion of luteinizing hormone and follicle stimulating hormone from anterior pituitary.

gonadotropins Hormones playing a role in the development and activation of the sex glands (testes and ovaries). It includes the mammalian hormones, FSH and LH (secreted by anterior pituitary), and hCG and eCG (secreted by placenta).

gonium

1. An undifferentiated germ cell.
2. A genus of colonial algae.

gonochorism Conditions of having sexesseparate; individualy having functional gonad of one type.

g

gonopoietic Producing gametes.

gonorrhoea The sexually transmitted disease, caused by bacterium *Neisseria gonorrhoeae.* It leads to inflammation and damage of epithelial cells of the reproductive system.

gonotome The portion of an embryonic somite that develops into gonads.

goose flesh The upright position of hair follicles and dimples on skin surface in response to cold, external temperature and fear. It is an important heat retention and protection mechanism in animals.

gorgonacea The horny corals, an order of cnidarian sub-class–Alcyonaria.

gout The inflammation of joints due to the accumulation of uric acid crystals.

G-protein An eukaryotic protein with 3 sub-units that occur on the surface of plasma membrane and transmit signals from outer cell surface to intracellular components.

Graafian follicle The fluid-filled spherical vesicle in mammalian ovary containing an oocyte attached to its wall. A Graafian follicle matures periodically from one of the enormous follicles present in ovary at birth. It enlarges and migrates through ovary to lie at the surface. Finally, it bursts and releases the ovum to Fallopian tube. The follicle then becomisd solid body, the corpus luteum. Its growth is under FSH influence from pituitary.

grade A group of organisms that share certain morphological characteristics but are not necessarily related in evolutionary terms. For example, all organisms that possess a coelom can be regarded as a grade.

graft The transplantation of an organ or tissue in animals (including humans). It is used to replace faulty or damaged parts of the body. It can be graft on the same individual (autograft) or different individual (allograft) of the same species. (Also see isograft and heterograft)

graft hybrid A type of plant chimaera that may be produced when a part of one plant (the scion) is grafted onto another plant of a different genetic constitution (the stock). Shoots growing from the point of union of the graft contain tissues from both the stock and the scion.

grafting A horticultural practice used to propagate plants, especially certain bushes and fruit trees, artificially. A shoot or bud of the desired variety (the scion) is grafted into a rootstock of either a common or a wild related species (the stock). The scion retains its desirable characteristics (e.g. flower form or fruit yield) and supplies the stock with food made by photosynthesis. The stock supplied the scion with water and mineral salts and affects only the size and vigour of the scion.

Gram's Stain A stain containing crystal violet and safranin used to differentiate bacteria. Gram positive bacteria retain the deep purple colour of crystal violet. Such bacteria have a thick layer of peptidoglycan in their cell walls.

In Gram-negative bacteria, the acetone-alcohol washes out of the violet dye and the counterstain safranin is taken up, the cells appearing red. The cell walls of these bacteria have an outer layer of lipoprotein overlying a thin layer of peptidoglycan.

granulocyte A white blood cell (leukocyte) that has granules in the cytoplasm. Their three types are–neutrophils (70%), eosinophils (1.5%) and basophils (0.5%). Sometimes called polymorpho nuclear leukocytes because of lobed nucleus.

granulosa cells (granular cells) Secretory cells in a Graafian follicle that surround the oocyte. They have a function analogous to that of the Sertoli cells in the testis, supplying nutrients to the developing oocyte. After ovulation they form the corpus luteum.

granum (pl. grana) One of the stack of thylakoids within a chloroplast that convert light into chemical energy. It bear the light receptive pigment chlorophyll and the enzymes for light dependent reaction of photosynthesis.

grassland A terrestrial biome where natural vegetation is grass; used as pasture. e.g. Savanna tropical grasslands, steppes of Asia (temperate grassland).

gravid

1. Pregnant.
2. Swollen with embryos or eggs.

gravitropism A gravity-stimulated response in an animal or plant. *See* geotropism.

gray crescent A cortical region of the newly fetilised eggs of frogs and some salamanders that forms just after fertilisation on the side opposite sperm penetration egg, grey in colour, future blas to pore and anus.

gray matter The nerve tissue of vertebrate central nervous system. It contains numerous cell-bodies and dendrites of nerve cells and unmyelinated terminations of nerve-fibres forming synapses with these; together with ganglia and blood vessels. The coordinating work of the CNS is done in grey matter by the large numbers of synapses present.

green algae *See* chlorophyta.

Green Fluorescent Protein (GFP) A naturally fluorescent protein obtained from the jellyfish *Aequorea victoria* and used as a marker for identifying cells containing recombinant DNA or for localizing specific proteins in cells.

green gland *See* antennal gland.

green manuring The ploughing of legume (containing nitrogen-fixing bacteria in their root nodules) plants in the field to enrich the nitrogen in soil.

green revolution The spectacular increase in the yield of crop plants, especially cereals beginning in late 1960's by the use of modern techniques of agriculture.

greenhouse effect The warming of atmosphere due to the accumulation of certain gases that absorb infrared radiation. The major greenhouse gases are methane, carbon dioxide, nitrous oxide and other fluorinated gases (sulphur hexafluoride, CFCs and HFCs).The greenhouse effect is a natural phenomenon, without which the earth's climate would be much more hostile to life. The overall effect is that the average temperature of the earth and its atmosphere is increasing (global warming).

grooming Actions of an animal of rearranging fur or feathers and cleaning the body surface by biting, scratching, licking, etc., which is important for removing parasites and spreading oils over the body surface.

In many mammals, especially primates mutual grooming between individuals (allogrooming) has an important role in maintaining social cohesion.

ground meristem The region of the apical meristem from which the ground tissues of pith, cortex, medullary rays and mesophyll differentiate.

ground tissue All tissues in plants except epidermis and vascular bundles. Consist of simple tissues such as parenchyma, collenchyma and sclerenchyma.

g

Group selection A postulated mechanism of evolution whereby characters disadvantageous to individuals bearing them, but beneficial to group of organisms they belong to, can spread through the population counteracting the effects of selection at individual level. It was originally proposed to account for the evolution of altruism in social groups of animals. It was suggested by the British ethologist VC Wynne-Edwards (1906-97) in 1962.

growth The increase in size and dry weight of an organism or cell; which cannot be reversed. It does not include the increase in size due to the uptake of water.

growth cone The enlarged motile tip of an elongating axon (nerve fibre) that determines the direction of axon growth. It has numerous finger-like projections, called filopodia which bear surface receptors.

growth factor Any of various chemicals, particularly polypeptides, that have a variety of important roles in the stimulation of new cell growth and cell maintenance, e.g. epidermal growth factor.

Growth Hormone (GH, somatotrophin) A hormone secreted by anterior pituitary. It stimulates growth and metabolism by acting on cells of the body, particularly those of bone. (See dwarfism, gigantism)

growth ring (annual ring) A cylinder of secondary wood laid down in a single season in a stem or root. It is seen as a circle in transverse section.

growth substance A term, used particularly for plant hormones, to include all natural and artificial substances with powerful and diverse physiological and morphogenetic effects in plants.

GTP (Guanosine 5′- Triphosphate) The purine nucleoside triphosphate. It is involved in tissue responses to various hormones and essential for chain elongation during protein synthesis.

guanine (G) A purine base present in nucleic acids; pairs with cytosine in DNA and RNA.

guano The accumulation of the droppings of birds, bats, or seals, usually formed by a long-established colony of animals. It is rich in plant nutrients and some deposits are extracted for use as fertiliser.

guanosine A nucleoside consisting of one guanine molecule linked to a D-ribose sugar molecule. The derived nucleotides, guanosine mono-, di-, and triphosphate (GMP, GDP and GTP respectively), participate in various metabolic reactions.

Guanosine Monophosphate (GMP) A monophosphate ester of phosphoric acid with the nucleoside guanosine. GMP is composed of a ribose, phosphate group attached to the carbon C5′ of the ribose and guanine attached to the carbon C1′.

guard cells Crescent-shaped epidermal cells occurring in pairs surrounding the stomata. Changes in shape of guard cells, due to the changes in their turgidity, control opening and closing of stomata and hence affect loss of water vapour and gaseous exchange.

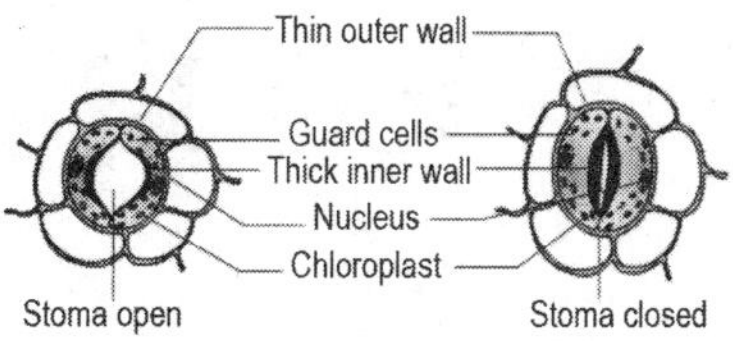

Position of guard cells in close and open condition

guide RNA (gRNA) The RNA that guides the insertion or deletion of uridines into mitochondrial RNA in trypanosomes.

guidepost cell A differentiated neuron that influences the development of the axons of other neurons so that they connect ultimately with their target cells, thus ensuring that the nervous system is correctly bound.

guild The group of organisms, often species within a higher taxon, having the same broad feeding habits. for example, different species of snakes and lizards may belong to a guild by virtue of occupying the same types of underground shelters and different species of seed-eating birds may constitute a guild.

Gullet *See* oesophagus.

gum Any of a variety of substances obtained from plants. Typically they are insoluble in organic solvents but form gelatinous or sticky solutions with water. Gum resins are mixtures of gums and natural resins. Gums are produced by the young xylem vessels of some plants (mainly trees) in response to wounding or pruning.

gustatory receptors Taste receptors.

gut *See* alimentary canal.

Gut-Associated Lymphoid Tissue (GALT) A tissue that protects digestive tract from infectious organisms and other ingested antigens.

guttation The exudation of water from hydathodes, which are commonly present at the end of the main veins of leaves. It happens especially in high humidity.

gymnosperms Vascular plants bearing naked seeds. The sporophyll is not infolded to form an ovary and each embryo may have several cotyledons.

gynandromorphs The animal, usually an insect that has both male and female cells. It also appears in the birds and mammals it is often due to the loss of an X-chromosome in a stem cell of a female (XX), so that all tissues derived from that cell are phenotypically male.

gynandrous (of flowers) Having stamens inserted on the gynoecium.

gynecology The study and medical treatment of the diseases of women, in particular those affecting the genitals and urinary tract.

gynobasic A style which arises near the base of ovary as in Labiatae.

gynodioecious A species in which female and hermaphrodite flowers are borne on separate plants.

gynoecium (gynaecium) All of the carpels of a flower taken together as a whorl.

gynogenesis The condition whereby a female animal must mate before she can produce parthenogenetic eggs.

gynogenone A diploid embryo with two maternal genomes produced by pronuclear transplantation.

gynomonecious Having both female as well as hermaphrodite flowers on the same plant.

gyrogonites Lime encrusted, fossilised oogonia encircling sheath cells of the charophyta.

gyrus (pl. gyri) One of the convolutions on the surface of cerebral hemisphere caused by infolding of cerebral cortex.

H

habitat The place where an organism tend to live (e.g. stream bank, forest, river, pond, land, sea and oceans). It is characterised by its physical features or by dominant plant types.

habituation

1. The learning, in which an animal's response to a stimulus declines with repetition of stimulus at same intensity.
2. A condition of being psychologically, dependent on a drug, with a desire to continue its used but not to increase the dosage.

hacking Part of the wildlife conservation, rehabilitation programme, where the animals or birds released in the wild for the first time, are periodically provided with food until they become independent.

Hadean An earliest eon in the history of the earth, from the time of the accretion of planetary material, around 4600 million years ago, to the date of the oldest known rocks. It marks the beginning of the geological record-about 3900 million years ago.

haem (heme) One of a group of iron-porphyrins, which are conjugated with proteins as prosthetic group, e.g. haemoglobin, myoglobin.

haemagglutination An agglutination of the red cells due to the mixing of blood of incompatible blood groups (e.g. group A and B).

haematology The study of blood, its constitution and the diseases connected with it.

haematophagous Organisms that feed on blood. e.g. horseflies and certain leeches.

haematoxylin A compound used in its oxidised form (haematein) as a blue dye in optical microscopy, particularly for staining smears and sections of animal tissue. It stains nuclei blue and is frequently used with eosin as a countrestain for cytoplasm.

haemerythrin A red iron-containing respiratory pigment protein. It contains two iron atoms to which two oxygen atoms can bind reversibly. It occurs in the blood of certain worms, especially sipunculids and priapulids and in lamp shells.

haemocoel The body cavity of arthropods and molluscs, containing blood. It is an enlarged blastocoel that reduces the coelom. Often functions as a hydrostatic skeleton.

haemocyte Blood cell, especially a red blood cell.

haemocytometer A device used to estimate the concentration of cells in blood samples.

haemodynamics Study of the flow of blood through the circulatory system.

haemoglobin (hemoglobin) Red proteinaceous respiratory pigment with Fe^{2+} containing porphyrin. It is found in red blood corpuscles to carry oxygen in blood. It comprises two pairs of polypeptide chains, known as α and β-chains (forming the globin protein), with each chain folded to provide a binding site for a haem group.

haemoglobin S An abnormal form of haemoglobin produced in sickle-cell disease.
See polymorphism.

haemoglobinic acid A very weak acid formed inside red blood cells when hydrogen ions combine with haemoglobin.

haemolymph Refers to the blood-like fluid of animals with open circulatory systems.

haemolysis The destruction of red blood cells.

haemophilia Sex linked recessive disease, which is transmitted from unaffected carrier female to some of the male offsprings.
In this disease, a single protein that is a part of cascade of proteins involved in the clotting of blood is affected. Due to this, in an affected individual, a small cut results in non-stop bleeding. The heterozygous female (carrier) may transmit the disease to sons. The possibility of a female becoming a haemophilic is extremely rare because mother of such a female has to be atleast carrier and father should be haemophilic.

haemopoiesis The blood formation in vertebrates includes both plasma and cells.

haemopoietic tissue The tissue that given rise to blood cells in the process of haemopoiesis. The haemopoietic tissue of the embryo and fetal stage of vertebrates is the bone marrow, lymph nodes, yolk sac, liver, spleen and thymus but after birth haemopoiesis occurs in the red bone marrow.

haemostasis The prevention of blood loss following rupture of blood vessels, which is effected by several physiological processes.

hair (pili)

1. Epidermal out growths of the Malpighian layer in mammalian skin contained within hair follicles. It consist mainly of hard keratin.
2. A single or many celled outgrowth from epidermal cell in plants with various functions (e.g. absorption) such as a trichome.

hair cell A mechanoreceptor ectodermal cell with one long cilium and a tuft of microvilli. Hair cells are found in the inner ears of vertebrates and the lateral lines of fish. Hair cells are important in our ability to hear and maintain balance.

hair follicle A narrow tubular depression in mammalian skin containing the root of a hair, lined with epidermal cells and extending down through the epidermis and dermis to its base in the subcutaneous tissue. The ducts of sebaceous glands empty into them.

hallucinogen A drug or chemical that causes alterations in perception (visual mostly), mood and thought, e.g. LSD.

hallux An innermost (first) digit on the hindlimb of a tetrapod vertebrate. In humans it is the big toe and contains two phalanges. It is absent in some mammals and in many birds, it is directed backwards as an adaptation to perching. *Compare* pollex.

halophyte The plant growing in and tolerating very salty soil typically of shores of tidal river, estuaries, saltmarshes, or alkali desert flats. e.g. species of *Spartina*.

halosere A series of successional stages leading to a climax, originating in a saline area.

haltere Modified hind wings of Diptera, for flight. One pair, club-shaped, protruding from side of thorax.

haplodiploid Organisms where one sex is diploid and the other sex is haploid. The females are diploid, the males haploid in all such known cases. Honeybees are a good example.

The queen and workers are female and diploid and the males are usually haploid.

haploid

1. A single-celled eukaryotic organism, having a single complete set of chromosomes.
2. A multicellular eukaryotic organism, having a single complete set of chromosomes in each somatic cell.

haplont Organism representing the haploid stage of a life cycle.

h

haplontic The life cycle pattern in many algae in which haploid spores divide mitotically and form the gametophyte. It is a dominant, photosynthetic phase in which gametophyte is free-living.

haplophyte Having the number of chromosomes characteristic of the gametes for the organism.

haplostele A type of protostele having a solid strand of stele with xylem on inside encircled by phloem, pericycle and endodermis.

haplotype The collective genotype of a set of linked loci.

hapten A foreign molecule that can bind to an antibody but does not evoke an immune response unless combined with a carrier protein.

hapteron (holdfast) A cell or cellular organ attaching a plant to the substrate, may be discoid or root like in structure.

haptonasty Refers to the movement induced by touch, though not influenced by the direction of the stimulus.

haptotropism (thigmotropism) The tropism in which stimulus is a localised contact. For example, a tendril in coiled form with a solid object. A twig, for instance; curves in that direction and coils round it.

Hardy-Weinberg equilibrium The situation in which the proportion of dominant and recessive genes remains constant from one generation to next, over a period of time. It is assumed that: (1) mating is random (2) no natural selection; (3) no migration (4) no mutation. In such *a* stable population, for a gene with two alleles, *A* (dominant) and *a* (recessive), if the frequency of *A* is *p* and the frequency of *a* is *q*, then the frequencies of the three possible genotypes (*AA*, *Aa* and *aa*) can be expressed by the **Hardy-Weinberg equation**

$$p^2 + 2pq + q^2 = 1,$$

where, p^2 = frequency of *AA* (homozygous dominant) individuals, $2pq$ = frequency of *Aa* (heterozygous) individuals and q^2 = frequency of *aa* (homozygous recessive) individuals. The equation can be used to calculate allele frequencies if the numbers of homozygous recessive individuals in the population is known.

harvest The process of gathering mature crops from the fields.

hatchling Young one that has just been hatched from an egg.

Hatch-Slack pathway
See C_4 pathway.

haustorium (pl. haustoria) A specialised organ of a parasite, which penetrates the host tissue and absorbs nutrients and water.

Haversian canals Channels carrying blood vessels and nerves, which ramify throughout bone, communicating with its surface and marrow. Sheets of bone and bone cells are arranged concentrically (lamellae) around the canals, known as Haversian system. It is the anatomical unit of a compact bone.

hearing

1. The faculty of perceiving sounds.
2. An opportunity to state ones's case.

heart A muscular organ, responsible for blood circulation. In mammals, it has two sides that each consist of an atrium and a ventricle. The right side pumps deoxygenated blood from the body to the lungs. The left side pumps oxygenated blood from the lungs to the body. It has a thick wall, composed of specialised muscle (cardiac muscles) and is surrounded by pericardium.

The contractions of the heart are initiated and controlled by the sinoatrial node (also called pacemaker). In an average adult human the heart contracts about 70 times per minute. The heart of other vertebrates are similar except in the number of atria and ventricles (there may be one or two) and in the degree of separation of oxygenated and deoxygenated blood. A great variation seen in the form and functioning of the heart of invertebrates.

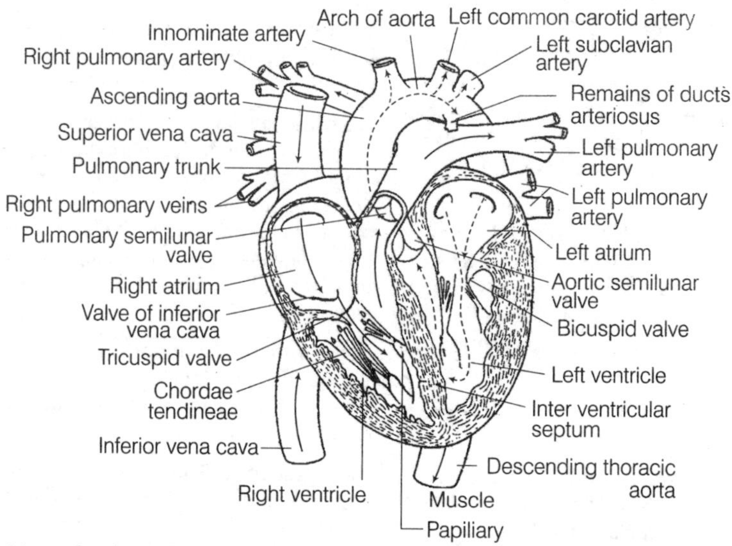

Mammalian heart showing direction of oxygenated (→) and deoxygenated (→) blood

heart failure The state of heart when it is not pumping blood effectively enough to meet the needs of the body. Also called congestive heart failure.

heartwood (duramen) Central layers of xylem tissue in tree trunk with no living cells; cease to conduct water and minerals. Provides only mechanical support. Darker in colour than sapwood, due to the blockage by substances such as resins.

heat sensitive pit The organ located on each side of the head, below a line from the eye to the nostril of some snakes, especially the vipers. It helps the snake detect their prey.

Heat-Shock Protein (HSP) Proteins synthesised by living cells in response to increased temperature or other forms of stress. They occur in both eukaryotes and prokaryotes and function mainly as molecular chaperones, protecting the cell's proteins as they become unfolded due to the heating and enabling them to refold correctly.

heater cell The specialised heat generating muscle cell found in certain fishes (e.g. marlin).

heliotrope An organism affected by heliotropism.

heliotropism The phototropism in which sunlight serves as the stimulus.

helix-turn-helix (helix-loop-helix) The first DNA binding protein motif. These were originally identified in bacteria but have been found now in both prokaryotes and eukaryotes.

helminthiasis The infestation with parasitic intestinal worms. e.g. filariasis, fascioliasis, etc.

h

helotism A form of symbiosis in which one partner benefits more than the other. For example, the fungus in a lichen thallus benefits more than the alga.

helper T-cells ($T_H$cell) A kind of T-cell without which some B-cells cannot make antibodies. They also help other T-cells to respond to antigens. Mature T_H cells express surface protein CD_4 one its surface.

hemal Pertaining to the blood or blood vascular system.

hemangiosarcoma A sarcoma arising in the cells that line the blood vessels, occurring mostly in dogs and cats (rarely).

hematocolpos A blockage of the vagina by an imperforate hymen resulting in retention of menstrual blood.

hemerythrin A red oxygen-carrying blood pigment similar to haemoglobin in containing iron and found in various invertebrates.

hemicellulose Carbohydrate related to cellulose, but is a mixed polysaccharide, containing more than one type of sugar. It is easily hydrolysed by dilute acids. Forms the part of plant cell walls and may function as reserve food in seeds, etc.

Hemichordata A small phylum of soft bodied marine invertebrates, with unsegmented body with three regions, i.e. proboscis, collar and trunk. Includes two class–Enteropneusta (acorn worms) e.g. *Balanoglossus* and *Pterobrachia* (e.g. *Cephalodiscus*).

hemicryptophyte A perennial plant, usually non-woody, with its over-wintering buds at soil level. The buds are protected by the soil or surface litter.

hemicyclic A flower in which some parts are in spirals and others are in whorls.

Hemimetabola Super-order of insects in which the larva closely resembles the adult insect throughout its entire period of development (Hemimetabolous). There are several molts and during the later stages (instars), wings appears. During the last molt, the fully developed adult (imago) emerges.

hemiparasitic (semiparasitic) A parasitic plant capable of producing some of its nutrients *via* photosynthesis, e.g. mistletoes.

hemiplacenta The chorion, the yolk sac and generally allantois which together serve as an organ of nutritional supply to the uterine young of marsupials.

Hemiptera Insect order that includes aphids, bugs, leaf hoppers, etc. The members have oval flattened bodies with two pairs of wings, folded back across the abdomen at rest. The forewings are hardened, either at their bases or uniformly. The mouthparts are modified for piercing and sucking, with long slender stylets forming a double tube.

hemizygote The individual hemizygous for a gene.

hemizygous gene A gene for which there is only one copy present in an organism. For example, in mammals, males are hemizygous for genes on the Y-chromosome.

hemoclasia (also haemoclasis) The disintegration of erythrocytes.

hemocyanin (also haemocyanin) A blue copper containing blood pigment found in many molluscs and arthropods.

hemoglobinometer (also haemoglobinometer) An instrument for determining the percentage of haemoglobin in the blood.

hemoptysis (also haemoptysis) The bloody discharge expelled from mouth or nose; derived from haemorrhage of lungs, trachea or larynx.

hemorrhage (also haemorrhage) The abnormal discharge from the blood vessels, whether internal or external.

hemostat (also haemostat) A medicine or surgical tool used to stop blood flow.

hemotoxin (also haemotoxin) A toxin that acts by lysing erythrocytes.

Hensen's node (primitive node; primitive knot) A knob like accumulation of cells, located immediately anterior to the primitive streak and is analogous to the dorsal lip of the blastopore in the blastula of an amphibian embryo. It acts as an organiser of development in the early embryo of birds and mammals. It consist of mesoderm cells that give rise to notochord.

heparin The substance which prevents blood clotting by stopping conversion of prothrombin to thrombin and by neutralising thrombin.

hepatic Of or pertaining to the liver.

hepatic portal system System of veins leading from the digestive tract to capillaries in the liver of a vertebrate. Thus, the absorbed products of digestion (except fat) go straight to liver and not to heart.

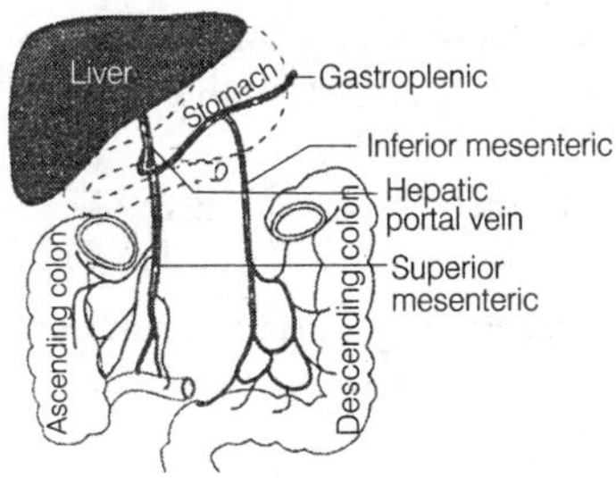

Hepatic portal system

Hepaticae (Liverworts) A bryophyte class placed in a separate phylum-Hepatophyta. It contain prostrate thallose dichotomously branching plants with unicellular rhizoids, e.g. *Marchantia*.

hepatitis Inflammation of the liver.

hepatitis A Refers to water-borne and food-borne disease of the liver caused by a highly resistant RNA virus and accompanied by abdominal pain, degeneration of the liver tissue and jaundice.

hepatitis B Refers to the blood-borne disease of the liver caused by a fragile DNA virus and accompained by jaundice, abdominal pain and damage of the liver tissue.

hepatocyte The most abundant type of specialised epithelial cell in liver. They are arranged around a central vein as lobules and functions in detoxification, bile production and metabolism.

hepatomegaly The enlargement or swelling of the liver.

hepatorrhexis Rupture of the liver.

hepatotoxin A toxic chemical that can cause liver damage, e.g. ethanol.

herb Plants without wood in their stems and roots as distinct from shrubs and trees.

herbaceous Having the characters of a herb.

herbarium Collection of preserved plant specimens for identification and reference purposes.

herbivore An animal that feeds on plant products, e.g. cow and rabbit.

heredity The transmission of traits from one generation to another, *via* chromosomes.

heritability The proportion of the total variation in phenotype of a population due to genetic factors, as opposed to environmental influences.

hermaphrodite (bisexual)

1. An animal with both male and female reproductive organs (in zoology).
2. A plant having both stamens and carpels in same flower (in botany).

hernia The protrusion of part of an organ usually a portion of the small intestine, through the abdominal wall.

heroin (diacetylmorphine) A highly addictive narcotic compound that is a synthetic derivative of morphine. It is easily absorbed by the brain, due to its lipid-like nature, and is used as a sedative and powerful analgesic. It is often abused by drug users.

herpesvirus One of a group of DNA containing viruses, about 100 nm in diameter that cause diseases like herpes and chicken pox.

herpetology Study of reptiles and amphibians.

hesperidium A segmented berry with leathery rind. The rind contains oil glands and is lined by white mesocarp (commonly called pith) e.g. citrus fruit.

heterobaric leaf anatomy The anatomical arrangement in leaf where smallest veinlets span the upper and lower surfaces of leaf blade and project as ribs on leaf surface, e.g. sunflower.

heterocercal tail The type of fish tail (caudal fin), characteristics of Chondrichthyes, in which vertebral column extends into dorsal lobe of fin, which is larger than the ventral lobe. Compare homocercal.

heterochlamydeous A flower having two kinds of perianth segments (sepals and petals) in distinct whorls.

heterochromatin The chromatin that is condensed (visible with a light microscope) during interphase. Heterochromatin is late replicating, transcriptionally inactive and is composed mostly of repetitive DNA.

heterochrony A difference in timing during the course of development of characters, which already present in ancestors, sometimes proposed as a mechanism of evolutionary change.

heterocyst Cyanobacterial specialised cell that carry out nitrogen fixation. They are enlarged cells with no chlorophyll but house the nitrogenase enzyme.

heterodont Animals possessing teeth in the jaws which are not similar in shape and structure, e.g. cows, humans, etc.

Heteroduplex DNA Double stranded DNA formed by annealing of two single stranded DNA from different DNA molecules. Occur as a result of genetic recombination and can be produced by DNA hybridisation (*in vitro*).

heteroecious Denoting rust fungi that require two host species to complete their life cycle. An example is *Puccinia graminis*, the stem rust of cereals and grasses.

heterogametes Two morphologically distinct male and female gametes. Opposite to homogametes.

heterogametic sex The sex that produces two types of sex cells. e.g. in mammals, the Y-chromosome in male being shorter than X-chromosome.

heterogamy The alternation of generations in which parthenogenesis alternates with bisexual reproduction in the life cycle. It occurs in certain aphids.

heterogeneous nuclear RNA (hnRNA) The assortment of RNA molecules assembled around newly transcribed RNA or pre-*m*RNA) in the cell nucleus. The *hn*RNA associates with proteins to form **heterogeneous nuclear ribonucleoprotein** (*hn*RNP). After release from the chromosome and processing of the pre-*m*RNA by the spliceosome. The mature messenger RNA associates with proteins to form messenger ribonuleoprotein (*m*RNP).

heterograft (xenograft) A type of graft from one organism to another of a different species.

heterokaryosis The presence of two or more nuclei with differing genotypes within a single cell. Fusion of fungal hyphae of differing genetic complements leads to heterokaryosis. Induced fusion of different animal or plant cells is also effective in causing heterokaryosis.

heterokonts An assemblage of organisms with motile cell or spore that possesses two flagella of unequal length. e.g. diatoms.

heteromerous The lichen thallus in which the layer of algal cells lie between two layers of fungus hyphae.

heteromorphism The existence of more than one form, used especially with reference to life cycles in which the alternating generations are markedly different morphologically, as in ferns and jellyfish.

heterophylly Denoting plant species that have more than one form of foliage leaf on the same individual. Where young leaves differ from adult leaves, as in ivy.

heteroplasmy The presence of more than one type of genome from organelle, within a cell or organism. e.g. *mt*DNA.

heterosis (hybrid vigour) In a hybrid, traits falling outside the range of parental variation. e.g. a hybrid might be larger than either of its parents.

heterosporous Pertaining to pteridophytes. Having a sporophyte that produces two types of gametophytes of different sexes; opposite to homosporus.

heterostyly Dimorphism in which the styles of flowers of the same species are of different length, thus dividing the species into groups. An example is the primrose, which is divided into pin-eyed and thrum-eyed types. Pin-eyed have long styles and short stamens and thrum-eyed have short styles and long staments.

heterothallism A condition in algae and fungi in which sexual reproduction occurs only through participation thalli of two different mating types each of which is self-sterile.

heterotherm An organism that is intermediate of endotherm and an ectotherm, in its ability to regulate body temperature. Some small birds and mammals can reduce their metabolic rate when conservation of food energy supplies is necessary.

heterotrichy The condition occurring in algae in which the algal body is composed of both prostrate filaments and protruding upright filaments common in filamentous forms.

heterotroph An organism that cannot produce all of the organic molecules it requires from an inorganic substrate. It instead obtains organic molecules, by eating other organisms or substances produced by them. Used in opposition to autotroph.

heterotropic bacteria Most abundant in nature. Majority are important decomposers. Helpful in making curd from milk, production of antibiotics, fixing nitrogen in legume, roots, etc. Some are pathogens causing disease in humans.

heterozygosity Presence of different alleles at one or more loci on homologous chromosomes.

heterozygous Having two different alleles at a given locus on a homologous chromosomes. Usually, the dominant allele from the two is expressed in phenotype.

hexacanth Having six hooks, thorns or barbs, e.g. onchosphere.

hexaploidy A condition in which a cell, or individual cells of an organism have, six complete sets of chromosomes.

Hexapoda (Insecta) The arthropod class. The members are 0.5-300 mm in length with body consisting of a head, a three segment thorax and 3 pairs of legs and 1-2 pairs of wings and abdomen of 11 segments.

hexose A carbohydrate sugar (monosaccharide) with six carbon atoms per molecule (such as glucose or fructose).

hibernation A physiological state, characterised by reduced metabolic activity that allows an animal to survive cold season with little or no food. Stored body fat supplies energy.

hidradenoma Adenoma of the sweat glands.

hidroschesis Non-secretion or suppression of sweating.

hierarchy A system in which members of an organisation or society are ranked according to relative status or authority.

High-Performance Liquid Chromatography (HPLC) A sensitive technique for separating or analysing mixtures, in which the sample is forced through the chromatography column under pressure.

Hill reaction Occurs in chloroplasts. Light induced transport of electrons from water to acceptors such as potassium ferricyanide (Hill reagents) which do not occur naturally, accompanied by the release of oxygen. The electron acceptors are reduced during the process.

hilum The scar on seed coat, at the point of attachment of seed to funicle.

hindbrain (rhombencephalon) Hindmost of the three divisions of vertebrate brain. Comprises cerebellum and medulla oblongata.

hindgut

1. Posterior part of alimentary canal of vertebrates, consisting of posterior section of colon.
2. The last part of alimentary canal of arthropods. It consist of ileum, colon and rectum and is lined with epithelium secreting protective chitin lining.

hip girdle *See* pelvic girdle.

hippocampus A ridge extending over the floor of each lateral ventricle of the vertebrate brain. It is linked together by a band of nerve fibres, the hippocampal commissure. It functions in short-term memory and the expression of instinct and mood.

Hirudinea An Annelida class that contains the leeches. They have flat body, with each segment subdivided externally into annuli, and anterior and posterior suckers for attachment to host.

histamine A local signalling molecule released by mast cells during an allergic reaction and as part of the normal inflammatory responses of the body. It causes the initial symptoms of an allergic response and also causes dilation and increased permeability of small blood vessel, which results in symptoms such as localised swelling, itching, sneezing and runny eyes and nose.

histidine A heterocyclic essential amino acid.

histiocyte A wandering amoeboid cell, capable of ingesting foreign particles, found in the matrix of connective tissue.

histochemistry The study of distribution of particular chemicals by chemical reactions. It makes use of specific staining methods electron microscopy, chromatography and autoradiography.

histocompatibility The extent to which an organism's immune system will tolerate tissue grafts from another organism. Two classes of proteins are important in histocompatibility and immune responses–class I MHC (Major Histocompatibility Complex) proteins and class-II MHC proteins. MHC-I occur on most body cells and MHC-II occur only on certain immune system cells, such as macrophages and B-cells.

histocompatibility antigen An antigen (often cell surface glycoprotein) that initiates an immune response resulting in rejection of a homograft.

histogenesis The differentiation of tissues.

histology The study of the microscopic structure of cellular tissues.

histone Positively charged eukaryotic proteins that bind DNA, (which has a negative charge). They form the nucleosome, the basic subunit of chromatin. Histones are organised to form a unit of eight molecules called histone octamer.

HIV (Human Immunodeficiency Virus) Retrovirus that causes AIDS in humans by infecting and ultimately destroying cells such as T_H and macrophages. (*See* AIDS).

HLA system (Human Leukocyte-Antigen system) A group of histocompatibility antigens found on the surface of human body cells and are encoded by the major histocompatibility complex of genes. It is the most important system of self-antigens in humans.

hnRNA The heterogeneous nuclear RNA.

holandric Describing a heritable trait that is inherited only by the males. Such traits are determined by genetic loci that occur only on the Y-chromosome, and hence are present only in males.

holarctic The zoogeographical region amalgamating the Palaearctic and Nearctic.

holdfast (hapteron) The anchoring process in algae that attaches the plant to the substratum. It is often dislike.

Holliday intermediate A heteroduplex formed by complementary strands of a pair of parental DNA molecules. These are joined at the holliday junction and are an intermediate in genetic recombination.

holoblastic Describing a cleavage type seen in certain eggs in which the first cleavages divide the egg cytoplasm and yolk completely (e.g. sea urchins and frogs).

holocarpic Fungi having mature thallus converted entirely into a reproductive structure, on maturation.

Holocene (recent) The most recent geological epoch, being the second epoch of quaternary period which began at the end of the last ice age, about 100 years ago, to present day.

holocrine secretion Secretion that involves the disruption of the entire cell to release its accumulated secretory vesicles.

holoenzyme A catalytically active complex made up of an apoenzyme and a coenzyme. The former is responsible for the specificity of the holoenzyme while the latter determines the nature of the reaction.

hologamy The fusion of two mature cells, each of which has been completely changed into a gametangium.

holometabolous Belonging to the insect super-order-Holometabola; comprised of all insects that undergo a complete metamorphosis. A larva bears no resemblance with the fully developed adult insect. During development an individual molts repeatedly, and eventually changes into a pupa. During the pupal stage, the original larval organs are destroyed and the organs of the adult stage (imago) develop. The imago then emerges from its chrysalis or cocoon.

holophytic Type of nutrition in which complex organic molecules are synthesised from inorganic molecules using light energy. Another term for photoautotrophic.

Holothuroidea A class of Echinodermata that contains the sea cucumbers. Body cylindrical, with mouth at one end and anus at the other; soft, muscular body wall with skeleton of scattered, minute plates; no, spines or pedicillariae; suckered tubefeet bottom-dwellers, often burrowing and bear tentacles (modified tube feet) around mouth for feeding.

holotype (type specimen) An individual organism, upon which naming and description of new species depends.

holozoic Describe organisms feeding on the solid complex food by ingestion, which is subsequently digested and absorbed. All vertebrates and many invertebrates show this mode of nutrition.

homeobox A conserved sequence found within the homeotic genes of *Drosophila*; also found in embryonically expressed genes of mammals and amphibians. Homeobox are about 180 base pairs long and sequences of bases is similar across a wide range of species. The sequence encodes the amino acids of a peptide sequence that enables parent protein to bind on DNA.

homeologous Chromosomes or genes that are only partly homologous.

homeosis The formation of a normal part or organ in place of another at an abnormal site.

homeostasis A tendency towards the maintenance of a steady physiological state.

homeothermic (homoiothermic) Refers to the ability to maintain a constant body temperature.

homeotic gene A gene that controls overall development (morphogenesis).

hominid Recent humans and extinct ancestral and related forms of family-Hominidae.

Hominae Family of the mammalian order-primates. It includes not only humans, but also the great apes (chimpanzees, orangutans and gorillas).

hominoid A member of the super family-Hominoidea containing gibbons, apes and hominids.

Homo A genus of primate family-Hominidae that includes modern humans (*H. sapiens*, as the only living representative) and various extinct species. The oldest first appeared in Africa perhaps 2.2-2.4 million years ago and used simple stone tools.

homobaric leaf anatomy A form of leaf anatomy in which there are few or no internal barriers to lateral movements of air inside the leaf.

homocercal tail The type of tail found in adult bony fish, in which the vertebral column does not extend to the end of the tail and the caudal fin has two lobes of equal size.

homodont Describing animals in which all the teeth in jaws are similar in shape and structure, e.g. frog, fish, etc.

homogametes The male and female gametes which are so similar in appearance that it is not possible to categorise them separately, e.g. in some algae.

homogametic sex The sex producing a single type of gamete. In mammals, XX.

homogamy The condition in flowers in which the anthers and stigmas ripen at the same time, to encourage self-pollination. Occurs in closed cleistogamous flowers that appear late in the season in certain plants.

homograft (allograft) A type of graft between individuals of same species.

homoiothermy (homeothermy) The maintenance of a constant body temperature higher than that of the environment. Characteristic of mammals and birds.

homokaryon (homocaryon) The cell, fungal hypha or mycelium with more than one genetically identical nucleus in its cytoplasm; in fungi such nuclei commonly are haploid.

homologous chromosomes Chromosomes that pair at meiosis. The members of each pair are similar to one another in size and shape.

homologous organs Organs in plants and animals which possess same fundamental structure and origin though they differ in their external appearance as well as in their function, e.g. forelimbs of land vertebrates.

homologous recombination The exchange of DNA between synapsed chromosomes.

homologous traits Traits that are presumed to be similar due to them, being inherited from a common ancestor (and not because of convergence).

homology

1. Regions of chromosomes having the same loci, in the same order.
2. Any similarity between organisms considered to reflect relationship (the term homology is often used to refer to what should be more properly called similarity between DNA or protein sequences).

homomorphic chromosomes Another name for homologous pair of chromosomes.

homonomy Characters of two or more taxa, found on different parts of organism.

homophyllous Having foliage-leaves all of the same kind.

homoplasmy The presence within a cell or organism of genetically identical mitochondria.

h

homoplastic (homoplasy) The same structure and manner of development, but not descending from a common source. Such a similarity may arise due to the convergent or parallel evolution.

homosporous Similar kinds of spores present in majority of pteridophytes.

homostyly The general condition found in plants in which the styles of all flowers of the same species are of about the same length.

homothallism A condition in algae and fungi, when sexual reproduction takes place in a colony derived from a single spore. Each thallus is self-compatible.

homotropous In a plant, embryo having the radicle turned toward the hilum.

homotypy Being of the same basic type.

homozygote A homozygous individual.

homozygous Having two identical alleles at a given locus on homologous chromosomes.

honey guides Colourful spots or stripes in petals of some flowers which lead to nectaries.

hormogonium A filament produced by filamentous cyanobacteria that becomes detached from the parent organism and acts as a means of vegetative propagtion.

hormone

1. A chemical messenger, secreted in small quantities into the blood stream by endocrine gland or group of cells. These regulates the growth or functioning of a specific tissue or organ in a distant part of the body. For example, the hormone insulin controls the rate and manner in which glucose is used by the body.
2. *See* plant hormone.

hormone-receptor Specific proteins located in target tissues to which hormones bind to produce their effects. Receptors present on the cell membrane of target cells are called membrane bound receptors and the receptors present inside the target cells are called intracellular receptors.

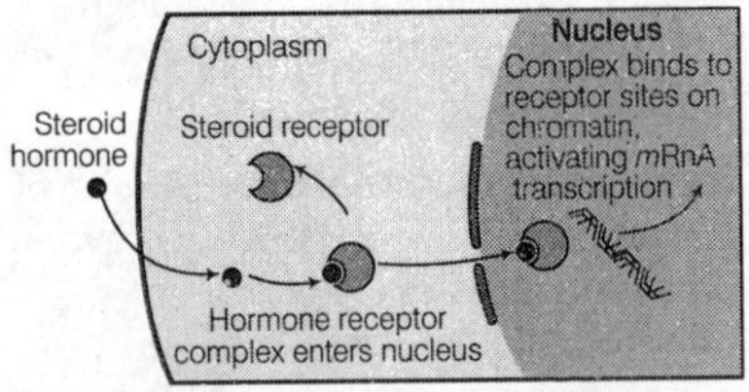

Hormone receptor complex at cellular level

host

1. An organism used as a nourishment source by another organisms (a parasite).
2. A bacterium harboring a plasmid or bacteriophage.
3. The recipient of a graft.

Human Artificial Chromosome (HAC) A mini chromosome that is constructed artificially in human cells.

Human Chorionic Gonadotropin (HCG) The hormone secreted by placenta in women during pregnancy. It prolongs the period of active secretion of oestrogens and progesterone by corpus luteum.

Human Genome Project (HGP) A mega project sequencing human genome launched in 1990. It is a 13 year project coordinated by the US Department of Energy and National Institute of Health. Project was completed in 2003.

Human Genome Therapy Insertion of normal DNA directly into cells to correct a genetic defect. The treatment of disease is done by replacing, altering, or supplementing a gene that is absent or abnormal and whose absence or abnormality is responsible for a disease.

Human Placental Lactogen (hPL) The hormone produced by placenta in women after about 5 weeks of pregnancy. Most important effect is to switch material metabolism from carbohydrate to fat utilisation.

humeroulnar Relating to both the humerus and the ulna.

humerus In humans, the bone of the upper arm. In tetrapods animals, the bone of the upper forelimb.

humidity The amount of water vapour in the air.

humor The fluid serving a function within the body.

humoral Pertaining to the various fluids functioning within the body and their contents.

humoral immune system A portion of the immune system that produces antibodies that circulate in the blood and lymph.

humus A dark-coloured amorphous substance in layer of soil obtained from the decomposition of organic matter by microbial action.

Huntington's disease A progressive and fatal disorder of nervous system which is caused by an expansion of a trinucleotide repeat and inherited as a dominant trait and develops between age of 30-50 yrs.

hyaline cartilage A cartilage consisting largely of glycosaminogly can, giving it a shiny glass like appearance. It gives flexibility and support at the joints.

hyaloid membrane The transparent membrane enclosing the vitreous humour of the eye.

hyaloiditis The inflammation of the hyaloid membrane.

hyaloplams *See* cytosol.

hyaluronic acid An organic acid that has lubricative properties. It is a major component of articular cartilage and skin, helping in tissue repair and the lubrication of joints. It also plays a role in cell movement and proliferation and may contribute to the growth of certain tumours.

hyaluronidase Enzyme that breakdown hyaluronic acid, thereby decreasing its viscosity and increasing the permeability of connective tissue. Used in medicines.

hybrid Offspring of genetically distinct parents.

hybrid dysgenesis Infertility, increased incidence of chromosome mutations and other defects observed in the progeny of laboratory-reared organisms crossed with wild organisms. Notably studied in strains of *Drosophila* fruit flies, when laboratory females are crossed with wild males.

hybrid infertility The reduction in the quality and/or number of gametes produced by a hybrid (in comparison with its parents).

hybrid inviability The impairment of a hybrid's ability to survive (in comparison with its parents).

hybrid swarm Very variable series of organisms resulting from the continual crossing, recrossing and back crossing of the hybrid- generations of two species.

hybrid vigour *See* heterosis.

hybrid zone A geographic region where two distinct populations come into contact and produce hybrids.

hybridisation

1. Interbreeding between distinct genetic types, including types treated as distinct species.
2. The process of joining two complementary strands of DNA, or one each of DNA and RNA, to form a double-stranded molecule.

hybridoma A type of cell used in the artificial production of monoclonal antibodies. It is a hybrid of two types of cells, a normal antibody-producing lymphocyte and a myeloma cell. The hybrid cell has the capability of indefinite cell growth, as well as produces the desired antibodies.

hydathode Water pore of water secreting gland found on the edges and tips of leaf veins of many plants.

hydatid cyst A sexual multiplicative phase of some tapeworms (e.g. *Echinococcus*) within the secondary host (e.g. man, sheep and pig). It consists of a fluid-filled sac that produces thousands of secondary cysts (brood capsules), each of which buds off a dozen or so retracted scolices. In humans the, cysts may become malignant and send **metastases** around the body with sometimes fatal results.

Hydra A tiny freshwater animal related to the sea anemones. It has a tubular body ending in a ring of tentacles containing stinging cells that are used to capture preys like water fleas and other small marine insects.

hydranth A polyp of a colonial hydrozoan specialised for feeding. It has a mouth surrounded by tentacles.

hydrocephalus An abnormal retention of cerebrospinal fluid in the brain. It results from blockage of normal circulation of the fluid. In children, affected by this condition, the cranium is enlarged and thin.

hydrocortisone (cortisol) A steroid hormone produced by the adrenal cortex having glucocorticoid activity.

hydrogen carrier (hydrogen acceptor) A molecule that accepts hydrogen atoms or ions and itself becoming reduced in the process. The electron transport system, involves a series of hydrogen carriers, including NAD and FAD, which pass on the hydrogen to the next carrier in the chain.

hydroid A specialised elongated water-conducting cells found in the cortex of certain mosses. These are aligned end to end and lose their cytoplasm and their end walls partially breakdown, becoming permeable to water and forming a rudimentary vessel, when mature.

hydrolase An enzyme that catalyses a hydrolytic reaction. Digestive enzymes are good examples. Hydrolases play an important role in changing insoluble food materials into a soluble form, which can then be assimilated.

hydrological cycle (water cycle) A constant exchange of water between air, land, ocean and the living organisms. The whose process by which water evaporates and falls on the land as rain and later flows back into ocean *via* rivers is called the **water cycle**. All of the water that falls on the land does not immediately flow back into the sea. Some of it seeps into the soil and becomes a part of the underground reservoir of freshwater. While, some of the underground water comes up to the surface through springs or wells. Water is also released by plants through the process of transpiration.

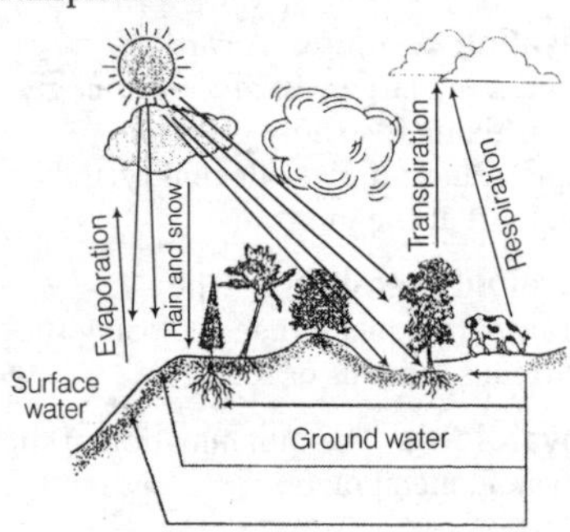

Water cycle in nature

hydrophily The pollination in which water carries the pollen from anther to stigma. It occurs in some pondweeds.

hydrophyte Plant adapted to live in water or extremely wet areas. Show certain adaptations that include reduction of mechanical and supporting tissues and vascular tissue, absence or reduction of a root system and specialised leaves (either floating or finely divided, with little or no cuticle) e.g. waterlilies and certain pondweeds.

hydroponics Growth of plants in culture solutions rather than soil. The solutions contain the correct balance of all the essential mineral requirements. The method has been used commercially for determining effects of mineral deficiencies though is not widespread.

hydrosere A sere beginning in a wet habitat.

hydrosphere Refers to the part of the physical environment that comprises all the liquid and solid-water at or near the Earth's surface.

hydrostatic skeleton The support system found in soft-bodies invertebrates, which relies on the incompressibility of fluids contained within the body cavity. For example, in earthworms the coelomic fluid is under pressure within the coelom and therefore, provides support for internal organs.

hydrotropism A tropism (growth movement) in response to water.

Hydrozoa The class-Cnidaria containing hydroids, corals, siphonophores, etc. Usually there is alternation of generation in the life cycle between a sessile polyp (hydranth) phase and a pelagic medusoid phase; but one or other may be suppressed. Most polyp forms (but not *Hydra*) are *Colonial*, showing division of **labour** between feeding and reproductive individuals. Gonads **ectodermal**.

hygrophilous Plants which need large supply of moisture to grow.

hygroscopic Readily absorbing and retaining moisture.

hylidae The anuran family comprised of the true tree frogs (frogs with well-developed gripping discs on the tips of the fingers and toes).

hymen A fold of membrane sealing the vaginal opening. The occlusion need not be complete even in a virgin.

hymenium Layer of regularly arranged spore-producing structures found in the fruit bodies of many fungi (e.g. Ascomycete, Basidiomycete).

hymenolepis A genus of small tapeworms parasitising human.

Hymenoptera An order of insects with two pairs of wings and biting mouth parts. The females have a well-developed ovipositor and the males are usually haploid. In all cases, hymenoptera have a larval stage which does not resemble the adult. Ants, bees, wasps are contained in this order.

hyoepiglottic Pertaining to the hyoid bone and the epiglottis.

hyoid arch The second visceral arch, lying between the jaws and spiracle. It forms the hyoid apparatus through its ventral elements in tetrapods. In fishes its dorsal element modifies to form hyomandibular.

hyoid bone U-shaped bone at the base of the tongue that supports the muscle of the tongue.

hyoscyamine An alkaloid obtained from Solanaceae plants, e.g. henbane, used as a sedative and antispasmodic.

hyostylic jaw suspension A type of jaw suspension (e.g. in most fishes). The attachment between cranium and upper jaw is made by a ligament at the front end and by the hyomandibular at the rear end Compare Amphistylic Jaw Suspension; Autostylic Jaw Suspension.

hypercalcemia Excessive calcium in the blood.

hyperglycaemia Excessive glucose in the blood.

h

hypermorph A mutant gene, having a similar, but greater effect than the non-mutant allelomorph.

hypermorphosis A form of heterochrony in which, during the course of evolution, the total development time is extended, permitting the addition of new stages to the end of the ancestral development sequence. However, the rate of development remains unchanged.

hyperostosis Excessive, abnormal bone growth.

hyperparasite A parasite whose host is itself a parasite.

hyperphoria A condition in which visual axis of one eye tends to turn upward.

hyperplasia An increase in size or an organ or tissue due to cell division (i.e. increase in number of cells).

hyperpolarisation An increase in the polarity of the potential difference across a membrane of a nerve or muscle cell. Caused by pumping of ions across membranes so that differential concentrations are created on either side. Stronger stimulus are needed to evoke a response.

hypersensitivity

1. Increased or abnormal sensitivity to compounds, which can elicit a specific immune response accompanied by tissue damage. Reactions include allergies and anaphylaxis. (in immunology)
2. A reaction by the tissues of a plant in response to invasion by pathogenic viruses, bacteria, fungi or other components. (in botany)

hypertension The term used for blood pressure that is higher than normal (120/80). In this measurements 120 mm Hg (millimetres of mercury pressure) is the systolic, or pumping pressure and 80 mm Hg is the diastolic or resting pressure. 140/90 or higher, shows hypertension.

hyperthermophile An organism that thrives in extremely hot environments from 60°C (140°F) upwards.

hyperthyroidism Disease caused by excessive production of thyroid hormone. It leads to weight loss and overactivity.

hypertonic solution The solution with a higher concentration of solute than some other specified solution (and which, therefore has a higher osmotic pressure than the other solution). A hypertonic solution, which has a higher concentration of salt than blood, causes cells in contact with it to swell, or even burst.

Hypertrophy An increase in the size of a tissue or organ due to increase in the size of its component cells. It often occurs in response to an increased worked in an organ, which may result from malfunction or disease.

hyperventilation An increase in the amount of air taken into the lungs caused by an increase in the depth or rate of breathing.

hypha (pl. hyphae) The individual filament of a fungal thallus. It forms a loose network (mycelium) or aggregate into fruiting bodies. It may be branched or unbranched and may or may not cell have cross walls.

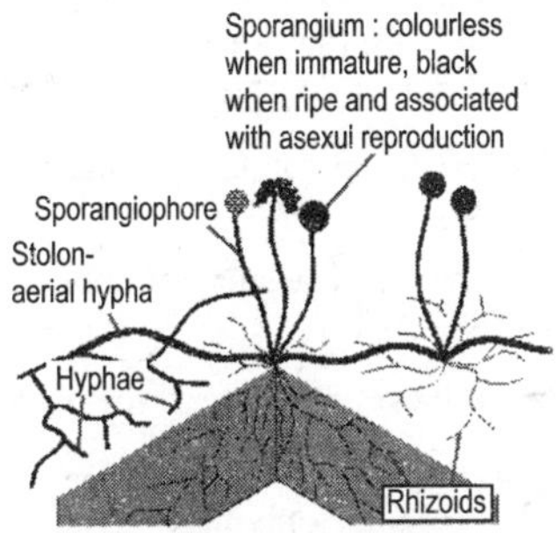

Hyphae growing on a fungus

hypochondriac A person affected by an abnormally heightened and unreasonable fear of disease.

hypochromia

1. Lack of or a reduced degree of colouration.
2. Deficiency of haemoglobin.

hypocotyl The part of a seedling axis between the radicle and the cotyledons.

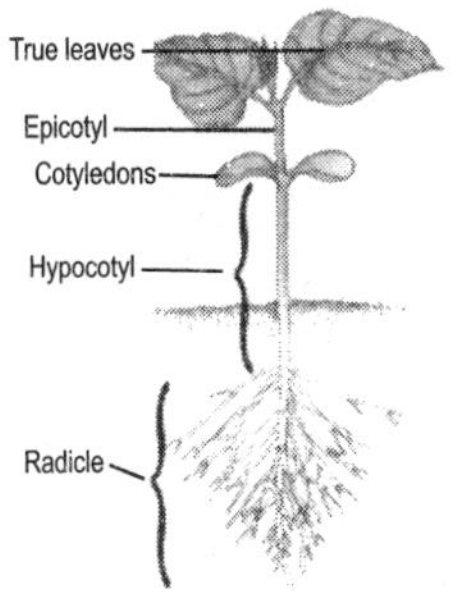

Position of hypocotyl in plant

hypodermis (exodermis) The outermost layer of cortex, next to epidermis in plants.

hypogeal Developing or growing beneath the soil. Hypogeal germinations as seen in oak, cotyledons remains below ground.

hypoglossal nerve (Cranial nerve XII) One of the pair of nerves that arises from posterior region of medulla oblongata in the brain of higher vertebrates. Carries motor nerve fibres to tongue muscles.

hypoglycaemia The abnormally low level of glucose in the blood.

hypogynous Of a flower, in which the other parts remain below the gynoecium. Gynoecium superior.

hypolimnion The lower layer of water in a lack.

hyponasty The lower side of an organ growing more quickly than the upper resulting in an upward bending.

hypophysectomy The surgical removal of pituitary gland.

hypoplasia A state of having incomplet5deficiency.

hypoploids Diploids lacking a piece or pieces of chromosomes from their complement.

hypostasis Situation in which expression of one gene (the hypostatic gene) is prevented in the presence of another non-allelic gene (the epistatic gene).

hypostomatic Having stomata on the lower surface of leaf.

hypothalamus A region of the vertebrate force brain involved in emotions, regulation of parts of the autonomic nervous system and providing the link between the endocrine system and the nervous system by its influence on the pituitary gland.

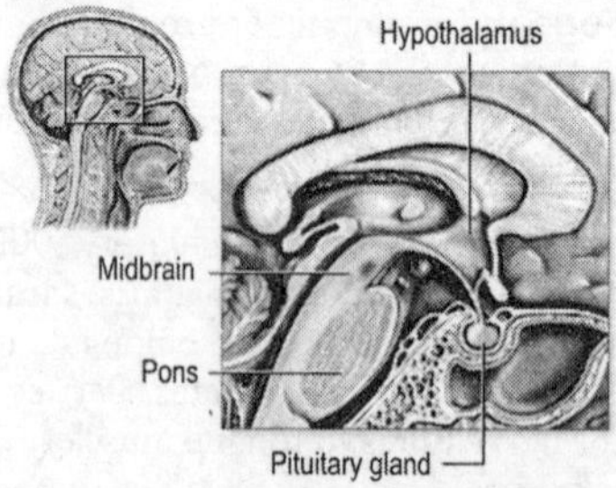

Position of hypothalamus in human brain

hypothyroidism Disease caused by insufficient hormone production by thyroid gland. Patient fails to attain normal physical or mental development. In children, disease results in cretinism and in adults causes myxoedema.

hypotonic solution A solution with lower solute concentration than another. Hypotonic solutions, with lower salt concentration than blood, cause cell shrinkage.

hypoxanthine A purine occasionally found in nucleic acids (in the anticodon of *t*RNAs in the form of its nucleoside inosine). It has a tautomer known as 6-hydroxypurine.

hypoxia The deficiency of oxygen in body tissue.

hysterectomy The surgical removal of the uterus.

hysterorrhexis Rupture of the uterus.

H-zone Region of a striated muscle fibre that contains only myosin, i.e. thick filaments. The H-zone appears as a lighter band in the middle of the dark. A band at the centre of a sarcomere.

I

IAA (Indole Acetic Acid) Auxin, one of the most common growth substance in plants. *See* auxin.

iatrology Medical science.

I-band One of the two types of bands present in striated and cardiac muscle fibres. It differs from the other type of band (the A-band) in containing only actin. It is visible as light bands at either side of a sarcomere.

ICAM (Inter Cellular Adhesion Molecule) Family of cell adhesion molecule (CAM) occurring on the surface of endothelial cell lining blood vessels. They help to fixate passing leucocytes at inflammation sites.

ice age The prolonged period of extensive glaciation. The last major ice age (of Pleistocene epoch) ended about 10,000 years ago.

ichnite (also ichnolite) A fossilised footprint or track of animal.

ichnology The scientific study of the traces of animal activity such as footprints, burrows, trails and borings.

ichthyic Of, pertaining to or like fish.

ichthyology A branch of zoology dedicated to the systematic study of fishes.

ichthyosaur These are extinct reptiles belonging to order Ichthyosauria, which lived during Mesozoic era. Vertebral column curved downwards to form reverse heterocercal tail; legs modified into paddles, with addition of extra digits. Fleshy dorsal fin and upper tail lobe lacked skeletal support. Jaws with homodont dentition. Became increasingly stream lined for aquatic locomotion, convergence with porpoises, etc. ICSH Interstitial Cell Stimulating Hormone. *See* luteinising hormone.

icteritious Colour of the skin when it is affected by jaundice.

icterohematuria Disease of sheep caused by the protozoan blood parasite *Babesia ovis*; a jaundice condition results from the destruction of erythrocytes.

identical twins (monozygotic twins) Genetically identical individuals that develop from a single fertilised egg.

idioblast Any cell that varies in cell content and wall thickness from its neighbours in the same tissue, e.g. lystolith containing parechyma cells.

idiogram (or karyogram) A photograph or drawing of the chromosomes present in an individual in which the chromosomes are arranged in homologous pairs.

idiopathic Pertaining to disease or condition that arise spontaneously without known cause.

idioplasm Equivalent to germplasm as described by Weismann.

idiosyncrasy An abnormal reaction to a drug or any foreign substance shown by individual which is usually genetically determined. Such individual is said to be **hypersensitive** to a particular substance, agent, etc.

idiotype An antigenic constitution of the variable (V) region of an immunoglobulin molecule.

Ig *See* immunoglobulin

IgA Monomeric or polymeric immunoglobulin, often dimeric (composed of two polypeptides). Most abundant in secretions such as saliva, breast milk, tear and other mucous secretions. Its role is to neutralise virus and bacteria as they enter the body.

IgD Low titre immunoglobulin found attached to B-cell membranes. Its function is unknown.

IgE Low titre immunoglobulin located on basophil and mast cell surfaces. It plays a crucial role in allergic reactions.

IGF *See* growth hormone

IgG (Gamma globulin) Main immunoglobulin in humans. It binds to microorganisms promoting their engulfment by macrophages and to viruses and bacterial toxins, thereby neutralising them. It also binds complement, which results in lysis of the target cell and protect the fetus by crossing the placenta.

IgM An immunoglobulin found in all the vertebrates. It is of much higher molecular weight than IgG. Chief role is to activate complement.

igneous rock Rock formed from the cooling and solidification of magma. Because igneous deposits can be accurately dated, they are useful for determining the age of fossils (the approximate age of a fossiliferous deposit can often be determined by evaluating the age of beds of igneous rock that over or underlie it).

ileal Of or pertaining to the ileum.

ileocecal valve One-way valve between the ileum and the caecum, which prevents the contents of the large intestine from returning to the ileum.

ileum Lower, third portion of the small intestine lying between the jejunum and the caecum. It is about 4 meters in length and is the site of digestion and absorption. Its internal lining bears numerous villi.

ilium A pair of bones that form the dorsal part of the tetrapod pelvic girdle; they are the fan-shaped part of the hip bones in humans. Each ilium is fused at its posterior side to the processes of the sacral vertebrae.

imaginal disc A group of undifferentiated cells in an insect larva of holometabolous insects. It develops into a specific adult structure. Imaginal discs are derived from the blastoderm and distributed mainly in larva thorax.

imago Sexually mature adult stage in the life of certain insects.

imbibition Uptake of water by substances that do not dissolve in water, so that the process results in swelling of the substance.

imbricate Closely overlapping leaves or petals, etc.

immigration Number of individuals of the same species that have come into the habitat from elsewhere during the time period under consideration.

immune clearance Fast removal of an antigen from the bloodstream, that follows the initiation of an antibody response by the immune system. It results in formation of antigen-antibody complexes, which are ingested by macrophoages and other phagocytic cells.

immune complex The small soluble cluster of bound antigen and antibody.

immune response The response to the introduction of antigen into the body involving production of specific antibodies or lymphocytes which combine with the antigen. It is the basic mechanism of active immunity and hypersensitivity, e.g. allergies.

immune system System that defends an organism against infection. The cellular components of this system are all derived from the same cells (haematopoietic stem cells), residing in the bone marrow that can alternatively mature into erythrocytes.

immunity The overall ability of an organism to fight the disease-causing organisms and effect of other harmful agents. *See* adaptive-active, humoral and passive immunity.

immunisation Process of making an individual less susceptible to infection by pathogens, toxins, etc. It may involve the use of vaccine (active immunisation) or injection of preformed antibodies (passive immunisation).

immunoassay Technique for measuring biological substances that depend on the substance acting antigenically and binding with a specific antibody, e.g. ELISA, radioimmunoassay.

immunoelectrophoresis An analytical technique used to identify antigens that have been separated by electrophoresis. The gel medium used is impregnated with specific antibodies for the required antigens.

immunoelectron microscopy A form of transmission electron microscopy in which specimen is treated with gold-labelled antibodies specific for the protein component of interest. The gold labels are then revealed as electron-dense dark areas on the image.

immunofluorescence Use of antibodies with a fluorescent marker dye attached in order to detect specific antigens, (e.g. enzymes, glycoproteins) by formation of antibody-antigen complexes which show up on appropraite illumination.

immunogenicity Degree to which an antigen can produce an immune response.

immunoglobulin (Ig) Group of proteins that take part in the immune responses of higher animals. They are produced by B-lymphocytes and act as antibodies by binding to foreign antigens such as bacteria, viruses or toxins. An antibody is typically a Y-shaped structure consisting of four polypeptide chains– two heavy chains and two light chains. Each arm of the 'Y' bears an antigen-binding site. There are five classes of immunoglobulins, IgA, IgD, IgE, IgG and IgM.

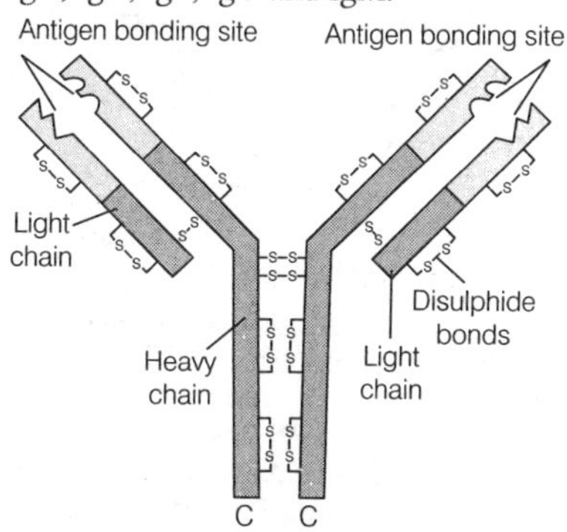

Structure of an antibody molecule

immunological memory Ability of the immune system to respond faster and more effectively to subsequent exposures to an antigen following an initial immune response to the same antigen. It may be due to an infection or immunisation and depends on the formation of memory cells, induced by the primary immune response. Memory T-cells appear typically only five days following initial immunisation, whereas memory B-cells may take about a month to reach maximum levels.

immunological synapse Highly organised region of contact between an effector CD4 T cell and its target antigen-presenting cell.

It involves accumulation of T-cell receptors within a local area of the T-cell plasma membrane and of complexes of the antigen-MHC class II protein in the close proximity of the target cell.

immunological tolerance Phenomenon by which cells of immune system are restricted from ilicitating an immune response against self tissues.

immunologically privileged sites Sites in the body where foreign tissue grafts do not induce immune responses, e.g. eye, brain, testis and unborn fetus.

immunology Study of the structure and function of the immune system, innate and acquired immunity, the distinction of self from nonself and laboratory techniques involving the interaction of antigens with specific antibodies.

immunosuppression Suppression of an immune response. It is necessary after organ transplants in order to prevent the host rejecting the grafted organ and is artificially induced by radiation or chemical agents that inhibit cell division of lymphocytes. However, it can occur naturally in certain diseases, e.g. AIDS.

immunotherapy

1. Use of the immune system to treat or prevent disease (e.g. vaccines).
2. Therapy of diseases caused by the immune system.

immunotoxin A therapeutic agent that consists of a monoclonal antibody linked to a toxin. It is used to target, for example, a particular tumour in a patient. The antibody selectively binds to cancer cells expressing the corresponding antigen and is taken into the cell, where the toxin is cleaved, allowing it to kill the cell.

imperfect flower Flower which lacks either anthers or carpels, or if present, they are non-functional.

implantation (nidation) Attachment of the fertilised ovum to the uterine wall, where it will continue its development following fertilisation in the Fallopian tube, the egg passes into the womb in the form of blastocyst. Its outer cells destroy cells of the uterine wall, forming a cavity into which the blastocyst sinks.

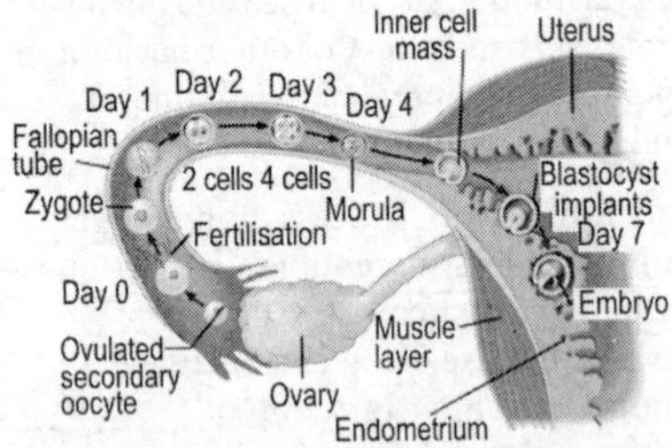

From ovulation to implantation

imprinting

1. Kind of learning that takes place during certain critical periods of an organism's life during which rapid and often irreversible receival of stimuli happens. For instance, many birds respond to the first organism they see upon hatching as they would to their mother.

impulse Electrochemical message conducted along a nerve fibre. It is marked by the flow of ions across the axon membrane caused by changes in the permeability of the membrane, producing a reduction in potential difference (action potential). The strength of the impulse produced in any nerve fibre is constant (*see* all-or-none response).

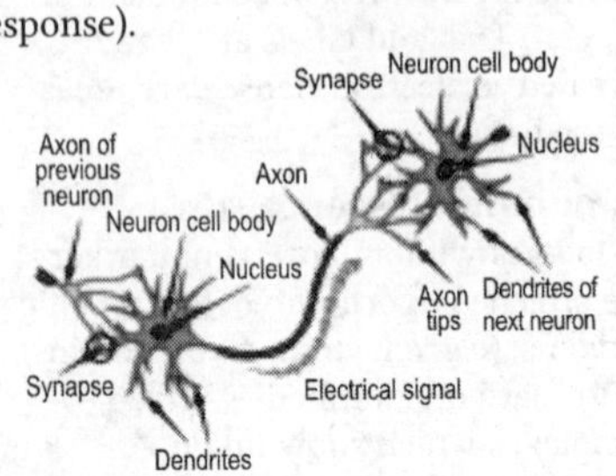

Conduction of impulse through nerve cell

inbreeding Mating between closely related individuals. The most extreme form of inbreeding is self-fertilisation, which occurs in some plants and primitive animals. It increases homozygosity.

inbreeding depression Reduction in fertility and productivity due to continued close inbreeding.

incisor One of the slicing chisel-edged teeth at the front of the mouth, in most mammals, e.g. Gnawing teeth of rodents and tusks of elephants. In humans, there are two pairs (central and lateral) in each jaw.

inclusion body Various visible bodies produced in plant and animal cell, as a result of viral infection.

inclusive fitness Quality that is tried unconsciously by organisms to maximise as the result of natural selection acting on genes that are influential in controlling their behaviour and physiology.

incompatibility Phenomenon or condition when pollen of one flower fails to fertilise other flowers on the same plant or on other genetically similar plant. Similarly, the rejection of foreign graft or blood by immune system.

incomplete dominance Mode of inheritance in which the heterozygote has a phenotype that is intermediate between the phenotypes of either homozygotes. For instance, in snapdragon flowers, if one homozygote is red, the other is white, then the heterozygote will be pink. Incomplete dominance usually arises, if one gene does not express itself and its other allele expresses an insufficient amount of character to mask the first allele.

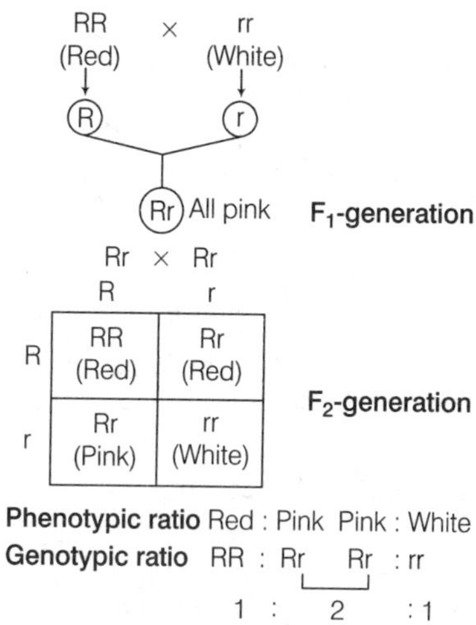

Results of monohybrid cross in the plant snapdragon, where one allele is incompletely dominant over the other allele

incomplete flower Flower which lacks one or more of its floral parts (i.e. lacking sepals, petals, stamens or carpels).

incubation

1. In reptiles, birds and egg laying mammals, the maintenance of a constant egg temperature.
2. The time spent by a premature human infants in an incubator.
3. The time between first infection and emergence of symptoms, in case of a disease.
4. The period of culture development in bacteriology.

incubation patch Featherless patch developed on the abdomen of certain brooding birds that becomes thick due to high levels of vascularity. This patch comes under direct contact with the eggs during incubation period.

incubator

1. An apparatus for maintaining eggs at a constant temperature.
2. An apparatus for rearing premature babies, in which factors such as temperature and oxygen levels can be closely controlled.
3. A device for culturing bacteria.

incus (anvil) One of the mammalian ear ossicle, that is homologous with quadrate bone of other vertebrates.

indefinite inflorescence
See racemose inflorescence.

indehiscent Fruits, fruit bodies, etc., which do not open on ripening to disperse their contents.
e.g. hazelnut.

independent assortment Random alignment and subsequent separation of pairs of homologous chromosomes at the equator of the cell during metaphase-I and anaphase-I of meiosis.
This process is important in recombination of the genetic material found in the gamete or spore producing organism.

indeterminate growth Growth that continues throughout life.

indeterminate inflorescence Type of inflorescence in which number of flowers can increase even if some are opening. In this type, most proximal flower will open first.

index fossil (Zone fossil) Animal fossil of a group that existed continuously during a particular span of geological time and can therefore be used to date the rock in which it is found. Found chiefly in sedimentary rocks, they are an essential tools in stratigraphy for comparing the geological ages of sedimentary rock formations. e.g. ammonities and graptolites.

indicator species A plant or animal that lives under special conditions of climate, or in a particular soil or in a particular community. Thus, its presence indicates the general nature of habitat, e.g. *Tubifex* worms indicate low levels of oxygen and stagnant water.

indigenous Originating or occurring naturally in a particular place.

indigestion Condition in which food is not properly digested leading to a feeling of fullness. Causes are inadequate enzyme secretion, anxiety, food-poisoning, over eating and spicy food.

induced fit model Induced fit model says the exposure of an enzyme to a substrate causes the active site of the enzyme to change shape in order to allow the enzyme and substrate to bind. This is in contrast to the old lock and key model of enzyme activity, which held that the enzyme active site had a specific rigid shape. The induced fit model is important because it suggests that enzymes do not have to perfectly 'fit' with the substrate in order to operate and thus evolution has plenty of ways in terms of new enzyme variants evolving.

induced mutation The mutation caused as a result of man-made factors.

inducer An agent which can start transcription from an operon.

inducible enzyme An enzyme synthesised only when its substrate is present.

induction

1. Ability of natural stimuli to cause unspecialised embryonic tissue to develop into specialised tissue (in embryology).
2. The initiation of child-birth by artificial means; e.g. by injection of the hormone oxytocin (in obstetrics).

indusium Membranous outgrowth on the underside of leaf which in ferns covers and protects a group (sorus) of developing sporangia.

industrial melanism Increase in melanic (dark) forms, e.g. in peppered moth (*Biston betularia*), in areas darkened by industrial pollution. Natural selection against normal pale forms by predators results in dark offspring being at selective advantage in such environments. This results in an increase in number of dark forms, that were better camouflaged.

infarct Area of dead tissue resulting from ischemia due to diminished or failure of blood circulation in the area.

infection A condition in which an organism or a part of an organism, is invaded by microorganism, fungus or animal parasite. When pathogenic, the invader causes damage to its host.

inferior ovary When receptacle completely encloses ovary and other flower parts arises above the receptacle.

inferior vena cava A large vein that carries deoxygenated blood from lower half of the body to right auricle.

inflammation Response of an animal tissue to injury, infection, etc characterised by increased blood flow, redness, accumulation of leucocytes, swelling and pain.

inflorescence A cluster of flowers sharing a common salk. Two major types of inflorescence are racemose and cymose.

influenza Commonly known as the flu. It is an infectious disease caused by the influenza virus.

infundibulum

1. Funnel-shaped process that connects the pituitary gland with hypothalamus in the floor of brain.
2. Other funnel-shaped structures, e.g. ciliated funnel of oviduct.

ingestion Swallowing (in order to introduce something into the digestive tract). The term ingestion can also refer to a cell engulfing or surrounding food or some other material.

inguinal Of pertaning to, or situated in the groin.

inhalation *See* inspiration

inheritance Sum of all characters that are transmitted by the germ cells from generation to generation.

inhibition

1. Reduction in the rate of an enzyme-catalysed reaction by substances called inhibitors. It can be competitive, non-competitive or feedback inhibition (in biochemistry).
2. The prevention or reduction of the activity of neurons or effectors (such as muscles) by means of certain nerve impulses. e.g. impulse to stimulate contraction of a voluntary muscle may be accompanied by an inhibitory impulse to prevent contraction of its antagonist (in physiology).

Inhibitory Postsynaptic Potential (IPSP) A localised hyperpolarisation of the postsynaptic membrane at an inhibitory synapse or a neuromuscular junction. It tends to inhibit production of an action potential in postsynaptic nerve fibre. The size of IPSP depends on amount of neurotransmitter released.

initiation codon (AUG) Three-base sequence on the messenger RNA that codes for the amino acid methionine. It is the start command for protein synthesis. So, it is called as initiation codon.

Initiation Factor (IF) Group of proteins that are required for initiating the translation stage of protein synthesis. It catalyses the binding of an **initiator** *t*RNA molecule.

This recognises the start codon of the mRNA, to the smaller subunit of the ribosome. The sub-unit then binds to the *m*RNA and the IFs dissociate, allowing the larger ribosomal subunit to bind to the smaller one. IFs also control the rate of translation in some cells.

innate behaviour Behaviour that is performed instinctively.

innate immunity (inherited or natural) Body's first line of defence. It plays a vital role in controlling invading organisms during early stages of infection. e.g. macrophages, dendritic cell.

inner ear The innermost sensory region of the vertebrate ear, situated in the auditory capsule of the skull. It is filled with a fluid (perilymph), in which is suspended the membranous labyrinth, responsible for hearing and balance.

innervation

1. Number, type and distribution of nerves that supply an organ or part of the body.
2. The nervous stimulation of an organ or part.

innominate artery The artery in mammals and some birds that arises from the arch of the aorta and divides to form the right carotid and right subclavian arteries. In fishes, it arise from the ventral aorta and divide into first and second pairs of afferent branchial arteries.

innominate bone Mass of bone forming each half of the pelvic girdle in reptiles, birds and mammals. It results from fusion of the ilium, ischium and pubis.

inoculate To inject or introduce a pathogenic (usually weakened) microorganism into the body in order to stimulate the production of antibodies.

Inositol A cyclic alcohol, component of vitamin-B complex and required for growth in certain animals and microorganisms. Its isomer myo-inositol is constituent of animal membrane, muscle and brain.

inquilinism Association between members of two different species in which one (the inquiline) lives on or in the other (the host) or inside the host's home obtaining shelter and in some cases, some of the host's food, e.g. certain mosquitoes live and breed in the fluid contained in the pitchers of pitcher plants. They take protection afforded by the pitcher and also make use of nutrients from prey trapped by the plant.

insect growth regulator Any of various substances that interfere with the normal development or growth of insects. Some of which are used to control insect pests, e.g. Juvenile hormones.

insecta *See* Hexapoda.

insecticide A toxin effective against insects.

Insectivora An order of small, mainly nocturnal, primitive insectivorous or omnivorous mammals. The members have long snout with stiff tactile hair and numerous teeth with pointed cusps. Includes hedgehogs, moles and shrews.

insectivore An organism that feeds chiefly on insects.

insectivorous plant Plant adapted to derive food by trapping, killing and digesting small animals.

insertion Chromosome abnormality in which a piece of DNA is incorporated into a gene and thereby disrupts the gene's normal function.

insertion sequence (IS) The mobile piece of DNA capable of inserting from one chromosome to another non-homologous chromosome.

insight learning Modification of an organism's behaviour brought about by combining information received from environment to solve problems. For example, characteristic of human beings and relatively small group of other vertebrates.

in silico A biological process or experiment simulated by a computer program.

***in situ* hybridisation** Use of a DNA or RNA probe to detect complementary DNA sequences present in cultured eukaryotic cells or bacterial clones.

inspiration (inhalation) The process of drawing gas into the lungs through the trachea. In mammals, the rib cage is raised by contraction of the external intercostal muscles and the muscles of the diaphragm, enlarging the thorax. As a result pressure in the lung cavity is reduced below atmospheric pressure, which causes an influx of air until the pressures are equalised.

Inspiratory Capacity (IC) Total volume of air a person can inspire after a normal expiration. It includes tidal volume and inspiratory reserve volume.

Inspiratory Centre Most important subcentre of the ventilation centre in the brain, which controls the regularity of breathing. It is situated in the medulla oblongata and is composed of a group of neurons (the dorsal respiratory group).

Inspiratory Reserve Volume (IRV) Additional volume of air, a person can inspire by a forcible inspiration. About 2500-3000 mL.

instar The period between consecutive molts in insect development.

instinct (instinctive behaviour) An instinct is a simple or complex set of behaviour that are innate rather than learned.

insulin Hormone that causes blood sugar levels to fall. Individuals with diabetes mellitus have a deficiency of this hormone. It is normally produced by the β-cells of the islets of Langerhans in the pancreas.

Insulin-Like growth Factor (IGF) Group of polypeptides that are structurally similar to insulin and act as growth promoting agents. Their production is stimulated by growth hormone and include IGF-I and IGF-II.

Integrated Pest Management (IPM) An approach to control insects and other crop pests. It combines various physical, chemical, and biological methods, with an aim to reduce dependency on chemical pesticides, and hence minimise pollution and harmful residues in the product. It also helps to deal with the threat posed by increasing resistance of insect pests to conventional insecticides.

integrin The adhesive extra cellular matrix protein found in animals.

integument

1. Membrane(s) enclosing the nucelus, finally forming the testa (seed coat) in plants.
2. Skin and structures and glands derived from it, such as hair, nails, sweat glands, sebaceous glands and mammary glands.

inteins An internal segment of amino acids that is excised from a protein precursor to generate a new protein.

interactome Interactions that occur between the various molecules that are produced in all cells, tissues, life stages of an organism.

intercalary meristem Meristematic tissue situated between the mass of permanent (non- dividing) tissue.

Occur at internode and leaf sheath bases of many monocots (such as grasses)

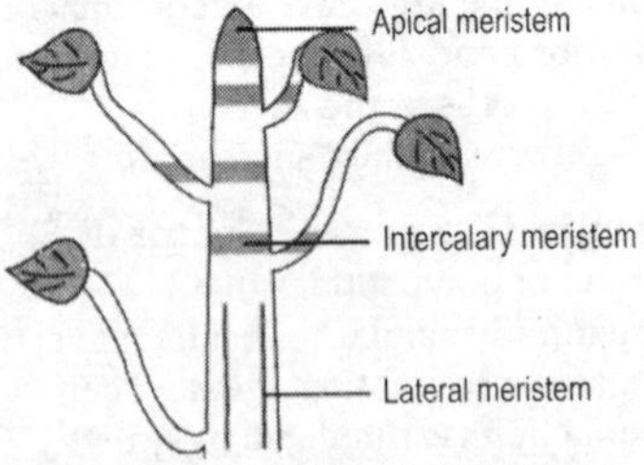

Position of different meristems

intercellular Materials found and the processes occurring between cells.

intercostal muscles A set of muscles connecting to adjacent ribs of reptiles, birds and mammals. Responsible for rotating the ribs to increase and decrease the volume of the rib cage, thus effecting breathing movements.

intercropping The special type of mixed cropping in which two or more crops are grown simultaneously in the same field following a definite row pattern.

interfascicular cambium The cambium between the vascular bundles in stem. It is formed when parenchyma cells resume meristematic activity.

interference Measure of independence of one cross-over even from other and is calculated by subtracting the coefficient of coincidence from one.

interferon (INFs) Protein produced in animal cells as a defence against virus. It inhibits replication of virus within cell.

intergenic suppressor Point mutation that lessens or contradict the effects of another mutation in the same gene, e.g. the effects of a primary frameshift mutation may be suppressed by a second mutation in which deletion, etc., of a nucleotide occurs.

interkinesis Refers to the abbreviated form of interphase that occurs between meiosis-I and II in some organisms.

Interleukin (IL) A class of cytokines that act as chemical messengers between various types of white blood cells (leukocytes). All are proteins or glycoproteins. for e.g. IL-4 activates β-cells.

intermediate filament Any of numerous microscopic protein fibres that form part of the cytoskeleton of eukaryotic cells. They are intermediate in size compared to the narrower microfilaments and the wider microtubules with diameter of about 100 nm.

intermembrane space The region in the mitochondria that is between the outer and inner membranes of the organelle. The intermembrane space is important, because the mitochondria concentrates hydrogen ions in this area as part of electron transport phosphorylation.

internal environment Medium surrounding body cells of multicellular animals.

internal fertilisation Which takes place inside the body of females.

interneurons The intermediate neurons that process signals from one or more sensory neurons and relay them to motor neurons.

internode

1. The part of plant stem between two successive node (in botany).
2. The region of a medullated nerve fibre between two nodes of Ranvier. It is covered by myelin sheath (in zoology).

interoceptor A receptor that detects stimuli from the internal environment of an organism. Chemoreceptors that detect changes in the levels of oxygen concentration in the blood are examples. *Compare* exteroceptor.

interphase Period in the eukaryotic cell cycle, when cell growth occurs and DNA is replicated in the nucleus. It follows the completion of cell division when the nucleus is not dividing.

interpolation hypothesis A hypothetical account for the emergence of a sporophyte generation during the evolution of vascular plants. It postulates that early plants were exclusively haploid gametophytes.

intersex An organism displaying characterstics that are intermediate between those of a typical male and typical female of its species. e.g. a human intersex may have under developed testes. Thus, he is technically a man, but has the external appearance of a woman.

interspecific hybridisation The hybridisation between organisms treated as distinct species.

interstitial cells Testosterone-secreting cells, which lie between the seminiferous tubules in the testes.

interstitial fluid A fluid that is in the tissues or in the spaces that are in between the cells.

interstitial fauna (meiofauna) Invertebrate animals that are small and live between individual substrate particles, such as sand grains, in the bed of a sea, lake, or river.

interstitial skin Skin found between the scales of a snake.

intervarietal A cross between two individuals of the same variety of the same species.

intervertebral discs Cartilage discs that separate the bones of the vertebral column. They allow a certain degree of flexibility to the vertebral column and also absrob shock.

intertidal zone The zone between the high and low tide marks, which is subjected to alternating flooding and drying.

interxylary phloem A strand of secondary phloem surrounded by secondary xylem.

intestinal juice A slightly alkaline liquid secreted into the small intestine from the cells that line the crypts of Lieberkuhn. Together with pancreatic juice, it provides an alkaline environment that helps in the absorption of digested food molecules entering the small intestine in chyme from the stomach. It also contain mucus.

intestine The portion of the digestive tract lying between the pylorus and the anus. Major functions are final digestion of food, absorption of water and production of faeces.

intine Inner layer of the wall of the pollen grain of gymnosperms and angiosperms. It is thin and composed of cellulose.

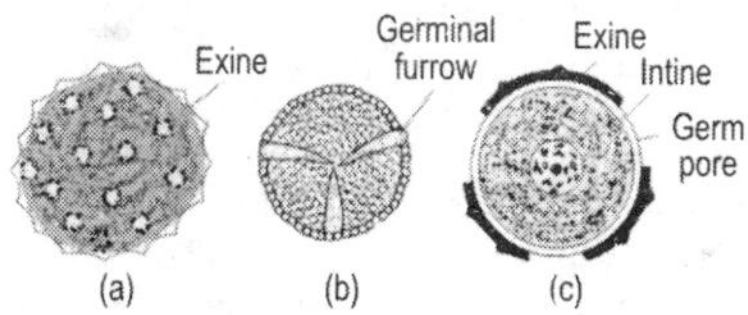

Different layers of pollen grain

Intra Cytoplasmic Sperm Injection (ICSI) Specialised procedure to form an embryo in the laboratory in which a sperm is directly injected into the ovum.

intracellular Inside the cavity of the cell.

intracellular immunisation Refers to any forms of gene-transfer-based cellular resistance to viral infection.

intrafusal Describing a type of muscle fibre that occurs in the muscle spindles in voluntary muscles. Two to twelve intrafusal fibres enclosed in a capsule are found in each muscle spindle, lying parallel to the main (**extrafusal**) fibres.

Each intrafusal fibre consists of a non-contractile (extrafusal) fibres. Each interfsal fibre consists of non-contractile (equatorial) region connected to a contractile (polar) region at each end of the fibre. The equatorial region is connected to stretch receptors.

intramembranous ossification Process of bone formation in which bones are formed directly in connective tissue.

intramuscular Into or within a muscle.

i

intraspecific Occurring between members of the same species. The term is applied, to some types of competition and communication.

intraspecific competition Competition occurring between members of a species.

Intrauterine Insemination (IUI) Artificial insemination technique in which semen collected either from husband or a healthy donor is artificially introduced into the uterus of the female to facilitate fertilisation.

intrinsic factor A glycoprotein that is essential for the absorption of vitamin-B_{12} from the gut. Lack of this factor results in pernicious anaemia.

introduced species Organism that would not normally occur but have been introduced in the habitat.

introgression (introgressive hybridisation) The insertion of the genes of one species into the gene pool of another can happen during inbreeding of two species to produce fertile hybrids. These can then back-cross with individuals of one of the parent species.

intromittent organ The male copulatory organ of an animal.

intron (intervening sequence) A portion of a gene that does not code for a gene product. These occur primarily in eukaryotes.

introrse Denoting anthers in which dehiscence lines are towards centre of flower.

intrusive growth A method of plant growth in which cell enlarge and force their way between other cells causing their disruption.

intussusception Incorporation of cellulose molecules into existing cell wall, giving an increase in wall area (in botany).

inulin A carbohydrate compound, clinically used to measure glomerular filtration rate.

invagination An act or process of formation of a pocket of tissue during embryonic development by an infolding of part of the wall of the blastula.

inverse myotatic response The excitation of antagonistic muscles by signals that relay information to CNS, during contraction of muscle. It protects muscle from tearing or detachment due to excessive contraction forces.

inversion A chromosome segment that has been reversed with respect to the remainder of the chromosome.

invertebrate Any animal lacking a backbone and cannot regulate their body temperature. Invertebrates do not develop a notochord during embryogenesis.

in vitro Performed or occurring outside a living organism, such as in a laboratory.

***in vitro* fertilization** (IVF) A fertility treatment in which sperm and egg are combined in a laboratory. The resulting embryo is transferred to the woman's uterus where it grows into a baby.

in vivo Carried out or occurring in a living organism.

involucre A protective structure seen in some angiosperms and bryophytes. In flowering plants, it consists of a ring of bracts arising beneath the flower cluster of those species with a capitulum (i.e. members of the dandelion family) or an umbel (i.e. members of the carrot family). In mosses and liverworts, the involucre is a projection of tissue from the thallus that arches over the developing archegonium.

involuntary muscle (smooth muscle) Muscle whose activity is supplied by the autonomic nervous system, i.e. not under the control of the will. Involuntary muscle comprises long spindle-shaped cells without striations. These cells occur singly, in groups, or as sheets in the skin, around hair follicles, and in the digestive tract, respiratory tract, urinogential tract, and the circulatory system. The cells contract slowly.

involution

1. A decrease in the size of an organ or the body. It may be associated with functional decline, as seen in the ageing process or follow enlargement, as when the uterus returns to its normal size after pregnancy.
2. The turning or rolling inwards of cells that occurs during the development of some vertebrate embryos.

iodopsin Light sensitive pigment present in cone cells.

ionophore A small hydrophobic molecule that facilitates the transport of ions across lipid membrane, they may be channel former or mobile ion carriers.

iris Pigmented diaphragm directly in front of the lens of the eye in vertebrates and cephalopods. Its muscles by contraction adjust size of the pupil.

ischemia Reduced blood supply to an organ.

ischium A bone forming the posterior part of the ventral side of the tetrapod pelvic girdle. In humans, it is L-shaped, passing down from the acetabulum and turning forward to meet the pubis.

islets of Langerhans Small clusters of endocrine cells scattered throughout the pancreas. Each islet of Langerhans consists of three types of cells—A-cells, which secrete glucagon, B-cells which secrete insulin and D-cells which secrete somatostatin.

isoantigen An antigen type that induces antibody production in genetically different individuals of same species.

isobilateral (of a leaf) Capable of being divided into symmetrical halves along two different planes.

isoelectric point pH value at which a substance, e.g. a protein solution is electrically neutral.

isoform Different forms of a protein that may be produced from different genes, or from the same gene by alternative splicing.

isogamete One of a pair of uniting gametes of similar size and form, that may be physiologically different.

isogamy Fusion of two similar gametes. Such type of reproduction is called isogamous.

isogenic Propagating entirely by means of apogamy.

isogenic graft A graft involving a genetically identical donor and host.

isokontan Bearing two (or more) flagella of equal length.

isolating mechanism Prevention of breeding between species due to behaviour, morphology, genetics or geographical barrier, e.g. ovum of mammals.

isolecithal An egg in which yolk is not abundant and nearly uniformly distributed throughout, e.g. ovum of mammals.

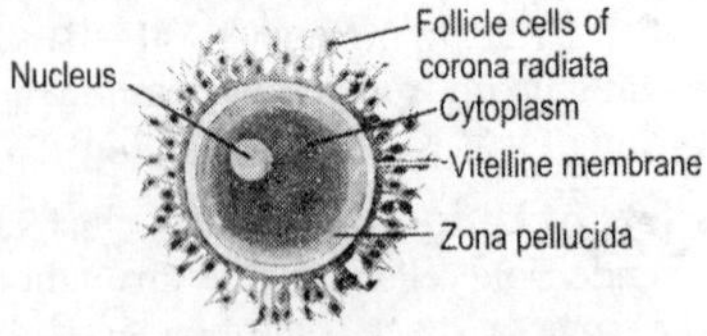

Internal structure of an isolecithal egg

isoleucine An essential amino acid.

isomerase A general class of enzyme which converts a molecule from one isomer to another.

isometry Growth of a part of the body at same rate as that of some other part.

isomorphic A condition seen in certain algae e.g. (*Ulva*, the sea lettuce) in which the alternating generations of the life cycle are morphologically identical.

isotonic Two aqueous solutions are considered isotonic, if they have the same concentration of solutes.

isozyme (isoenzyme) An enzyme performing the same function as another enzyme but having a different amino acid sequence. They catalyse the same reactions but differ in molecular structure and with respect to other characteristics such as isoelectric point and molecular weight.

itaconic acid A product of fermentation by the fungus *Aspergillus niger* and commercially used in production of paints and adhesives.

iteroparity Condition of producing offsprings repeatedly during the lifetime of an individual. It eroparous organisms include perennial plants and majority of animals.

J

Jacob-Monod hypothesis An influential theory to explain the prokaryotic gene expression. Jacob and Monod investigated the expressions of the gene that codes for the enzyme β-galactosidase. It breaks down lactose; the operon that regulates lactose metabolism is called the *lac* operon.

Jak *See* Janus kinase.

Janus kinase (JAK) Any of a family of receptor-associated tyrosine kinases that play a crucial role in many intracellular signalling pathways. Mainly the ones that relates to cytokines.

jasmonate (Jasmonic acid) An organic compound thought to play a role in plant's defensive response against attack by insects, fungi and other pathogens.

jaundice The condition in which skin becomes yellowish due to accumulation of bilirubin in blood. It is due to its failure to be excreted by the liver into bile.

jaws Bony structures in vertebrates enclosing the mouth. These contain teeth for grasping prey and chewing food. In invertebrates, grasping structure surrounding the mouth.

jejunum The part of small intestine between duodenum and ileum in mammals. It has numerous villi and is the main absorptive region.

joint Place of union of two or more bones or other hard plates. It allows or different degrees of movement for, e.g. hinge joint, in one direction.

J-shape growth form Type of population growth form, in which density increases rapidly in an exponential manner and then stops abruptly as environmental resistance or other factors limit the growth, more or less suddenly.

jugular vein A pair of vein returning blood from head and neck region in vertebrates. They unite with subclavian veins and lead into the anterior venae cavae.

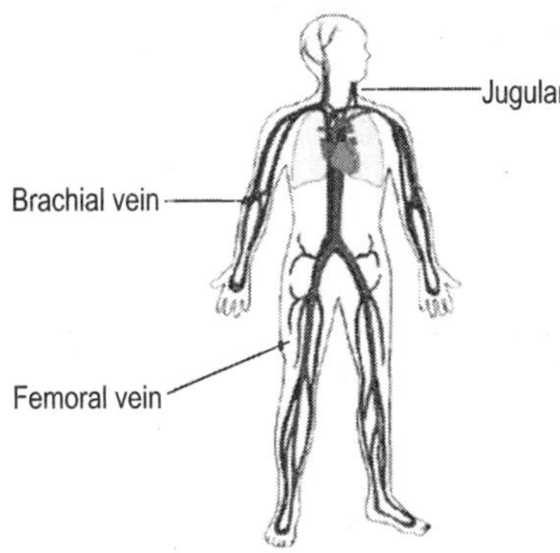

Veins in human body

jumping genes *See* transposon.

junctional complex A site of direct contact between neighbouring cells in animals, e.g. desmosomes.

Jurassic Middle period of the Mesozoic era, 215-135 million years ago. During this period, dinosaurs became large and abundant and bony fishes (teleosts) were also evolving rapidly.

juvenile hormone (neotenin) Hormone in insects that is secreted at certain stages of development by endocrine glands associated with the brain (corpora allata). It inhibits metamorphosis to the adult form and maintains the presence of larval features.

juvenoid A juvenile hormone mimics in insects, that disrupt the larval development's hormonal control. The met amorphosis either facts to occur or adults that emerge are weak and sterile, e.g. methoprene.

Juxta Glomerular Apparatus (JGA) Collection of modified smooth muscle cells in kidney, near the afferent arterioles entering the glomeruli which produce renin. It is important in regulating blood pressure and body fluids and electrolytes.

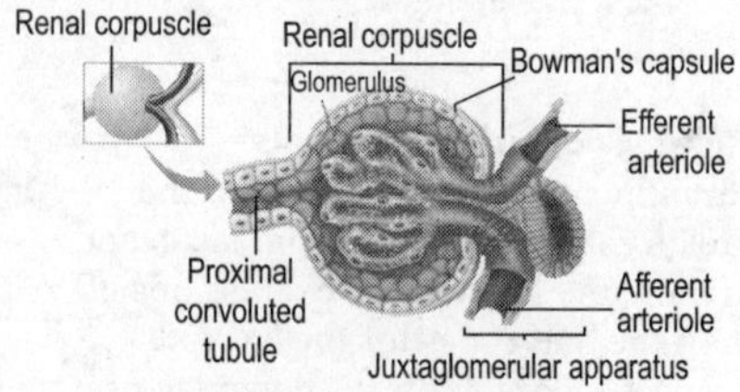

Structure of juxta glomerular apparatus in kidney

K

kairomone A chemical messenger emitted by an individual of a species which causes a response in an individual of another species. For example, many parasites are attracted to their hosts by an excreted kairomone.

kappa particles Gram-negative bacterial species present endosymbiotically (as commensals) within cytoplasm of ciliate *Paramecium aurelia*. Their maintenance requires activity of some nuclear genes of host cells. May be transferred from one host cell to another during conjugation and may produce toxins (e.g. paramecin) which kill sensitive *Paramecium* strains but not the producer cells (termed killer cells).

karyogamy Union of two nuclei, especially gametic nuclei during fertilisation.

Karyogram *See* idiogram

karyokinesis Division of nucleus as distinguished from the division of the cell or cytokinesis.

karyology Study of nuclear structure and function.

karyolymph The fluid or gel-like substance of nucleus.

karyomere A swollen condition sometimes seen in chromosomes towards the end of nuclear division.

karyoplasmatic ratio The ratio of the volume of the nucleus and that of the cytoplasm of the same cell.

karyorrhexis A stage of cell death that involves fragmentation of a cell nucleus. The nucleus breaks down into small dark beads of damaged chromatin. This usually leads to further dissolution of the nucleus (karyolysis), whereupon the damaged chromatin gradually fades.

karyosome Thickened mass of chromatin in a cell nucleus.

karyotype

1. Set of chromosomes present in a particular type of organism, viewed in terms of their characteristic size and structure. The karyotype is identical in all the diploid cells of an organism.
2. A photomicrograph of the chromosomes present in an individual's somatic cells, often arranged in a standard format showing the number, size and shape of each type of chromosome.

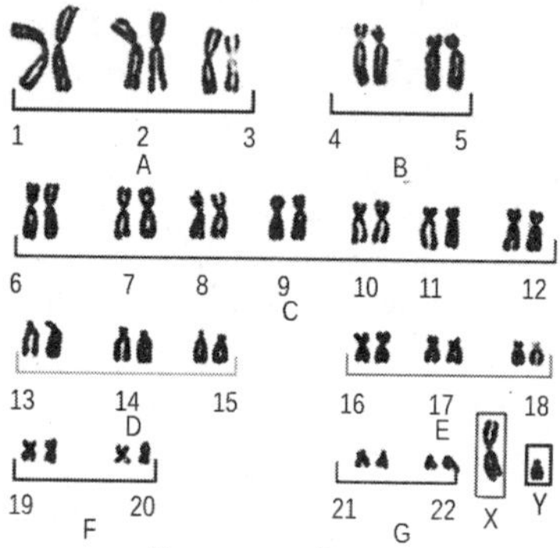

Karyotype of man

karyotypic Of or pertaining to a karyotype or karyotypes.

keel (carina) Thin plate-like projection of bone from ventral surface of breast bone (sternum) of birds and bats, to either side of which powerful wing muscles are attached.

kelp Sea-weed (brown) used as a source of iodine.

keratin A hard insoluble protein substance found in hair, nails and horns of animals. It is a fibrous sulfur containing protein that produces a strong but elastic structure.

keratinisation (cornification) Process in which the cytoplasm of the outermost cells of the mammalian epidermis is replaced by keratin. It occurs in the stratum corneum, feathers, hair, claws, hooves and horns.

keratoconjunctivitis An inflammation of the cornea and conjunctiva of the eye.

ketogenesis Formation of ketone bodies.

ketohexose Ketose sugar with six carbon atoms.

ketone body One of a group of organic substances formed in fat metabolism, e.g. acetoacetic acid and acetone. Normally used as energy sources by peripheral tissues.

ketonuria Presence of ketone bodies in the urine in excessively large amounts. It is generally associated with the disorder diabetes mellitus, or starvation, etc.

ketopentose Ketose sugar with five carbon atoms.

ketose A sugar containing a ketone (=CO) group.

ketosis Presence of ketone bodies in the body in excessive amounts.

kidneys Organs present in vertebrates, usually near the spine, that function in excretion of nitrogenous wastes and osmoregulation. In humans, they are bean-shaped. The interior surface of the kidney is lined with renal pelvis, which is continuous with the lining of the ureter. Each kidney is composed of numerous (nearly one million) complex tubular structure called nephrons, which are the functional units of kidney.

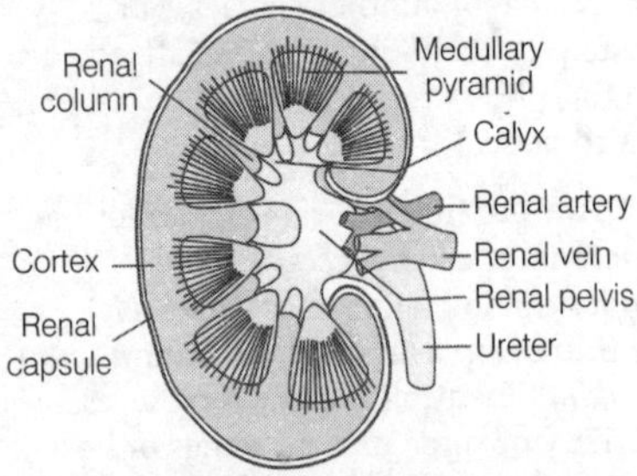

LS of kidney

kilobase Unit of length for nucleic acid, equal to 1000 base pairs.

kin selection Kind of natural selection in which individuals increase the number of their genes left to future generations, by providing aid to relatives who also share some of the same genes.

kinase An enzyme that catalyses the transfer of a phosphate group from ATP to another substrate.

kinesialgia Pain resulting from movement of the muscles.

kinesin An eukaryotic motor protein with the ability to bind separately to microtubules and organelles and then generate the force required to move organelles.

kinesiology Study of the movement of muscles.

kinesis An undirected change in the movement of a cell or organism in response to an environmental stimulus. Unlike a taxis, kinesis is not oriented towards or away from the stimulus.

kinetin An artificial cytokinin found in extracts of denatured DNA.

kinetochore A region of specialised proteins at the centromere that serves as an attachment point for the spindle microtubules, during late prophases in mitosis and meiosis.

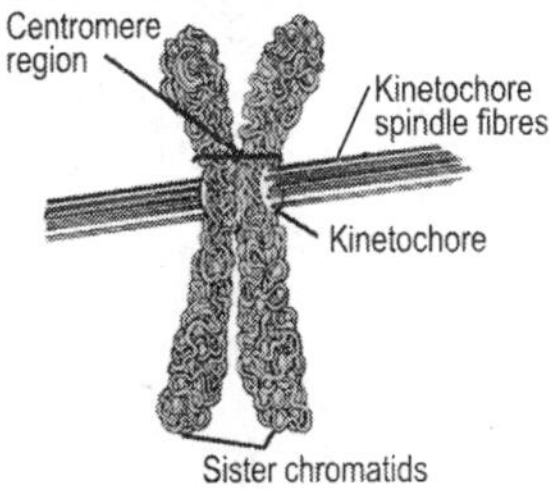

Position of kinetochore in a chromosome

kinetoplast A modified mitochondrion situated near the origin of a flagellum in some flagellate protozoans (kinetoplastida).

kingdom In taxonomy, the highest domain that itself contains one or more phyla.

kinin

1. A group of plant hormones which seem to be particularly important in influencing the rate of mitosis. Found in young developing tissues and embryos.
2. Any of a class of peptides occurring in blood and function as local hormones.

kinocilium Motile cilium of one of the hair cells of the sensory epithelium of vertebrate inner ear.

kinomere *See* centromere

kinoplasm Protoplasm which appears to be composed of fibrils and which in cell division composes the spindle fibres and astral rays.

Kleiber's law The relationship whereby metabolic rate (*R*) varies with body mass (*M*) of an organism raised to the power 3/4, i.e. $R \propto M^3/4$. Shown to apply to organisms ranging in size form bacteria to whales and for plants as well as animals.

kleptoparasitism A parasitic characteristic of opportunistically stealing food and/or nests from other organisms.

Klinefelter's syndrome A genetic disorder in males who have an extra X-chromosome. Instead of having usual XY-chromsomes, these males have an XXY-chromosome pattern. Symptoms include infertility, under developed testes and some female characteristics (such as breast development gynaecomastia).

klinostat A device used in experiments to test the influence of gravity on the growth movements of plants (see geotropism).

knee-jerk reflex *See* stretch reflex

knockin Technique related to gene knockout in which a gene is inserted into the genome of a cell, cell line or organism. Using the technique of Cre/*loxP* recombination, such as insertion can be targeted to a particular site in the genome and may replace or supplement existing genes.

kolypeptic Interfering with digestion.

koniology (also coniology) Study of dust in relation to its effects on environment.

koniometer A measurement device for estimation of atmospheric dust loads.

Kranz anatomy Found in C_4 plants.

It is characterised by the presence of mesophyll cells and bundle sheath cells with dimorphic chloroplasts wreath-like arrangement, palisade mesophyll cells around a layer of bundle sheath cells.

Krebs' cycle (Citric Acid Cycle/ TCA cycle) The fundamental step of metabolism of aerobic organisms. It is a complex cyclic series of biochemical reactions in which pyruvate, formed by glycolysis is oxidised to CO_2 and H_2O with large amount of energy. It requires oxygen and occurs in the matrix of mitochondria. The two-carbon acetyl coenzyme-A (acetyl Co-A) reacts with the 6 carbon oxaloacetate to form the 6 carbon citrate in a series of seven reactions, this is reconverted to oxaloacetate and produces two molecules of carbon dioxide.

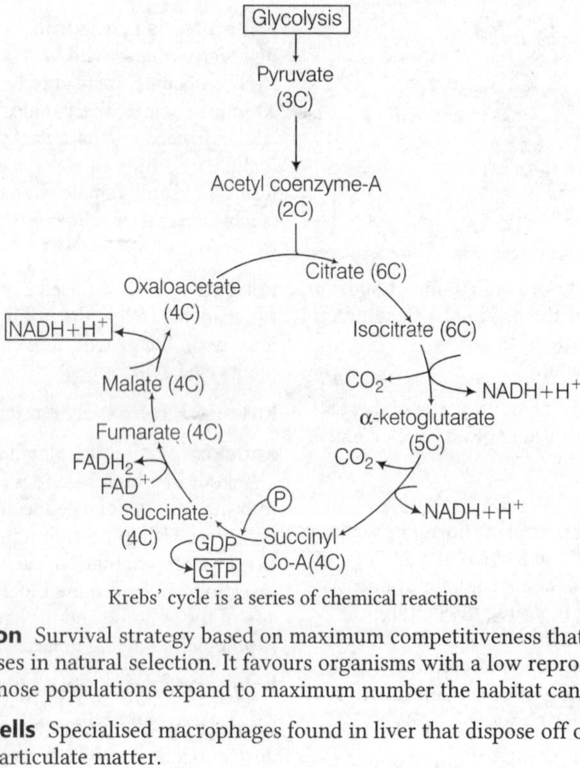

Krebs' cycle is a series of chemical reactions

***K*-selection** Survival strategy based on maximum competitiveness that is one of the processes in natural selection. It favours organisms with a low reproductive rate but whose populations expand to maximum number the habitat can support.

Kupffer cells Specialised macrophages found in liver that dispose off old blood cells and particulate matter.

kwashiorkor A type of malnutrition developed by lack of proteins in the diet.

kyphosis A curvature of the upper spine; commonly known as hunch back or hump back.

L

labial Of, or pertaining to the lips.

labium

1. Lower lip in the mouth parts of an insect.
2. Either member of two pairs of fleshy forms that form part of vulva.

labrum The upper border of the mouthparts of an insect or crustacean of vertebrates.

labyrinth Fluid-filled inner ear of vertebrates which consists of two parts, i.e. bony and the membranous labyrinths.

lac operon (lactose operon) A nucleotide sequence in the genome of *Escherichia coli,* that regulates lactose metabolism. It comprises an operator, promoter and three structural genes, transcribed into a single *m*RNA.

lacrimal gland Tear producing gland found in all land vertebrates. In human beings, it lies beneath the upper eyelid. The fluid (tears) from the gland washes and lubricates the exposed surface of eye.

lacrimal sac A small expanded chamber at the upper end of the nasolacrimal duct. Lacrimal fluid flows from the eyes through the nasolacrimal duct into the nasal cavity. This flow can cause a running nose during excessive tearing, as in states of high emotion or allergic reaction.

lactase (β-galactosidase) Enzyme which catalyses the conversion of lactose into glucose and galactose.

lactate (lactic acid) Organic acid produced from fermentation by a wide range of bacteria and skeletal muscles of our bodies under anaerobic conditions.

lactate dehydrogenase (LDH) Enzyme present in both plants and animals, that catalyses the interconversion of lactate and pyruvate (and at the same time, the interconversion of NADH and NAD^+).

lactation The production of milk by mammary glands. It involves hormone activity, notably prolactin and oxytocin.

lacteal The lymph vessels connected to the intestines and taking up some of the products of digestion.

lactic acid It plays a variety of biological roles. It is known to have three isomers

1. It is a dextrorotatory L-form found in muscle tissue and in the blood.
2. It is a laevorotatory D-form, a product of sucrose fermentation.
3. It is a DL-form found in foods that have undergone fermentation (such as beer, pickles and sour milk).

Lactobacillus (lactic acid bacteria) A group of bacteria that ferment carbohydrates in the presence or absence of oxygen forming lactic acid as a major end product. They have a high tolerence of acid conditions. Lactic acid bacteria are involved in the formation of yoghurt, cheese, etc.

lactogenic hormone *See* prolactin.

lactometer A cylindrical instrument to measure the density of milk. If the reading is less than 1.026, it indicates water in milk.

lacuna (pl. lacunae) A small open space or gap, as within a cell or bony tissue.

lag growth phase The period during which there is little increase in the number of organisms present in a population.

lagging strand Strand of the DNA double helix that is synthesised discontinuously and contains Okazaki fragments, as opposed to the leading strand.

lagomorpha The mammalian order comprsing of the rabbits, hares and pikas. Gnawing, herbivores differing from rodents. Cheifly in having two pairs of incisors as opposed to one pair in upper jaw.

lagomorphs A member of the mammalian order–Lagomorpha.

Lamarckism The name given to the idea that an alteration in the body, caused by 'use or disuse' of the affected part, will tend to pass on to the offspring of an individual. This idea is widely attributed to Jean-Baptiste Lamarck, but he did not originated this idea. It was accepted by all of his contemporaries and thereafter, by most naturalists, up to about the year 1900.

lambda phage A bacteriophage that infects cells of bacterium *E. coli.* It has been studied as a model of viral infection and replication and is extensively used in genetic research in genetic engineering. Modified forms are used as vectors in gene cloning, especially for packaging relatively large amounts of foreign DNA.

lamella (pl. lamellae) The thin layer or plate-like structure.

1. One of the spore-bearing gills in the fruiting body of a mushroom or related 'fungus, attached to the underside of the cap (pileus) and radiating from centre to margin.
2. Any of the paired folds of membranes seen between the grana in chloroplast of plants.
3. In bryophytes, a thin sheet of flap-like plates of tissue on dorsal surface of the thallus or leaves. One of the concentric layers of hard calcified material of compact bone forming part of a Haversian system (in zoology).

lamellibranchia *See* bivalvia.

lamellipodium A sheet-like extension or flowing pseudopodium of a vertebrate cell during locomotion, some forming attachments with substratum. It contains bundles of actin microfilaments which polymerise and extend at the leading edge. These become cross-linked to form supporting network for maintaining the shape of lamellipodium.

lamin Any of a group of proteins that are components of intermediate filaments and form the nuclear lamina on the inner surface of the nuclear envelope. They are classified into three types i.e. A, B and C.

lamina

1. Thin and flattened blade of a leaf in which photosynthesis and transpiration occurs. The bulk of it is made up of mesophyll cells interspersed by a network of veins.
2. The leaf-like part of the thallus of certain algae, notably kelps. *See* also stipe.

laminar placentation Attachment of ovules over surface of carpel.

laminin Protein abundant in the basal lamina of epithelial tissue. These help to create and stabilise the basic fibrous network.

lampbrush chromosome Long chromosome (up to 1 mm), covered by very fine hair-like projections which are actually loops. Each loop with a core of DNA found in the oocytes of animals, (newts and other amphibians).

lancelets *See* Cephalochordata.

Langerhans cell A type of macrophage found in epidermis. As part of the epidermal immune system, these cells act as antigen- presenting cell.

large intestine The last part of the alimentary canal consisting of the caecum, colon and the rectum. It is called 'large' because it has a much wider diameter than the small intestine. Receives the undigestible remains of the food and functions in absortion of water and formation of faces.

lariat The loop structure formed by bending of intron during RNA processing. *See* RNA processing.

larva A free-living, sexually immature form that occurs in the life cycle of many animals; a larval form is often quite distinct from the adult (e.g. maggot and fly, tadpole and frog) and the two usually differ in diet.

larviparous Laying larvae, as opposed to eggs.

laryngopharyngitis The inflammation of the larynx and pharynx.

laryngoscope The instrument for examining the larynx.

larynx Organ of speech (voice-box) that makes up the upper end of the trachea in tetrapod vertebrates.

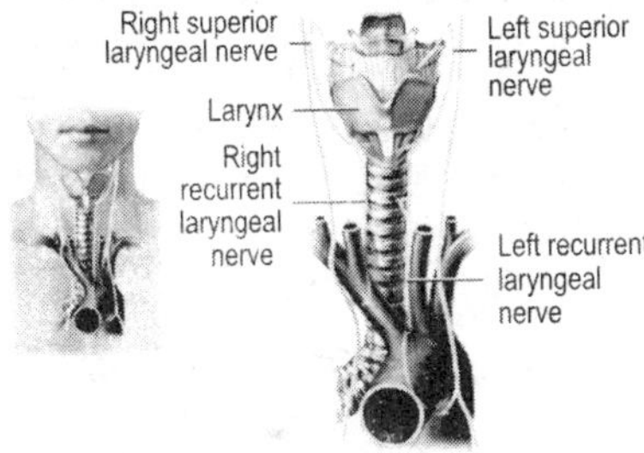

Internal structure of upper respiratory tract of human

lasso cell (colloblast) Cell types characteristics of comb jellies (phylum-Tenophora). These cells are embedded in the tentacles of the organism. Each has two protruding filaments, one straight and one spiralled, with adhesive heads used to catch prey.

latent learning A form of learning in which there is apparently no immediate reward for the animal and what is learnt remains 'latent'. Many insects learn the details of landmarks near their nest by making orientation flights. This process enables them to locate the nest when returning from distant sites.

latent period (latency)

1. Short time between the reception of a stimulus and the start of the response. For a contracting muscle, the latent period lasts about 0.02 seconds. Calcium ions are released from the sarcoplasmic reticulum in this time.
2. The period during which a virus persists in a latent state.

latent virus A virus that can remain inactive in its host cell for a considerable period after initial infection. Eventual replication inside the host cell may be triggered by factors such as radiation and chemicals. An example of a latent virus is herpes simplex virus.

lateral Toward, at or pertaining to the side.

lateral-line canal Sense organ in fish consisting of a groove on the head and along each side of the body that contains sensory hair cells. These are sensitive to variations in water pressure and are used to detect moving objects such as predators, and to monitor progress through water during swimming. The canals and the hairs within them collectively form the lateral-line system.

lateral root *See* tap root.

laterodorsal Denoting a position in the region where back and side meet.

latewood (also summer-wood) The xylem lignified during the summer.

latex A milky fluid of mixed composition found in some herbaceous plants and trees. It may assist in protecting wounds and may be involved in the nutrition of the plant. The latex of some species, notably rubber trees is collected for commercial purposes.

laticiferous tissue A very long, well-branched, thin-walled, tube-like tissue in certain families of angiosperms. It contains milky juice called latex which remains irregularly distributed between the parenchy- matous tissues.

latifoliate Broad-leaved.

latrodectism Poisoning caused by a widow spider bite.

latrodectus A cosmopolitan genus of spiders known as the widow spiders, responsible for most human fatalities from spider bites worldwide.

law of dominance (first law) States that characters are controlled by genes which occur in pair, when two alternate forms of a trait or character (genes or alleles) are present in an organism, only one factor expresses itself in F_1-generation. This factor is called dominant, while the other factor which remains hidden is called recessive. It explains expression of genes in a cross and 3 :1 ratio obtained at F_2.

law of independent assortment One of the Mendel's law. It states that pairs of 'factors' segregate independently of each other when germ cells are formed (see also independent assortment).

law of segregation A law on inheritance given by Mendel. It states that each hereditary characteristic is controlled by two 'factors' (now called alleles), which segregate (separate) and pass into separate germ (reproductive) cells.

layering Artificial method of vegetative propagation. Runners or stolons or a branch of a plant is made to bent towards the ground and covered with the moist soil. After sometime, roots develop where node touches and new plantlets develop.

L-dopa (I-dihydroxy- phenylalanine) An intermediate in the pathway from phenylalanine to noradrenaline and immediate precursor of the brain neurotransmitter dopamine.

leaching The removal of soil nutrients by water moving down the soil profile. It makes soil more acidic since cations (e.g. potassium and magnesium) are replaced by hydrogen ions. Leaching leads to the formation of podsolized and laterite soils.

leader sequence A segment at the 5′ end of *m*RNA. It precedes the start codon and contains the ribosome binding site and usually, regulatory sequences such as an attenuator region.

leading strand The strand of the DNA double helix that is synthesised without interruption, as opposed to the lagging strand, during discontinuous replication.

leaf Flattened appendage of the stem that arises as a superficial outgrowth from the apical meristem. Leaves are arranged in a definite pattern, have buds in their axils and show limited growth. See lamina, petiole. Leaves show considerable variation in size, arrangement of veins, type of attachment to the stem and texture. They may be simple or compound and may be modified for special purposes. e.g. leaves reduced to spines for protection.

Leaf Area Index (LAI) The total surface area of the leaves of plants in a given area divided by the area of ground covered by the plants. In an dense vegetation area e.g. forest, the LAI will be high.

Leaf Area Ratio (LAR) The total leaf area of a plant divided by the dry mass of the entire plant, measured in square centimetres per gram or square metres per kilogram. It indicates the efficiency with which a plant uses its leaves to produce plant material. It is equal to the product of the Leaf Mass Ratio (LMR) and the Specific Leaf Area (SLA).

leaf blade Especially a leaf of grass or the broad portion of a leaf as distinct from the petiole.

leaf buttress A leaf primordium appearing as a protuberance below the stem apex, produced by periclinal division of the tunica and corpus and associated with a procambium.

leaf–height–seed scheme (LHS scheme) A scheme proposed to describe the ecological strategy of plant species in terms of three parameters: specific leaf area (L), height of the plant at maturity (H) and seed mass (S). The strategy of a species is thus depicted by its position in the three-dimensional space bounded by the axes for the three variables.

Leaf Mass Ratio (LMR) Ratio of leaf dry mass to the dry mass of the entire plant, typically measured in grams per gram or kilograms per kilogram. *See* also leaf area ratio.

leaf trace A vascular bundle or group of vascular tissues carrying fluid connecting the vascular systems of leaf and stem. It stretches from the leaf base to the vascular ring in the stem axis.

leaf veins The vascular tissue of leaves, arranged in a net-like network in dicots and running parallel to each other in monocots known as reticulate venation and parallel venation respectively.

leaflet A leaf-like part in a compound leaf; a foliole.

learning Ability of an organism to modify its behaviour based on past experience. It allows an animal to respond more flexibly to the situations it encounters. Learning abilities in different species vary widely and are adapted to the species' environment. On a physiological level, learning involves changes in the connections of neurons in the central nervous system.

lecithin One of a group of phospholipids that contain glycerol, fatty acid, phosphoric acid and choline. Found widely in higher plants and animals particularly as a component of cell membranes.

lectin Group of proteins that bind to specific carbohydrate group. Found in plants, animals, fungi, algae and bacteria. e.g. collectins, galectins.

lectotype The specimen selected from original material to serve as the type-specimen, when no holotype was designated or as long as it is missing.

leghaemoglobin A monomeric haemoglobin synthesised in the root nodules of leguminous plants that are host of N-fixing bacteria. Has high affinity for dioxygen.

legume (pod) A dry fruit member of an economically important family-Fabaceae (of pod-bearing plants) that includes peas, beans, lentils, soy, alfalfa, clovers, peas, beans, lupins, peanuts, wisteria, locust tree. Legumes are notable for their ability to carryout nitrogen fixation, due to the presence on their roots containing nitrogen-fixing bacteria, which can carry out this process only in association with a legume.

leiomyoma Benign tumour of smooth muscle tissue, occuring most often in the uterus.

leiomyosarcoma Malignant sarcoma of smooth muscle cells. Such tumours are most common in the digestive tract, uterus and pelvis, but can occur almost anywhere in the body.

lek Area of ground divided into territories for courtship display of mating by certain animals, where males gather around the females for being selected for copulation.

lemma One of two bracts that protects the floret of grasses. The lemma is situated beneath the second (smaller) bract, known as the palea, which it surrounds. *See* spike.

lens Transparent biconvex structure in the eyes or analogous organs of animals. It is responsible for directing light onto light sensitive cells. In vertebrates, it is centred behind the iris and attached by suspensory ligaments to the ciliary body as a flexible structure.

lenticel Found on the stems of woody plants, that lack suberin. Gas exchange occurs through the lenticels (photosynthesis takes place even in leafless trees in winter, if the temperature rises above about 40°F). Developed in woody stems, when epidermis is replaced by cork.

lepidoptera Order of insects including moths and butterflies (lepidopterans). Adults possess two pairs of membranous wings, often brightly coloured and usually coupled together. The wings, body and legs are covered with minute scales. Adult mouth parts are generally modified to form a long proboscis for sucking nectar, fruit juices, etc.

leprology The scientific study of leprosy and its treatment.

leprosy (hansen's disease) A chronic disease caused by the bacteria, *Mycobacterium leprae*. Untreated, it causes progressive, permanent and disfiguring damage to the skin, nerves, eyes, and appendages. It primarily affects the peripheral nerves and upper respiratory tract.

leptin A protein hormone, that is secreted by adipose tissue in humans. It regulates adipose tissue mass and energy balance. It acts on leptin receptors in the hypothalamus and inhibits expression of neuropeptide Y, thereby countering the appetite-stimulating effects of the latter and inhibiting food intake. It also promotes synthesis of the appetite suppressant melanocyte-stimulating hormone. Deficiency of leptin or its receptors leads to severe obesity.

leptoid (leptome) Photosynthate conducting cells, similar to sieve cells of phloem, found in bryophytes.

leptospirosis (also Weil's disease) Disease caused by spirochetes of the genus-Leptospira. These micro-organisms have an exceptionally broad host range. Leptospira causes disease not only in humans and other mammals, but also in birds, amphibians and reptiles. The disease is transmitted when infected urine or semen comes in contact with the eyes, mucous membranes or broken skin. Although many different animals are carriers of this disease, it is most commonly spread to humans *via* the urine of infected rats and mice, either in contaminated foods or directly by skin contact. Leptospirosis begins with flu-like symptoms (vomiting, fever, intense headache). The patient seems to recover, then a more serious phase sets which can lead to meningitis, liver damage, kidney failure and death.

leptotene First stage of meiotic prophase-I. During it, chromosomes are thin and attached at both ends to the nuclear membrane.

lethal gene Mutant form of a gene which leads to death of the individual bearing it.

leucine One of the essential amino acid. Due to its hydrocarbon side chain, leucine is hydrophobic.

leucocyte (white blood cell) A colourless cell with a nucleus, found in blood and lymph. They can produce antibodies to fight against infection.

leucoplast (amyloplast) The organelle related to chloroplasts that stores starch. The white appearance in the ground tissue of potato is due to leucoplasts. In actuality, the leucoplast starch grains are typically clear, not white as implied by the name.

leukaemia Blood cancer; cancer characterised by an uncontrolled increase in the number (mitosis) of leukocyte in the blood.

leukotrienes Group of substances that are produced in tissues to serve as local hormones. They are derived from arachidonic acids and contain at least three double bonds.

levulose Alternative name for fructose.

Leydig cells or interstitial cells Present in the intertubular spaces of siminiferous tubules of vertebrate testis and produce a group of hormones called androgens mainly testosterone.

L-form bacteria Cell-wall deficient bacteria which are able to grow as spheroplasts or protoplasts.

LH *See* luteinising hormone

libriform fibre The xylem fibre having thick walls and greatly reduced pits.

lichens Organisms composed of a fungus and an alga (usually a green or blue-green alga), growing in an intimate, closely entangled, symbiotic relationship. Lichens are most commonly seen on rock crops and tree bark. As lichen grows and respires, it gives off carbon dioxide and water which combine to form carbonic acid. This acid decomposes rocks and releases nutrients essential for the growth of other living things. Lichens do not injure trees.

lienitis Inflammation of the spleen.

lienocele Hernia of the spleen.

life cycle Progressive series of changes in life that animals go through starting from fertilisation to birth, to sexual maturity, till death.

lift The airfoil that controls the movement of the birds when in flight, as well as the force with which the birds fly is termed as lift.

ligament A fibrous band of connective tissue serving to bind bones together at joints.It restrain the movement of bones at a joint and is hence, important in preventing dislocation.

ligand A molecule that binds to a protein with a high degree of specificity, e.g. substrate of an enzyme or hormone binding to cell receptor (in cell biology).

ligase Enzyme catalysing formation of a covalent bonds using energy released by ATP cleavage. e.g. DNA Ligase, that links two DNA molecules through the formation of a phosphodiester bond between the 3′ hydroxyl and 5′ phosphate of adjacent nucleotides.

ligate

1. To join two or more DNA fragments end-to-end.
2. To tie a ligature (in a medical context).

light dependent reactions Step in photosynthesis in which light energy is captured by and converted to ATP and NADPH. These in turn supply energy for the rest of photosynthesis. Oxygen is produced as a waste material by the light dependent reactions. It requires the presence of sunlight.

light independent reactions Calvin cycle and other related reactions in photosynthesis whereby carbon dioxide is fixed from the atmosphere and hydrogen ions and electrons are added to produce sugars.

lignin A hard material that joins with cellulose to form stiff cell wall in vascular plants; it also cements cells together providing structural strength to the plant as a whole.

ligule

1. A small membranous structure that develops on the upper surface of a young leaf base in certain club mosses.
2. A strap-shaped extension from the corolla tube in certain florets of a capitulum. They are termed ligulate (ray) florets.
3. A membranous scale-like outgrowth from the leaves of certain flowering plants. Many grasses have a ligule at the base of the leaf blade.

limb

1. Appendage of a vertebrate animal, such as the leg or arm of a mammal or the wing of a bird.
2. The expanded upper part of a sepal, petal or leaf.
3. The widened upper section of a gamopetalous corolla.

limbic system A group of regions in the brain. It is involved in the expression and control of mood and instinct and plays a major role in long-term memory. Hippocampus and the hypothalamus are included in it.

limbus Border of or margin of the cornea of the eye. It is continuous with the sclera.

limestone Sedimentary rock containing more than 50% calcium carbonate.

liming Application of lime (calcium hydroxide) to soils to increase levels of calcium and decrease acidity.

limiting factor Environmental factor that limits the growth, metabolic processes or distribution of organisms or populations. It may be in decreasing or increasing amount or may be its absence or presence. In a desert ecosystem, for example, low rainfall and high temperature will be factors limiting colonisation in desert. When a metabolic process is affected by more than one factor, the **law of limiting factors** states that its rate is limited by the factor that is nearest its minimum value.

limnology The scientific study of freshwater and its flora and fauna. It also includes study of physical and chemical aspects of inland water.

limnoplankton Freshwater especially lake plankton.

LINE (Long Interspersed Element) Any of group of non long terminal repeats found in eukaryotes. These are retro-transposons and can spread by reverse transcription. These consisting of numerous copies of relatively long (generally 6-8 kb) sequences scattered through out the genome, e.g. L1 element in humans and other mammals.

lineage Descent through a single line from a progenitor.

lingual Pertaining to the tongue.

linguiform Tongue-shaped.

linkage Increased tendency of two alleles on same chromosome, to be inherited together together because their loci are on the same chromosome. In general, two such alleles will segregate together more frequently when their corresponding loci are closer to each other on a given chromosome (i.e. the two alleles will be more closely linked). Linkage can be broken by crossing over or by a chromosome mutation, when sections of chromosomes are exchanged and new combinations of genes are produced.

linkage disequilibrium A phenomenon found in linkage mapping of genes whereby a certain combination of alleles at two or more marker loci occurs more frequently than expected.

linkage map A map based on linkage data, that shows the relative positions of genetic loci on a chromosome.

linked genes/markers Genes/markers that due to their occurrence on the same chromosome, are inherited together at rates above random expectation. Do not show independent assortment and results in parental allele combinations occuring greater than recombinant arrangements.

linker Double-stranded oligonucleotide containing a restriction site. Linkers are ligated to the ends of a DNA fragment.

Linnaean System *See* binomial nomenclature.

linoleic acid Liquid polyunsaturated fatty acid abundant in many plant fats and oils, e.g. linseed oil, groundnut oil and soyabean oil. It is an essential fatty acid.

lipase Enzyme that breaks down fats into fatty acids and glycerol in the small intestine. It is secreted by pancreas.

lipid bilayer The enclosing layer, composed primarily of phospholipids, that forms the boundaries of cells and cellular vesicles. Each lipid molecule comprises a hydrophilic 'head' (having a high affinity for water) and a hydrophobic 'tail' (having a low affinity for water). In the lipid bilayer, the molecules are aligned so that their hydrophilic heads face outwards, forming the outer and inner surfaces of the membrane, while the hydrophobic tails face inwards, away from the external aqueous environment.

lipids Wide variety of biological molecules, which are sparingly soluble in water, includes fats, fat-soluble vitamins, monoglycerides, diglycerides, phospholipids, waxes and sterols. These are generally soluble in non-polar solvents such as ether or chloroform and are important structural components of plasma membrane. They also serve in energy storage and as signaling molecules.

lipoblast A connective tissue cell that develops into fat cell.

lipocyte A fat cell.

lipoic acid Sulphur containing fatty acid found in a wide variety of natural materials. It is essential as a coenzyme for certain dehydrogenase enzymes notably pyruvate dehydrogenase. It is classified with water-soluble B-vitamins. Its sources are liver and yeast.

lipolysis Splitting of the component of fatty acids from a lipid. It occurs in the body, mainly in gut by the lipase enzymes.

lipopolysaccharide Conjugated polysaccharide in which the non-carbohydrate part is a lipid. They are the constituents of the cell wall of certain bacteria.

lipoprotein A complex molecule of lipids and protein. In the blood of humans and other mammals, cholesterol, triglycerides and phospholipids associate with various plasma proteins to form lipoproteins, so they can be carried in blood.

liposarcoma Malignant tumour arising from fat cells.

liposome Vesicle composed of phospholipids like those found in the lipid bilayer of a cell's plasma membrane. Liposomes occur naturally but can also be artificially constructed for various purposes, such as to deliver drugs to target cells. A liposome usually has an aqueous core.

lipotropin The of two polypeptide hormones produced in mammals by the pituitary gland that stimulate lipolysis. Both occur in human blood, but their precise physiolgical role is uncertain.

lithosphere The solid inorganic outer portion or mantle of the Earth, containing soil and rocks.

litter The dead organic matter present in soil that has to be decomposed consists of fallen leaves, etc.

littoral Refers to the marginal shallow-water zone of a sea or lake, especially between high and low tide lines in the sea. In this zone, enough light penetrates to the bottom to support rooted aquatic plants.

liver Large, dark-red organ, made up of several lobes, lying close to the stomach in vertebrates. Its main function is to regulate the chemical composition of the blood and to act as a depot where the nutrients obtained by digestion are converted to materials that are required for metabolism. (convert excess glucose to glycogen as food reserve). Other functions of the liver are production of bile, the break down (detoxification) of poisonous substances in the blood; removal of damaged red blood cells, the synthesis of vitamin A and the blood-clotting substances prothrombin and fibrinogen; and the storage of iron.

liverworts Type of bryophytes which grow usually in moist, shady habitats. Plant body of liverworts is thalloid, e.g. *Marchantia*.

living fossils Organisms (such as the horseshoe crab, dawn redwood, crocodilians, coelocanth or the inarticulate brachiopod lingula) whose closest relatives are extinct and that was once itself thought to be extinct.

lizards Vertebrates belonging to class reptilia, first found in the Jurassic period. They have long bodies, typically with four legs and a tail, have movable eyelids and external ear openings. These may be the most successful reptiles living today.

loafing platform A nest like structure built by few birds that can float on water. These birds keep their young ones in them to keep them out of water and afloat.

loam Fertile soil is made up of organic matter mixed with clay, sand and silt. Loams differ in their ratios of clay, sand and silt which influences which types of plants they can support.

lobopod

1. Blunt-ended pseudopodium found in certain protozoans.
2. A member of the phylum-Lobopodia.

local signaling molecules Molecules released by cells that generally travel through the interstitial fluid and have their effect locally. Examples of local signaling molecules include histamine and prostaglandins, as well as neurotransmitters.

lock and key mechanism Mechanism to explain binding between the active site of an enzyme and a substrate molecule. It was proposed in 1890 by Emil Fischer (1852-1919). The active site was thought to have a fixed structure (the lock), which exactly matched the structure of a specific substrate (the key). Thus, the enzyme and substrate interact to form an enzyme-substrate complex. The substrate is converted to products that no longer fit the active sit and are therefore released, liberating.

locomotion Ability of an organism to move in a particular direction in its environment.

locule (loculus) Small cavity in a plant or animal body. In plants the locule of the ovary is the cavity containing the ovules and the locules of the anther contain the developing pollen grains.

locus Position on a chromosome or within nucleic acid, occupied by a gene.

lodicules Two small scales situated at the base of the floret of grasses. May represent the reduced perianth and be involved in the opening of the floret at the time of pollination.

loins Region of the back and sides between the ribs and pelvis.

logistic growth Type of population growth in which the population starts growing exponentially but as resources become limiting, growth rate becomes zero and the population is at a long term equilibrium.

lomentum A dry dehiscent fruit formed from a single carpel and bearing more than one seed. It resembles a legume or siliqua but is divided by false septa into single seeded compartments.

long–day plant A plant in which flowering can be induced or enhanced by long days, usually of more than 12 hours of daylight, e.g. spinach

Long Terminal Repeat (LTR) Long nucleotide sequence that begins and ends the genome of retroviral DNA and certain retrotransposons.

loop of Henle Hairpin-shaped region of the kidney's nephron situated between the proximal and distal tubules. It is involved with the resorption of salt and water. Longer loops of Henle are associated with more efficient water conservation. Hence, desert animals have longer loops of Henle than their non-desert counterparts.

lophophore A ring of cilia-covered tentacles characteristic of brachiopods, phoronids, and bryozoans. The cilia create a current that carries food particles to the mouth.

Lophotrochozoa A major group of protostome animals. It includes the molluscs and annelid worms, together with the nemertoies, bryozoans and other worm like phyla.

lores Space between the eye and the bill of a bird or the eyes and tip of the mouth of any animal.

lorica A hard protective shell present in certain cells as of rotifers and microorganisms.

lower critical temperature The minimum body temperature that can be tolerated by an organism. Below this temperature, the biochemical properties of cell structures are altered and reactions are slowed such that the organism cannot maintain its usual bodily functions and death may follow.
LSD- Lysergic Acid Diethylamide.

luciferase An oxidising enzyme act on protein luciferin and causes luminosity in glow-worm and firefly and many other animals.

luciferin A compound found in bioluminescent organisms. It emits a nearly heatless light when it undergoes oxidation.

lucy The skeletal remains of the most famous specimen of the female hominid *Australopithecus afarensis,* found in Ehiopia in 1974. It is the most complete skeleton yet discovered.

lumbar

1. Pertaining to the loins.
2. With reference to vertebrae, those lying between the thoracic and sacral vertebrae.

lumbocostal Pertaining to the loin and ribs.

lumen

1. Space enclosed within a living cell or organelle.
2. Space within a tubular organ or other tube-like living structure.

lung Paired triangular bags that constitute the respiratory organ in vertebrates.

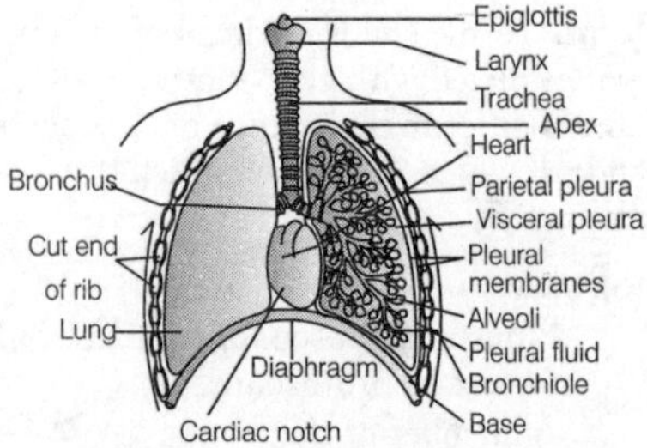

Human respiratory system

They lie in the thoracic cavity on the sides of the heart. Each lungs, in humans is enclosed in two membranes called the pleura.
Pleural cavity is a very narrow space that exists between the two pleura. It contains the pleural fluid secreted by the pleura, for reducing friction on the lung surface.
The left lung has two lobes, i.e., superior lobe and inferior lobe separated by oblique fissure. It has a cardiac notch, a concavity where the heart lies. It is longer and narrower than right lungs.
The right lung is bigger and has three lobes, i.e. superior lobs, middle lobe and inferior lobe separated by horizontal fissure and oblique fissure. There are about 300 million alveoli in the two lungs with a combined surface area of about $10m^2$.

lung book The respiratory organ of some arachnids, occur in pairs on the ventral side of the abdomen. They consists of leaf-like folds of ectoderm sunk into pockets having slitlike openings at the abdominal surface. Gaseous exchange occurs by diffusion through the ectodermal folds.

lungfish *See* dipnoi.

luteal phase Second phase of menstrual cycle in women, that occur after ovulation. It is marked by the production of corpus luteum.

luteinising hormone (LH, also ICSH) Glycoprotein hormone secreted by the anterior pituitary gland in mammals. It stimulates ovulation progesteron synthesis and corpus luteum formation in females and production of sex hormones by testes in males.

lyases Enzymes that catalyse removal of groups from substrates by mechanism other than hydrolysis leaving double bonds.

lymph A colourless liquid, circulating through the lymphatic system, similar to blood plasma and leucocytes, but has no erythrocytes.

lymph heart The series of muscular pumping structures that occur in the major lymphatic vessels of certain lower vertebrates and maintain a flow of lymph around the body. The animals which do not possess lymph hearts maintain the flow of lymph by a series of muscle contractions and valves, e.g. mammals.

lymph node (lymph gland) One of a large number of flat oval structures distributed along the lymphatic vessels and clustered in certain regions, such as neck, armpits and groins in humans. Lymph nodes are composed of lymphatic tissue and contain special white blood cells (lymphocytes). They provide defence against the spread of infections, by filtering out bacteria and other foreign particles.

lymphangiosarcoma Sarcoma arising in the cells that line the vessels of the lymphatic system.

lymphatic system A system of vessels and nodes containing lymph which returns protein and fluid from the tissues to the blood. It connects with the blood circulatory system in the large veins near the heart.

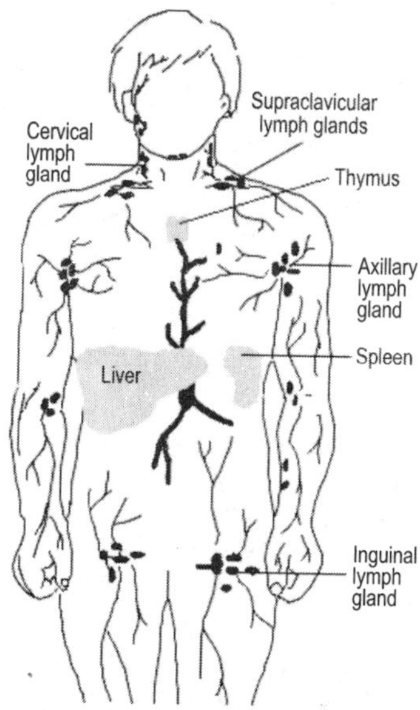

Lymphatic system of human

lymphectasia Dilation of lymphatic vessels.

lymphocyte Leukocyte formed in lymph, nodes, with large nucleus and little cytoplasm. There are two main types, the large granular lymphocytes and the small lymphocytes. Most large granular lymphocytes are Natural Killer (NK) cells that protect the body from tumour and virus-infected cells. Small lymphocytes are of two types, T-cells and B-cells.

lymphoid tissue The type of tissue found in the lymph nodes, tonsils, spleen and thymus. It is responsible for producing lymphocytes and therefore contributes to the body's defence against infection, e.g. BALT, MALT

lymphoidectomy The surgical excision of lymphoid tissue.

lymphokine Any of various cytokines that are released by lymphocytes when activated on encountering their specific antigen. They serve as chemical signals to other cell types involved in immune response, such as macrophages, neutrophils and basophils.

lymphoma Cancer of lymphoid tissue.

lysergic acid Substance obtained from ergot and used in the manufacture of the hallucinogen LSD (Lysergic Acid Diethylamide).

Lyseric Acid Diethylamide (LSD) A chemical derivative of lysergic acid that has potent hallucinogenic properties. It occurs in the cereal fungus ergot and acts as an antagonist at serotonin receptors.

lysigeny Localised disruption of plant cells to form a cavity in which secretions accumulate. Example the oil cavities in the leaves of citrus trees.

lysimeter An instrument used to measure the loss of water from an area of land covered with vegetation. Water that evaporates from both the soil and from the plants can be determined.

lysine One of an essential amino acids.

lysis

1. Dissolution of the cell.
2. In a medical context, refers to the gradual subsidence of a fever or acute disease.

lysogenic cycle A complex pattern of virus reproduction in which viral genetic material inserts itself into the host cell's genetic material. The host cell then reproduces, reproducing virus material along with its own genetic material. The result is a number of generations of host cells with the viral material incorporated into their own genetic material.

lysogenic repressor Protein that prevents a prophage from entering the lytic cycle.

lysosome A membrane bound organelle containing hydrolytic enzymes for intracellular digestion found in animal cells and in single-celled eukaryotes. In multicellular animals, lysosomes are important for programmed cell death that is a part of organism's development.

lysozyme An enzyme present in saliva, tears, egg white and mucous, discovered by Alexander Fleming in 1922. It destroys bacteria by hydrolysis of the cell wall.

lytic infection Infection with a bacteriophage that ends in lysis of the host cell.

lytic pathway Type of simple pattern of virus reproduction in which a virus invades a cell, causes the cell to produce more viruses. When the cell is full of viruses, it breaks open or lyses, releasing the new viruses.

lytic virus Virus that causes lysis of the host cell.

M

MAb (MoAb or Monoclonal Antibody) A specific antibody produced by one of numerous identical cells derived from a single parent cell. The parent cell (hybridoma) is obtained by the fusion of a lymphocyte with a cell derived from malignant tumour of lymphoid tissue. Used to identify a particular antigen within a mixture and can, therefore be used for identifying blood groups. They also enable production of highly specific and effective vaccines.

macroevolution The production during the course of evolution of new forms of life treated as distinct species, e.g. emergence of mammals.

macrofauna The collection of larger animals that do not need microscope for their observation.

macrofossil (also mega fossil) Fossil large enough to be inspected with the naked eye.

macrogamete The larger usually female gamete or egg cell.

macromolecule A large, complex molecule, such as DNA, RNA and proteins.

macronucleus (meganucleus) The larger of the two nuclei of ciliata. It is concerned with vegetative functions.

macronutrient Nutrient required in relatively large amounts by an organism.

macrophage A type of leukocyte produced by precursor cells (promonocytes) in bone marrow become monocytes in blood stream and settle as mature macrophage in tissues. Protect the body by engulfing microorganisms and debris.

macrophagus An organism feeding on relatively large particles of food.

macropinocytosis A form of pinocytosis by which dendritic cells engulf relatively large volumes of fluid from their surroundings. This enables them to take in antigens non-specifically in fulfilling their role as antigen monitors.

macropodid A member of the mammalian family–Macropodidae which includes the kangaroos, wallabees and wallaroos.

macroscopic Visible to the naked eye and measurable.

macula

1. A patch of sensory hair cells in the utriculus and sacculus of the inner ear that provides information about the position of the body in relation to gravity.
2. An area of increased visual activity in retina of vertebrate eye.

macula adherens *See* desmosome.

macula densa Thickening in the wall of distal tube of kidney nephron. It contains specialised cells, responsive to concentration in the luminal fluid and regulate the secretion of renin.

macula lutea The central region of retina surrounding the fovea.

maggot The worm-like, small legless larva with a pointed head-end and a truncated ear, e.g. of house flies, blue beetles, etc.

Magnetic Resonance Imaging (MRI) An imaging and diagnostic technique based on nuclear magnetic resonance. The body or part to be imaged in placed in an MR scanner. This creates a strong magnetic field, which causes the alignment of certain tissue nuclei, notably hydrogen nuclei in water and lipids. Its benefits is that it is non-invasive.

Major Histocompatibility Complex (MHC) A region composed of many separate genes that control the immune system. In humans, it is found on chromosome 6 and contains over 200 genes. The MHC is part of the major immunogene complex.

malacophily Pollinated by snails.

malaria A mosquito borne infectious disease caused by the parasitic protozoan *Plasmodium,* which requires two hosts, the bloodsucking female *Anopheles* mosquito and a human, in order to complete its complex life cycle.

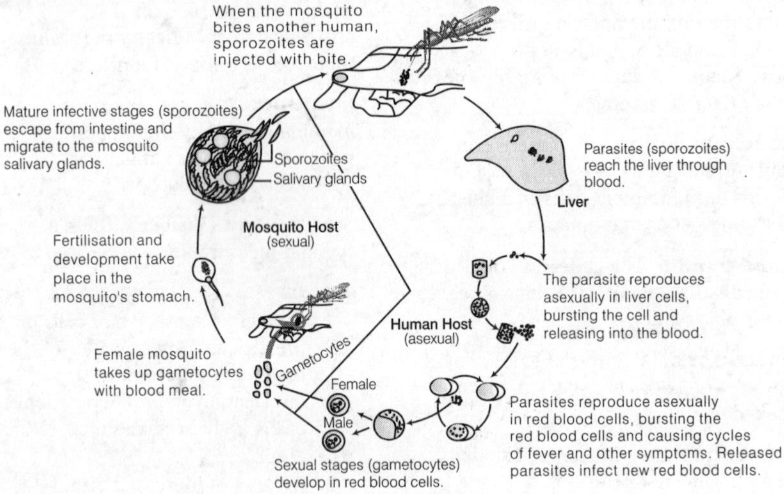

Spread of malaria disease

male

1. Denoting the gamete that, fuses with a female gamete in the process of fertilisation. Male gametes are generally smaller than the female gametes and are usually motile.
2. (Denoting) an individual whose reproductive organs produce only male gametes.

male haploid Having a single set of chromosomes. *See* haploid.

malic acid A colourless crystalline carboxylic acid which occurs in acid fruits, such as grapes and gooseberries. In biological processes, malate ion is an important part of the Krebs' cycle.

malignant A mutant cell or group of cells that can invade nearby tissue and spread elsewhere in the body at a fast rate.

malleus (hammer) An auditory ossicles, the fist of the three in mammalian middle ear.

Mallophaga An order of secondarily wingless insects comprising the bird lice, ectoparasites of birds, feeding on particles of dead skin, feather fragments and sometimes blood. These are minute with dorsoventrally flattened ovoid bodies, reduced eyes and biting mouth parts.

malnutrition A condition that occurs when a person's diet does not contain the right amount of nutrients, required to maintain health, e.g. Kwashiorkor.

Malpighian body (Malpighian corpuscle) The part of a nephron in the kidney that consists of its cup-shaped end together with the glomerulus that it encloses.

Malpighian layer (stratum germinativum) An innermost layer of dividing cells at the base of epidermis of vertebrates.

Malpighian tubule An organ involved in the excretion of nitrogenous waste in insects, arachnids and centipedes. They lie in the haemocoel. In insects Malpighian tubules open into the intestine and selectively extract uric acid from blood, together with water and salts it is deposited into the hindgut and excreted in the faeces. Some reabsorption of water and salts can occur in the tubule itself but most reabsorption occurs in the hindgut.

malt The product of the hydrolysis of starch by β-amylase that occurs during the germination of barley in brewing.

MALT *See* Mucosal-Associated Lymphoid Tissue.

maltase (α-glucosidase) A membrane-bound enzyme in the small intestine that hydrolyses the disaccharide maltose into glucose.

maltose (malt sugar) A sugar found in germinating cereal seeds. It is a disaccharide composed of two glucose units. Maltose is an important intermediate in the enzyme hydrolysis of starch. It is further hydrolysed to glucose.

mammal A member of the vertebrate class—Mammalia. Mammals nurse their young with milk.

Mammalia A class of vertebrates. The presence of mammary glands is most characteristics feature. Heart four chambered, excretion through kidney and internal fertilisation are other notable features parental care is highly developed. They feed their young ones on their milk and protect them from enemies, e.g. rabbits, whales, monkeys, etc.

mammalogy Study of mammals.

mammary gland The milk producing organ in female mammals. There may be one or more pairs depending on the species, situated ventrally. Each gland consists of fatty tissue in which are embedded lobules consisting of clusters of milk-producing alveoli.

mandible

1. The lower jaw of a vertebrate animal.
2. Either of the (upper or lower) part of the beak (bill) in birds.
3. One of a pair of horny mouthparts in insects, crustaceans, centipedes and millipedes, lie in front of the weaker maxillae. Their lateral movements assist in biting and crushing.

mandibular arch The first visceral arch; it is modified to form the upper and lower jaws in chondrichthyes (cartilaginous fish).

mangrove swamp A region of vegetation, found along tropical coasts, in which mangrove trees occurs in majority. The waterlogged soil is highly saline and mangroves are adapted to withstand these conditions. They possess aerial roots (pneumatophores) through which gaseous exchange occurs, to counteract effects of the badly aerated soil. Mangroves are complex ecosystems that play a vital role in protecting coastal regions from the effects of tropical storms and high tides.

mannan *See* mannose.

mannitol A soluble sugar alcohol found widely in plants and forming a characteristic food reserve of the brown algae. It is derived from mannose or fructose and is used as a sweetener in certain foodstuffs and as a diuretic in medicine to relieve fluid retention.

mannose An aldohexose sugar, that occurs naturally only in polymerised forms as mannans. Found serving as food energy reserves in plants, fungi and bacteria.

Mannose Binding Lectin (MBL) An acute-phase protein (that forms part of the innate immune response by binding to bacterial cells. It enhances their susceptibility to ingestion by phagocytic cells, particularly monocytes and triggering the complement cascade leading to target cell destruction. Secreted into blood plasma by the liver in response to infection.

mantle

1. A dense mass of fungal hyphae surrounding a root (in botany).
2. Part of the diatom valve that bends away at 90" (in zoology). Surface layer of visceral hump of Mollusca secreting shell. Its flap enclose the mantle cavity. Similar structure is found in brachiopods.

map distance A measure of the frequency of crossing over between two linked chromosome marker loci, equal to the percentage recombination in meiotic products. The further apart geographically two loci are on a chromosome the more will their apparent distance (as derived from cross-over values) be foreshortened by double and multiple cross-overs between them.

MAP kinase (Mitogen-Activated Protein Kinase or MAPK) An intracellular signal molecule that forms part of a key signal transduction pathway in eukaryotic cells. It enters the nucleus, so that transcription factors controlling gene expression are activated.

marginal meristem Located along margin of leaf primordium and forming the blade.

marginal placentation Attachment of ovules to carpel margin.

marine biome Refers to the aquatic biome comprising 3.5% salty water on an average. It includes the oceans and covers over 70% of the earth's surface and is divided into benthic and pelagic zones.

marker gene

1. A gene used for identification of a particular bacterial colony or bacteriophage plaque. Such genes are incorporated into cloning vectors to enable the isolation and replication of colonies containing a desired vector. Typically, marker genes confer resistance to specific antibiotics or produce colour changes.
2. (Genetic marker) a gene that acts as a tag for another, closely linked, gene. Such markers are used in mapping the order of genes along chromosomes and following inheritance.

marsupial A member of the infraclass-Metatheria of mammals. See Metatheria.

marsupium

1. A fold of skin or pouch, containing the mammary glands, and serving to hold and protect the young during early development; found in many marsupials.
2. A pouch present in certain non-mammals, such as fish, in which eggs are carried externally.

masquerade The resemblance of an organism to some inanimate object in its environment, so that it remains effectively 'hidden' from predators, e.g. insects have evolved colour or shape similar to leaves or flowers, so it is difficult to locate them.

mass extinction An extinction of a large number of species within a relatively short interval of the geological time scale. It has profound effect on course of evolution. The fossil record provides evidence for mass extinction as many as 20, since the start of the Phanerozoic eon about 570 million years ago. The biggest being at the end of the Permian period (about 248 million years ago), when over 80% of all marine invertebrate genera disappeared (including the trilobites) and at the end of the cretaceous (65 million years ago), when some 50% of all genera became extinct, including all dinosaurs.

mass flow hypothesis The pressure flow hypothesis given by Munch in 1930 is the well known and most accepted hypothesis to explain translocation of sugar from source to sink. According to this hypothesis, the organic substances flow in solution form in the seive elements due to the development of an osmotically generated pressure gradient between source end and sink end.

mass spectrometer An instrument that determines the chemical compound's composition on the basis of the mass-to-charge ratio of charged particles (used in DNA sequencing).

mast cell A type of white blood cell (leukocyte) with granular cytoplasm found within connective tissue, e.g. in skin and around blood vessels. Responsible for causing the symptoms of various allergies and other hypersensitive reactions. Also releases heparin that prevents clot formation.

mastication The process of chewing food. Mastication breaks up the food into small particles, providing greater surface area for digestion and enabling formation of bolus.

mastigoneme Exceedingly fine lateral projection (tinsel or flimmer) or certain flagella. Often numerous; Probably increase surface area and frictional resistance during movement.

mastoid process An outgrowth from the temporal bone of the skull. It contain air cavities that communicate with the cavity of the middle ear. In humans, it is a route through which infection may spread from the middle ear.

maternal effect (m. influence) Instances in which some aspect of phenotype of sexually produced offspring is controlled more by the genotype of maternal parent than by its own, e.g. distribution of cytoplasmic molecules in the egg, under the control of maternal genes, may have a marked effect upon cleavage and later development.

maternal gene Term sometimes used specifically for a gene responsible for a maternal effect.

maternal inheritance Characters inherited through the female line only. *See* maternal effect, cytoplasmic inheritance.

mating *See* sexual intercourse.

mating season *See* breeding season.

mating type Describes an equivalent in microorganisms, fungi and algae of the male and female individuals of higher organisms. These are morphologically same and difficult to distinguish. Thus, are designated as + and –. The mating types are distinct strains that are compatible in sexual reproduction; they give rise to a zygote, which can develop into a new individual.

matric potential (ψ_m) A component of water potential due to the adhesion of water molecules to non-dissolved structures of the system, i.e. the matrix, such as plasma membranes or soil particles. It is always negative and is significant only outside living cells in relatively dry system, e.g. soils, where much of the water is tightly bound to soil particles.

matrix The component of tissues in which cells are embedded.

maxicell A bacterial cell that has been irradiated with ultraviolet light to intentionally damage the cell's own DNA, so that plasmid genes which are relatively spared, can be preferentially expressed.

maxilla (pl. maxillae)

1. One of a pair of mouth part lying in insects, crustaceans, centipedes, and millipedes. Lying behind the mandibles and feeding assists in crustaceans have two pairs of maxillae but in insects the second pair are fused together forming the labium.
2. One of a pair of large tooth-bearing bones in the upper jaw of vertebrates.

maxilliped A paired appendage in crustaceans that contributes to the mouthparts. There may be up to three pairs of maxillipeds, which are located behind the mandibles and are used to move food to the mouth.

maxillula A paired appendage that contributes to the mouth parts in certain crustaceans. Lying between the mandibles and the maxillae, they may bear hairs or spines designed to move food particles deeper into the mouth.

M-band (M-line) One of the membrane bounding the sarcomere, which is the point of attachment of myosin filaments. See sarcomere.

M cells (Multifenestrated cells) Specialised epithelial cells in the follicle of Peyer's patches exposing to the gut lumen. These cells have microfolded luminal surface, and take up particles from the gut by endocytosis.

meatus A small canal, or passage, e.g. external auditory meatus.

mechanoreceptor A receptor that responds to a mechanical stimulus, e.g. touch, pressure, sound, etc.

meconium The first faecal material passed by a newborn mammal, which is usually dark green in colour.

median eye (pineal eye) An eyelike structure, with a lens and retina, found on the top of the head of some lizards, *Sphenodon*, and the Cyclostomata (lampreys) as well as in many fossil vertebrates. It is thought act as a photoreceptor, detecting changes in light intensity and modifying the physiology and behaviour of the animal. It corresponds to the pineal gland of other vertebrates.

mediastinum
1. A membrane in the midline of the thorax of mammals that separates the lungs.
2. The space between the two lungs, which is occupied by the heart and oesophagus.

medulla The central region of an animal organ, differs in structure or function from the outer regions kidney, e.g. renal medulla.

medulla oblongata A region of the hindbrain that is concerned with the functioning of the visceral organs, e.g. the stomach, lungs and heart. It is continuous with the spinal cord and controls respiratory rhythm, blood circulation, the reflex movements of the eye muscles and other involuntary functions. Many of the cranial nerves arise from the medulla.

medullary bundle A vascular bundle situated in the peripheral part of the pith of a stem and sometimes held to be an extension of a leaf trace.

medullary ray (pith ray) An undifferentiated parenchyma tissue found between the vascular bundles and connecting the pith and cortex in monocotyledons. In older dicotyledons and gymnosperms in which secondary tissues are formed, cells of the medullary ray differentiate to form interfascicular cambium.

medullated nerve fibre A nerve fibre that is characterised by a myelin sheath, which insulates the axon.

medusa The sexual stage free-swimming in the life cycle of a coelenterate, such as a jellyfish or a *Hydra.* Medusae are umbrella-shaped with tentacles round the edge and mouth in centre underneath.

megabase (Mb) A unit of nucleotide sequence equal to 10,00,000 base pairs.

megadont (macrodont) Having large teeth.

megalocephaly Having an abnormally large head.

megaphyll A type of foliage leaf in ferns and seed plants that has branched or parallel vascular bundles running through the lamina, e.g. fronds. A megaphyll was formerly called a macrophyll.

megasporangia In gymnosperms, cones bearing megasporophylls with ovules or megasporangia. Also called macrosporangiate or female strobili.

megaspore In an ovule of a flower, the four haploid cells produced by meiosis. Normally, only one goes on to produce a female gametophyte; the others degenerate.

megaspore mother cell (megasporocyte) A diploid cell in plants that divides by meiosis to give rise to four haploid megaspores. In flowering plants the megaspore mother cell (embryo mother cell) is situated in the ovule.

megasporophyll A leaf like structure that bears megasporangia.

megasporogenesis The process of formation of megaspores from the Megaspore Mother Cell (MMC).

Meibomian glands A sebaceous gland that lubricates the eyelids.

meiocyte The cell which can undergo meiosis. It is always diploid.

meiofauna Animals inhabiting the bottom of a river, lake or sea, e.g. small polychaetes, bivalves, nematodes.

meiosis (reduction division) The process of two consecutive cell divisions that produces haploid sex cells and spores from diploid progenitor cells. Meiosis results in four daughter cells, each with a haploid set of unreplicated chromosomes.

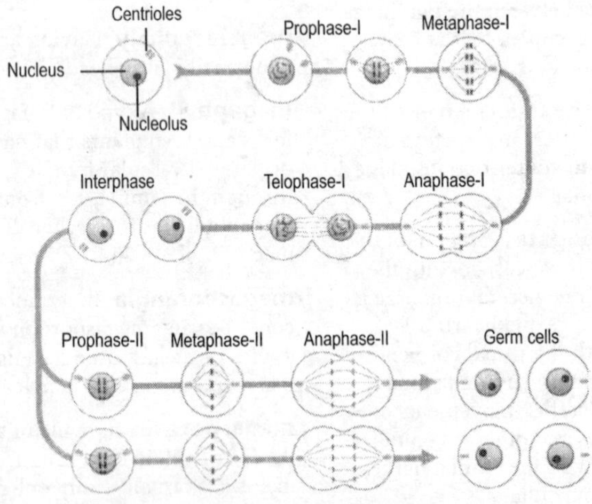

Different stages of meiosis

m

meiospore A spore produced by meiosis, which is haploid and gives rise by division to haploid daughter cells, or to a haploid individual such as the gametophyte of a seed plant.

Meissner's corpuscles Receptors for touch that are found in the papillary region of the dermis of the skin. Each corpuscle consists of a mass of dendrites encapsulated by connective tissue. They are found in abundance in the palms of the hands, fingertips, and soles of the feet.

Meissner's plexus The network of nerves in the submucosa of the small intestine.

melanin A group of tyrosine-derived compounds found causing pigmentation in eyes, skin and hair of vertebrates. Most common ferns are eumelanin, (brownish-black) and pheomelanin (reddish-brown).

melanism A characterised by a high level of concentration of melanin (black pigmentation) occurring in the coat, skin or plumage of the animals.

Melanocyte Stimulating Hormone (MSH) Several related peptide hormones that affect skin colouration and also act as anti-inflammatory cytokines and as neuropeptides in the brain secreted by the intermediate lobe of pituitary gland, it stimulates dispersal of melanin in the chromatophores of amphibian skin, causing the skin to darken; in humans it stimulates production and dispersal of melanin in the pigment cells (melanocytes) of the skin.

melanophore The chromatophore in skin containing pigment melanin.

melatonin The hormone secreted by pineal gland located on the dorsal side of forebrain. Plays vital role in the regulation of 24 hour (diural) rhythm of body.

membrane
1. Thin, flexible layer of tissue covering a bodily structure, or separating one structure from another.
2. The limiting surface of cellular cytoplasm, or of a cellular organelle.
3. Thin, flexible partition or covering of any type.

Membrane Attack Complex (MAC) The assembly to complement proteins that creates a pore in the plasma membrane of a pathogen, thus causing its destruction. It is one of the mechanisms of innate immunity that follows activation of complement cascade.

membrane bone (dermal bone) The bone formed directly in connective tissue by intramembranous ossification, rather than by replacing cartilage. Some face bones, skull bones, and part of the clavicle are membrane bones.

membrane potential The difference in charge between the cytoplasm and the fluid outside the cell. It due to the differences in the distribution of ions within and outside the cell.

membranous labyrinth The soft tubular sensory structures that form the inner ear of vertebrates and are housed within the bony labyrinth.

meme A self-replicating unit of cultural inheritance analogous to a gene. It is transmitted between individuals and across generations, so that it is inherited and (potentially) can change over time. Such a concept can provide information into nature or cultural development.

memory The means of information storage in brain.

memory cells *See* immunological memory.

menarche The onset of menses.

Mendelian disorders Genetic disorders determined by alteration or mutation in a single gene, e.g. colour blindness.

Mendelian trait The transmission of hereditary traits from parent organisms to their offsprings, the trait mainly emphasises on a single locus.

Mendelism The theory of heredity that forms the basis of classical genetics, proposed by Gregor Mendel in 1866 and formulated in two laws. Mendel suggested that individual characteristics were determined by inherited 'factors' and with technological advancements, the details of cell structure showed that behaviour of Mendel's factors could be related to the behaviour of chromosomes during meiosis.

Mendel's laws Two laws summarising Gregor Johann Mendel's theory of inheritance see and law of segregation law of independent assortment.

meninges Three membranes enclosing the spinal cord and brain. Outermost thick layer is **dura mater**. Inside is delicate, arachnoid layer. The innermost is pia mater which intimately covers the brain and spinal cord. Cerebrospinal fluid is between arachnoid membranes and pia mater in a space called subarachnoid space.

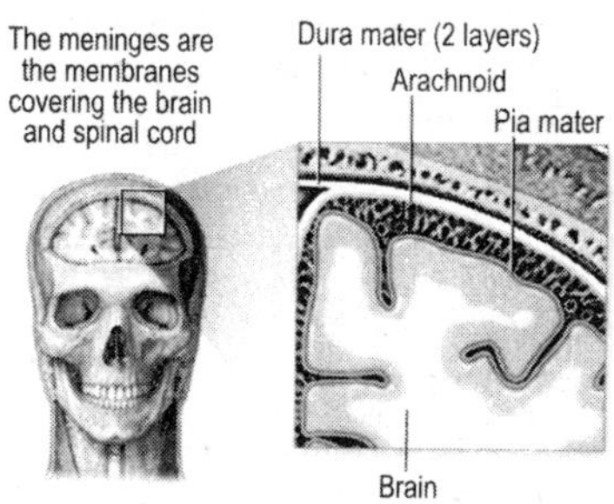

Different layers of meninges in human brain

m

meningitis Inflammation of the meninges.

menopause The permanent cessation of ovulation and menstrual cycle in human females that normally occurs between the age of 42-52.

menses The monthly blood discharge occurring in women after puberty.

menstrual cycle A rhythmic change in the reproductive organs of the female primates (monkey apes and humans). Average interval of menstruation in human female is about 28-29 days. The cyclical events starting from one menstruation till the next one constitute menstrual cycle. The four phases of menstrual cycle are menstrual phase, follicular phase, luteal phase and ovulation.

mericarp The one-seeded dehiscent or indehiscent portion of a schizocarpic fruit. It is found, e.g. in the Geraniaceae, whose fruit, the regma, splits into five dehiscent mericarps.

meristele A segment of a dicotystele formed by overlapping leaf gaps. A meristem contains all the tissues of a stele, usually arranged in a concentric pattern. The xylem is surrounded by a layer of phloem and bounded by a pericycle and endodermis. Seen in many club mosses and ferns.

meristem Plant cells that remain embryonic and that are therefore, mitotically active, allowing plants to go on growing indefinitely as long as they live (indeterminate growth).

meristoderm An outer meristematic layer of the thallus of certain brown algae (Phaeophyta). The cells of the meristoderm divide to increase the width of the thallus.

Merkel's disc A touch receptor, occurring in the outer layer (epidermis) of hairless skin, that responds to sustained pressure. These discs are found in large numbers in the fingertips.

meroblastic The type of incomplete cleavage that occurs in very yolky eggs (e.g. birds and sharks) in which the non-yolky portion engages in cell division.

merocrine gland A gland whose cells secrete their product while remaining intact: no portion is pinched off nor do cells have to disintegrate in order to release their product (see holocrine gland). e.g. include vertebrate salivary glands and exocrine cells of the pancreas.

merogenesis Reproduction by segmentation.

merogony (also schizogony) A mode of asexual reproduction seen in certain protozoan parasites. The feeding-stage form (trophozoite) of the parasite increases in size while its nucleus and other organelles divide repeatedly without cytokinesis. Then this enlarged, reproducing form (schizont) ruptures, releasing merozoites.

meropia The partial blindness.

meroplanktonic Term describing organisms that spend part of their life cycle in the plankton and part in the benthos as a resting stage.

Merostomata A class of aquatic arthropods. Includes the extinct *Eurypterida*. The only living forms (king crabs, e.g. *Limulus*) have a broad cephalothorax (prosoma) covered dorsally by carapace which covers limbs and in which is located a pair of eyes. Chelicerae chelate. These are pedipalps resemble walking limbs. Gnathobases (spiny basal segments of legs) function as mandibles, which are absent. Opisthosoma represented by fused tergites forming a single dorsal plate, with a long caudal spine hinged to its posterior border. Gill books serve for gaseous exchange.

merozoite The life cycle of certain protozoan parasites, a stage produced by merogony.

merozygote A bacterial cell that contains more than the haploid number of chromosomes but less than the full diploid number. Merozygotes arise when the genetic material from one bacterial cell is only partially transferred into another cell during conjugation, transduction or transformation.

mesarch A condition in which the protoxylem elements are surrounded by the metaxylem elements.

mesencephalon *See* midbrain.

mesenchyme A loose network of cells, usually underlying epithelial layers in animal embryos. That give rise to connective tissue, blood, etc.

mesenteric artery An artery that delivers blood to the intestines.

mesentery A thin sheet of tissue, bounded on each side by peritoneum, that supports the gut and other organs in the body cavities of animals. Vertebrates have a well-developed dorsal mesentery that anchors the stomach and intestine and contains blood vessels and nerves supplying the gut. The reproductive organs and their ducts are also supported by mesenteries.

mesocarp The middle layer of the pericarp in an angiospermic fruit which is situated between exocarp and endocarp. In peach and mango, mesocarp is fleshy edible part of the fruit.

mesoderm An early embryo's middle cell layer that lies between the ectoderm and endoderm. It develops into muscles, skeleton, kidneys, spleen, gonads and much of the circulatory system.

mesofauna *See* macrofauna.

mesogloea The layer of jelly-like material between ectoderm and endoderm of coelenterates. It is present as a non-cellular collagenous membrane in hydrozoans such as *Hydra* but thickened and fibrous in jellyfish (scyphozoans) and some other groups where it contain cells derived from the two tissue layers. It is not itself a germ layer.

mesoileum The mesentery supporting the ileum.

Mesopelagic zone *See* dysphotic zone.

mesophilic Describing an organism that lives and grows optimally at moderate temperatures be (10-40°C). The vast majority of organisms are mesophiles, occupying all the major biomes of temperate and tropical climates.

mesophyll The mid-layer of a leaf between the upper and lower epidermis; portion of a leaf where photosynthesis occurs.

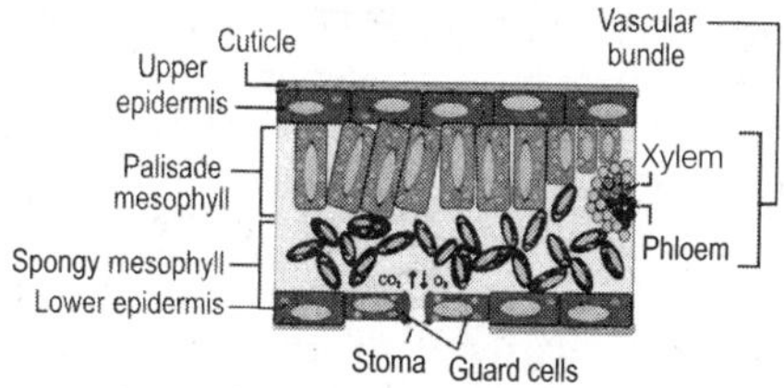

Internal structure of a leaf showing mesophyll cells

It consists of parenchyma cells and has two distinct forms palisade and spongy-mesophyll.

mesophytes Plants which grow in balanced availability of water and mineral salts, e.g. rose, sunflower, mango, etc.

mesosome The special membranous structure in prokaryotes. Formed by the infolding of plasma membrane into the cell. Helps in cell wall formation, DNA replication.

mesothelium The tissue consisting of one thin plate-like layer of cells that lines abdominal cavity and thorax. It is derived from mesoderm.

mesotrophic A water body such as a lake, that is intermediate between a eutrophic lake and an oligotrophic lake in the amount of nutrients contained within it.

Mesozoic era The second of the three geologic eras of the Phanerozoic Era. It lasted from 251.0-65.5 million years ago. It comprises the Triassic, Jurassic and Cretaceous periods. It is often known as the age of reptiles.

messenger RNA (mRNA) RNA that serves as a template for protein synthesis. See RNA.

metabolic pathway A series of chemical reactions catalysed by enzymes, occurring within an organism. The product of each successive reaction is the substrate of the next.

metabolic rate Measure of the energy used by an animal in a given time period. It is affected by several interacting factors, considering temperature and the level of activity. The metabolic rate of a resting animal is known as the Basal Metabolic Rate (BMR).

metabolic waste Waste products, collectively of metabolism.

metabolise To transform a chemical compound into some other compound(s) *via* a metabolic pathway.

metabolism Chemical processes, taken as a whole, occurring within an organism.

metabolite A product or substance of a metabolic pathway.

metabolome An entire complement of metabolites found within a cell under defined condition. It is determined using various forms of mass spectrometry. It excludes nucleic acids and other large molecules.

metabolomics Study of change in the pool of metabolites under various physiological or developmental conditions or in response to genetic modification. (e.g. mutation).

metabotropic receptor The cell receptor that, when activated by binding of a ligand, triggers changes in cell metabolism *via* intracellular second messengers.

metacarpal One of the bones in the metacarpus.

metacarpus The hand in terrestrial vertebrates consisting of a number of bones (metacarpals).

metacentric Chromosome with its centromere at the centre. *See* centromere.

metafemale Fruit fly (*Drosophila* sp.) with three instead of the normal two X-chromosomes. In such flies the disproportion of sex chromosomes to autosomes leads to impaired development and often failure to emerge from the pupae.

metamale Fruit fly (*Drosophila* sp.) with three (instead of the normal two) copies of each autosome. The disproportion of autosomes to the normal complement of one X-chromosome (plus one Y-chromosome. It plays little or no part in sex determination) but impairs development and such flies are typically weak and sterile.

metamerism Body is externally and internally divided into segments with serial repetition of atleast some organs, e.g. in earthworm. Each segment is called metamere.

metamorphosis The process of marked change in the appearance and habits of some invertebrates and amphibians, as part of their normal development. Metamorphosis of a caterpillar into a butterfly is one such example.

metanephridium An excretory gland in many annelids, arthropods and molluscs. It consists of a tubule that opens into the coelom by a ciliated funnel and conducts waste fluids to the exterior.

metaphase The stage of cell division during which the nuclear membrane breaks down, the spindle forms and centromeres attach the chromosomes to the equator of the spindle. In the first metaphase of meiosis, pairs of chromosomes (bivalents) are attached, while in mitosis and the second metaphase of meiosis, individual chromosomes are attached.

metaphloem Later formed phloem, having bigger sieve tubes that develops after stem or other part has finished elongating. First formed phloem called protophloem have narrow sieve tubes.

metaplasia The change of one tissue to another as seen in response to certain diseases. e.g. the epithelium of the respiratory tract may show metaplasia in response to irritants, e.g. smoke.

metapopulation A group of populations that are separated by space, but consist of same species.

metarteriole (or arterial capillary) A short vessel that links arterioles and venules.

metastasis The spread of cancer cells to new regions of the body.

metatarsal One of the bones in metatarsus.

metatarsus The collection of metatarsal bones forming part of the hind foot or lower hindlimb in tetrapods; the arch of the foot in humans.

Metatheria An infraclass of mammals containing the marsupials. The female bears an abdominal pouch (marsupium) into which the newly born young, which are in a very immature state, move to complete their development. The obtain nourishment from the mother's mammary teats, e.g. kangaroos.

metaxylem The later formed elements of xylem tissue. First formed xylem elements are called protoxylem.

metazoa (eumetazoa) The term including all multicellular animals other than sponges.

methanogens Bacteria which grow anaerobically on cellulosic material and produce large amount of methane along with CO_2 and H_2, e.g. *Methanobacillus*. These bacteria are important in production of biogas.

methionine An essential amino acid.

methylene blue Blue dye used in optical microscopy to stain nuclei of animal tissues. It is also suitable as a vital stain and a bacterial stain.

MHC *See* major histocompatibility complex.

MHC class I protein
Histocompatibility proteins that are expressed by all nucleated cells. After synthesis in the endoplasmic reticulum, these proteins bind peptide fragments derived from viruses or other pathogens invading the cytoplasm and are transported to the outer membrane of the cell, effectively flagging the cell for destruction by cytotoxic T-cell bearing CD8 proteins.

MHC class II protein
Histocompatibility proteins whose expression is restricted to certain immune cells, in particular B-cells, macrophages, other antigen-presenting cells, epithelial cells of the thymus, and (in humans) activated T-cells. They bind antigenic peptide fragment derived from pathogens or toxins. They are recognised on the cell surface by receptors of CD4 bearing T-cells.

micelles An aggregate of surfactant molecules dispersed in a liquid colloid. The products of fat digestion are dispersed into micelles by the action of bile salts, which facilitates their absorption in the small intestine.

microarray A glass slide or bead on which are deposited biomolecules or other material in a regular microscale pattern to enable automated simultaneous multiple assays of target substances or activity, Microarrays are powerful analytical tools with wide-ranging applications, e.g. DNA microarray.

microbiology The study of microorganisms (e.g. bacteria and viruses), including their interactions with other organisms and with the environment.

microbody Any of a class of cell organelles that are spherical vesicles, 0.2-1.5 μm, bounded by a single membrane. They contain enzymes responsible for the oxidation of various materials and they originate from the endoplasmic reticulum. They include peroxisomes and glyoxysomes (see glyoxylate cycle).

microcirculation The part of the blood circulatory system consisting of the networks of capillaries and associated vessels that supply blood to tissue cells.

Arteriole → Metarterioles →
Capillaries ↓ Post capillary venules

microclimate The local climate of a particular habitat or small area which is different from the macroclimate of the larger surrounding geographical area.

microdissection (micromanipulation) A technique used for the dissection of living cells under the high power of an optical microscope. It utilises instruments, such as needles, scalpels, micropipettes and lasers.

microelectrode A micropipette filled with a physiologically appropriate electrolyte solution (e.g. potasium chloride), whose tip can be inserted through the plasma membrane into cells without major disturbance to cellular function. It can be used to detect changes in electrical potential.

microevolution An evolution on a relatively small scale, involving the emergence of new species or of new groups below the species level, such as races and subspecies.

microfauna

1. Animals that cannot be seen with the naked eye. They are normally observed with the aid of a microscope.
2. Animals that live in a particular microhabitat.

microfibril A microscopic fibre. Cellulosic microfibrils appear in plant cell walls. These are about 5 mm diameter.

microfilaments Strands of actin present in almost all eukaryotic cell. Microfilaments are present in undulipodia and cilia, and make up the part of cytoskeleton. They act alone or with myosin to bring about contraction of the cell.

microflora

1. Plants and algae that cannot be seen with the naked eye. They are normally observed with the aid of a microscope.
2. Plants and algae that live in a particular microhabitat.

microfossil Fossil too small to be inspected without the aid of a microscope (less than about 4 mm in diameter), e.g. are microorganisms and pollen grains.

microgamete Smaller, actively motile male gamete (particularly in protozoans).

microglia A class of glial cells occurring in nervous tissues.

microhabitat The local habitat of an organism, e.g. bird's nest.

microinjection An injection of a DNA or other substance into a cell, cell organelle, etc., by means of a microcapillary pipette.

micronucleus One of the two nuclei cell of certain protozoans that is involved in sexual reproduction.

micronutrients Dietary minerals needed in very small quantities throughout life (generally less than 100 micrograms/2day). In animals, they include iron, cobalt, chromium, copper, iodine, manganese, zin selenium and molybdenum. In plants, they include B, CI, Cu, Fe, Mn, Mo and Zn.

microorganisms (microbe) Organisms that too small to be seen with the naked eyes. Include bacteria, viruses, protists and fungi.

microphagous The method of feeding of heterotrophic organisms that take in their food in the form of tiny particles, e.g. Filter feeding and ciliary feeding.

microphyll A type of foliage leaf with single unbranched midrib, e.g. club mosses and horsetails. These are generally no more than a few millimetres long.

micropipette A glass pipette with an ultrafine tip, typically less than 1 μm in diameter. It can be inserted into single cells or other microscopic structures and used, e.g. to inject materials (see microinjection). It is usually held by a micromanipulator a mechanical device that allows precise movement of the tip.

micropropagation The *in vitro* propagation of plants by cloning. It involves culturing excised meristematic tissue on a special medium that encourages axillary bud development. The new shoots are then separated and cultured and the cycle is repeated until finally the shoots are transferred to a medium that promotes root development, to produce plantlets. Micropropagation is used in agriculture, horticulture and forestry and special genotypes can be bred and maintained, the process is rapid and plants can be kept disease-free.

micropyle

1. Small opening in the surface of a plant ovule through which the pollen tube passes prior to fertilisation.
2. Small pore in some animal cells or tissues, e.g. in insect eggs (*see* chorion).

microRNA (miRNA) A small RNA molecule that is encoded by a cell and can 'silence' the expression of a particular target gene within the cell *mi*RNAs bind to target messenger RNA (*m*RNA) molecules and suppress translation of the *m*RNA into protein. *See* RNA interference

microsatellite DNA A form of highly repetitive DNA that consist of large number of short base sequence repeated a variable number of times in a tandem array at any given locus. throughout the eukaryotic genome.

microscope Any device used for forming a magnified image of a small object. The simple microscope consists of a biconvex magnifying glass or an equivalent system of lenses, either hand-held or in simple frame. The compound microscope uses two lenses or systems of lenses, the second magnifying the real image formed by the first see electron microscope; nomarski microscope; phase-contrast microscope; scanning probe microscopy; ultraviolet microscope.

microsome A fragment of endoplasmic reticulum formed when cells or tissues are disrupted. commonly used to investigate the functional properties of endoplasmic reticulum, such as enzymic activity and protein synthesis.

microsporangia Anther is a four-sided structure consisting of four microsporangia located at the corners. Microsporangia develop further and become pollen sacs.

microspore Haploid cells formed by microspore mother cell in plants. They develop into pollen grain.

microspore mother cell (microsporocyte) A diploid cell in plants that divides by meiosis to give rise to four haploid microspores. In flowering plants, microspore mother cells are formed within the pollen sacs of the anthers by mitosis.

m

microsporophyll Sporophyll bearing male spores in gymnosperms. Strobili bearing microsporophylls are called microsporangiate or male strobili.

microtome The instrument for cutting thin sections of biological material for microscopic examination.

microtubules Hollow strands of tubulin present in all eukaryotic cell and cilia and make up part of the cytoskeleton. Present in the cytoplasm below the cell membrane.

microvilli Hair-like projections on the surface of the epithelial cells, e.g. on the villi in the small intestine. They increase the surface area of the intestine to improve absorption of digested nutrients.

micturition An act of passing urine.

midbrain (mesencephalon) One of the three sections of the brain of a vertebrate embryo. It does not undergo further subdivision to form additional zones. It becomes part of the brainstem in mammals but in amphibians, reptiles and birds, the roof of the midbrain becomes enlarged as the tectum, a dominant centre for integration and may include a pair of optic lobes.

middle air (tympanic cavity) An air-filled cavity in the skull, between the outer and inner ear in most tetrapods. It is connected to the back of the throat by the Eustachian tube. In mammals, it contains three small bones, the ear ossicles (malleus, incus and stapes).

middle lamella A thin cementing layer holding together neighbouring plant cell walls. It consists mainly of pectin substances (e.g. calcium pectate). The middle lamella is laid down at the cell plate during cell division.

midgut The middle part of the alimentary canal of arthropod and in vertebrate animals, responsible for digestion and absorption.

midrib The strengthened central vein of a leaf or flower petal.

migration Seasonal movement of complete populations of animals to a more favourable environment. It is usually a response to lower temperatures common in mammals, fish, some insects and birds.

milk

1. A whitish opaque nutritious fluid that is produced by the mammary glands of the female mammals for feeding their young. It contains carbohydrates, fats, proteins, certain mineral salts, vitamin and water. It is rich in calcium and phosphorus, vitamins-A, D and riboflavin but a poor source of iron and vitamin-C and K.
2. In plants, any of various milk-like fluids, such as coconut milk or the latex of certain flowering plants.

milk teeth *See* deciduous teeth.

millipedes *See* Diplopoda.

mimicry Resemblance of one animal to another by which the mimic gains advantage from its resemblance to the model, e.g. Batesian mimicry.

mineral salts Inorganic salts that need to be ingested or absorbed by living organisms for healthy growth and maintenance. They comprise the salts of the trace elements in animals (see essential element) and the micronutrients of plants.

mineralocorticoid A type of steroid hormone produced by the adrenal cortex. Mineralocorticoids (e.g. a ldosterone and deoxy corticosterone) control salt and water balance by their action on the kidney.

minicell

1. A small roughly spherical cell produced by the abnormal division of a rod-shaped bacterium that contains no chromosomal DNA and is unable to grow or divide. It is used experimentally to express nucleic acids and proteins encoded by DNA introduced to the cell, e.g. in the form of a cloning vector.
2. (Micro-cell) an experimentally produced body consisting of a nucleus surrounded by a thin layer of cytoplasm and a plasma membrane and containing variable numbers of chromosomes. It is used as a vector for transferring one or several chromosomes into a normal cell following cell fusion.

minisatellite DNA *See* variable number tandem repeats.

miocene The first epoch of the Neogene. It lasted from 23.03-5.33 million years ago. It saw the appearance of modern mammal groups.

miracidium The first larval stage of trematode worms (flukes), which hatches from eggs excreted in the faeces of the primary host. Its leaflike body is covered with cilia, enabling the larva to swim towards a secondary host, in which it continues its development.

miRNA *See* micro RNA.

mismatch repair A form of DNA repair that replaces mismatched bases in newly replicated DNA in living cells.

missense mutation A type of point mutation that converts one codon to another, specifying a different amino acid. This results in incorporation of the wrong amino acid in the polypeptide chain during translation and consequently, a potentially faulty protein.

mites *See* acarina.

mitochondria (sing. mitochondrion) Organelles that are the site of respiration in eukaryotic cells. Reaction of Krebs' cycle and electron transfer chain take place in mitochondria. A mitochondrion contains two membranes, of which outer membrane is smooth and forms the continuous boundary of the organelle. The inner membrane is semipermeable to some metabolities. It is infolded into the matrix as incomplete partitions called. cristae. The cristae are responsible for increasing the physiological active area or surface area. The density of cristae determines the intensity of respiration.

mitochondrial DNA (mtDNA) The DNA found in mitochondria. It codes for ribosomal and transfer RNA.

mitochondrial theory of ageing A senescence theory which proposes that over an individual's lifetime, mutations accumulate in mitochondrial DNA to the extent that there is an age-related decline in the formation of ATP by oxidative phosphorylation.

m

mitosis The process of eukaryotic cell division that produces daughter cells genetically identical to each other and to the parent cell. No change in chromosome number occurs.

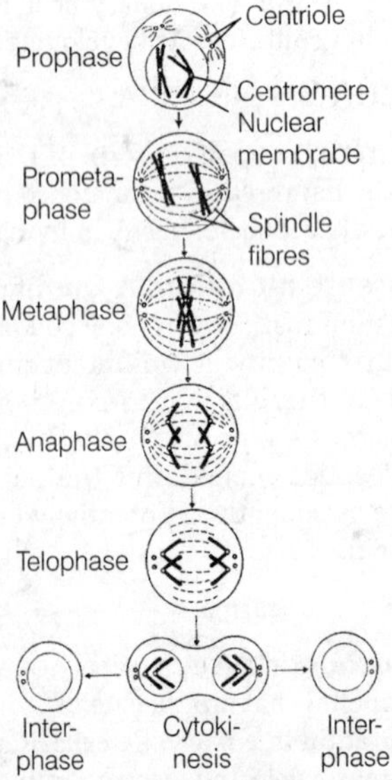

Diagram showing the behaviour of one pair of homologous chromosomes in animal mitosis

Mitosis Promoting Factor (Maturation Promoting factor MPF) A protein complex responsible for triggering mitosis in somatic cells and for maturation of oocytes into egg cells.

mitral valve (bicuspid valve) The valve consisting of two membranous flaps situated between the atrium and ventricle of the left side of the heart in mammals and birds.

mobbing An aggressive stand taken by some birds to ward off an intruder or predator from the area.

model organism An organism studied as representative, either with respect to the course of some disease that also occurs in humans or with respect to the general category of organisms to which the organism belongs.

modern synthesis *See* neo-Darwinism.

modifier gene A gene that influences the expression of another gene, e.g. one gene controls whether eye colour is blue or brown but other (modifier) genes can also influence the colour by affecting the amount or distribution of pigment in the iris.

MOET (Multiple Ovulation Embryo Transfer Technology) The programme for cattle improvement. This method is applied in cattle, rabbits, mares, etc. Cow is administered with hormones, like FSH to induce follicular maturation and super ovulation. Instead of one egg which they normally yield per cycle, 6-8 eggs are produced. The animal is either mated with an elite bull. Fertilised eggs at 8-32 celled stage are recovered non-surgically and transferred to surrogate mothers. The genetic mother is available for another round of ovulation.

molar One of the grinding teeth in jaw of mammals, behind the premolars. There are two or more molars on each side of both jaws; their surfaces are raised into ridges or cusps. In humans the third (and most posterior) molar does not appear until young adulthood (17-25 years) these molars are known as wisdom teeth.

molecular biology The study of the formation, structure and function of macromolecules found in living organisms, particularly nucleic acid and proteins.

molecular chaperone A group of proteins in living cells that assist newly synthesised or denatured newly synthesised or denatured proteins to fold into their functional three-dimensional structures. There are several unrelated families of chaperones, including five classes of heat-shock proteins chaperonins, calnexin and calreticulin.

molecular clock

1. Rate of fixation of mutations in DNA and thus a measure of the rate of genetic diversification.
2. A biological system capable of maintaining a timing rhythm or pulse.

molecular farming The development of transgenic animals to produce human proteins.

molecular fossil A substance found in geological deposits that is indicative of a particular species or group of organisms that lived when the deposits were formed, e.g. isoprene residues indicative of ancient archaeans and oleanane, a defence chemical found only in flowering plants.

molecular genetics The study at the molecular level, of the function and structure of genes.

molecular marker The site (locus) in the genome of an organism at which the DNA base sequence varies among the different individuals of a population. Used for chromosome mapping, DNA profiling and genetic screening, these markers have no effect on phenotype of individual.

molecular mimicry The close resemblance of an antigenic molecule of an invading pathogen to a host tissue molecule. This can cause autoimmune diseases.

molecular systematics (biochemical taxonomy; molecular phylogenetics) The use of amino-acid or nucleotide-sequence data in determining the evolutionary relationships of different organisms. Essentially it involves comparing the sequences of functionally homologous molecules (e.g. ribosomal RNA subunits) from each organism being studied, and determining the number of differences between them.

Mollusca Very diverse phylum, including animals, such as gastropods, bivalves, octopuses, squids, cuttlefish and the chambered nautilus. The molluscs have marine, freshwater and terrestrial representatives. Body differentiated into head, a ventral muscular foot (used in locomotion) and a dorsal visceral hump covered by a, respiration is by means of gills (ctenidia) or a lung like organ and the feeding organ is a radula.

molluscs Animals in the phylum–Mollusca.

monandry The condition of having a single anther.

Monera The kingdom of bacteria. All prokaryotes belong to this kingdom according to Whittakar system of classification.

Monoamine Oxidase (MAO) An enzyme that breaks down monoamines (e.g. adrenaline and noradrenaline) in the body by oxidation.

monobiontic Describing a life cycle in which there is no alternation of generations, i.e. only a single independent generation.

monocarpic Bearing fruits only once and dying after fructification, as beans, maize, mustard, etc.

monocistronic A eukaryotic *m*RNA that contains information coding for only one protein.

monoclinous Having stamens and carpels in same flower.

Monocotyledoneae (monocot) An angiospermic class with one cotyledon generally have parallel leaf veins, scattered vascular bundles within the stems and flower parts in threes or multiples of three, e.g. cereals, tulips, palms, etc.

monoculture Growing of a single kind of crop plant.

monocyte A type of leukocyte in vertebrates that functions in the immune system by migrating to tissues and then differentiating into macrophages and dendritic cells.

monoecious Having both pistillate and staminate flowers on the same individual, e.g. maize.

monogenic disorder A disorder caused by mutation of a single gene.

monohybrid cross The cross in which inheritance of only one pair of contrasting character is studied at a time. All the offspring (called monohybrids) have one dominant and one recessive allele for that gene (i.e. they are hybrid at that one locus). Crossing between these offspring yields a characteristics 3 : 1 (monohybrid) ratio in the following generation of dominant: recessive phenotypes (see illustration overleaf). Compare dihybrid cross.

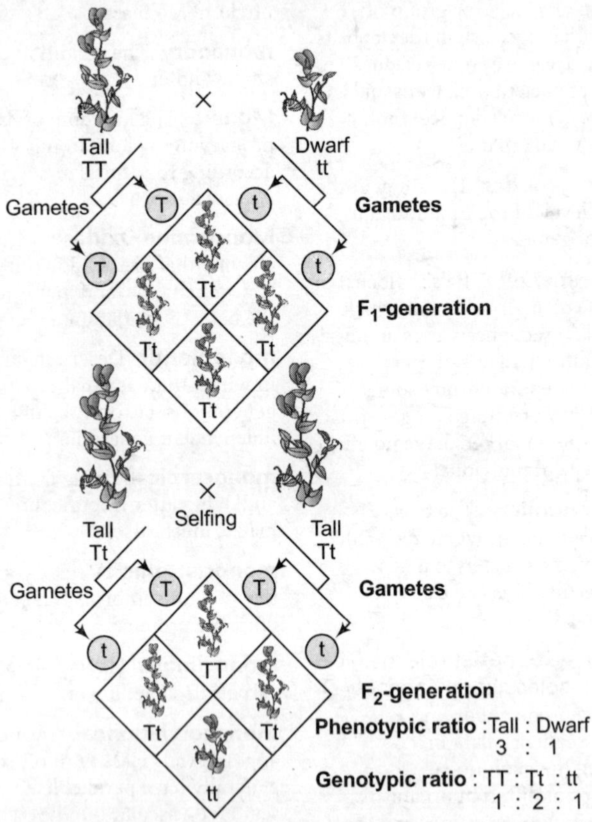

Monohybrid cross conducted by Mendel between true-breeding tall plants and true-breeding dwarf plants

monolayer culture A type of cultural true in which cells are grown in a single layer on a flask or petridish containing the culture medium.

monomer An atom or single molecule that can join with other, like atoms or molecules to form a polymer.

mononuclear phagocyte system (reticuloendothelial system) The network of macrophages and their precursors (monocytes) in the body's tissues, which is concentrated in the bone marrow, liver, spleen and lymph nodes.

monooxygenase (mixed function oxidase; mixed function oxygenase) Any of a large group of enzymes that perform oxidation-reduction reactions in which one atom of the oxygen molecule is incorporated into the chemical donor substrate and the other oxygen atom is combined with hydrogen ions to form water.

monophyletic Describing a group of organisms that contains all the discendants of a particular single common ancestor in taxonomy.

monophyodont A type of dentition in which an animal has only one set of teeth during its lifetime, which are not replaced, if fall out.

monoploid An individual that contains one half the normal of chromosomes is a monoploid and exhibits monoploidy.

monopodium The main axis of a plant (such as trunk of a spruce) that maintains a single line of growth, giving off lateral branches.

monosaccharides (also monosaccharoses) Soluble simple sugars that cannot be broken down into simpler sugars of lower molecular weight; they are the simplest carbohydrates, acting alone or as monomers in disaccharides. Monosaccharides are unaffected by enzymes and are absorbed directly by the body without change, e.g. glucose, fructose.

monosomy Having only one copy of a given chromosome instead of the two normally present in diploid cells.

monospermy The fertilisation accomplished by only one sperm. Opposite of polyspermy.

monosporic The method of embryo sac formation from a single megaspore.

monosynaptic reflex The simple reflex that involves transmission of information from sensory neuron to the appropriate motor neuron across a single synapse in the spinal cord, e.g. knee jerk reflex action.

monotremes A member of the major division–Monotremata (egg-laying mammals) of class–Mammalia It includes the platypus and the *Echidna.*

monotypic Describing a species whose members show only minor variation throughout the entire geographical range of the species, so that there are no recognised races or subspecies.

monozygotic twins *See* identical twins.

mons pubis The female external genitalia. Cushion of fatty tissue covered by skin and pubic hair.

montane Occurring in the mountains.

morphine An alkaloid present in opium. It is an analgesic and narcotic, used medically for the relief of severe pain.

morphogen A substance that can direct the differentiation of cells, as along a gradient of morphogenetic substances.

morphogenesis The development, through growth and differentiation, of form and structure in an organism.

morphology The scientific study of the structure and form of organisms.

mortality *See* death rate

morula A spherical mass of blastomeres, without blastocoel, 16-celled in humans, but usually 32 or 64 celled–in other animals.

mosaic A hybrid organism whose cells differ genetically, although they have all arisen from a single zygote. The genetic difference is usually created by irradiating the zygote during early stages of cleavage. A viral disease in plants that causes yellow patches to develop on the leaves, giving these a variegated appearance, e.g. tobacco mosaic caused by tobacco mosaic virus.

mosaic evolution An evolution of different parts of an organism at different rates. For example, some proteins evolve very rapidly while others remain unchanged over million of years.

mosses *See* Bryophytes.

moths *See* Lepidoptera

motile Capable of movement (as a gamete or microorganism).

m

motility The characteristic of being motile.

motor cell A type of plant cell that acts like a hinge at joints to enable the movement of plant parts, such as the closing and opening of leaflets in response to light intensity. These cells adjust their internal concentration of potassium ions (K^+) to alter their turgidity and hence the cell shape.

motor nerve An afferent nerve that carries impulses from the motor end plate muscles and glands.

motor neuron A nerve cell (neuron) that transmits impulses from the brain or spinal cord to a muscle or other effectors.

mould A fungus that grows in the form of multicellular filaments called hyphae.

moulting The shedding and replacement of an outer covering, such as feathers in birds or exoskeleton of chitin as in insects.

mouth An opening of the alimentary canal. It is used in most animals for the ingestion of food and leads to the buccal cavity.

mouth cavity *See* buccal cavity

mouthparts Modified paired appendages on the head segments of arthropods, used for feeding. A typical insect has a labium (lower lip) one pair each of mandibles and maxillae and a labrum (upper lip). The mouthparts in many are modified to form piercing stylets or a sucking proboscis. Crustaceans, centipedes and millipedes have one pair of mandibles and two pairs of maxillae used for cutting and holding the food. Crustaceans also have several pairs of maxillipeds Arachnids have chelicerae and pedipalps.

M-phase (mitosis phase) The phase when the actual cell division or mitosis occurs. Divided into prophase, metaphase, anaphase and telophase.

MTP (Medical Termination of Pregnancy) induced abortion It is the intentional termination of pregnancy before full term. MTPs are essential in certain cases where pregnancy can be harmful and even fatal either to the mother or the foetus or both. It is considered relatively safe during the first trimester (up to 12 weeks) of pregnancy. The Government of India legalised MTP in 1971 with some strict conditions to avoid its misuse. These are important to check indiscriminate and illegal female foeticides which are reported to be high in India.

mucigel (slime) A mixture of plant secretions, bacteria and soil particles that surrounds the tip of plant roots. Contain mucilages secreted by the root cap and help to lubricate passage of the growing root tip through the soil.

mucigen The substance, present in mucous cells, that is converted into mucin upon extrusion from the cell.

mucilage Any of a large group of complex polysaccharides present in the cell walls of aquatic plants and in the seed coats of certain other species. Hard when dry and slimy when wet. They probably have a general protective function or serve to anchor the plant.

mucin The glycoprotein that gives wet mucus its slippery texture. Mucin is also present in various other fluids that protect the body from abrasion.

mucoprotein A glycoprotein in which the carbohydrate component is a relatively large polysaccharide. These readily form gels with water and constitute the mucin in mucus. The mucoprotein chains are joined end to end by disulphide bonds, forming very long mucin strands. When released into water, the network of strands expands rapidly to produce a large volume of mucus.

mucosa Refers particularly to the cellular lining of the digestive tract derived from endoderm.

Mucosal-Associated Lymphoid Tissue (MALT) Any of various lymphoid tissues or organs that protect the mucosal surfaces, including the digestive tract respiratory tract eyes and reproductive tract. This network constitutes the mucosal immune system and overall contains about half of all the body's lymphocytes, emphasizing its importance in immune defence. Plasma cells of MALT produce chiefly IgA-type antibodies, which are found widely in mucosal secretions.

mucous A slimy fluid secreted by goblet cells on the surface to mucous membrane. In addition to water, it contains mucin, leukocyte, epithelial cells and inorganic salts.

mucous membrane A tissue which lines organs of an organism that are open to the outside, e.g. the alimentary canal, lungs, vagina.

Muller cell Supporting cells of the neural retina in vertebrates. They also regulate the concentrations of certain physiologically important substances within the retina, such as potassium ions, glutamate and gamma-aminobutyric acid, thereby affecting the activity of other cells.

Mullerian duct An oviduct of female vertebrates. It develops in both sexes from embryonic mesoderm, in association with the Wolffian duct but becomes vestigial in the male.

Mullerian mimicry A type of mimicry in which the model and mimic are distasteful, inedible or harmful.

multiaxial Having more than one axis; developing in more then a single line or plain.

multicarpellary The gynoecium having more than one pistil.

multicellular Composed of multiple cells of various types; said of a single organism.

multienzyme system An enzyme complex within a cell that form a reaction sequence of a biochemical pathway, so that the product of the first enzyme reaction is transferred directly to the next enzyme and immediately undergoes a second reaction and continue in same manner. The enzyme reaction rate often depends on the concentration of the enzyme and the substrate, both being required in relatively high amounts.

multifactorial inheritance *See* quantitative inheritance

multimer (in molecular biology) An aggregate molecule that consists of two or more (identical or non-identical) held together by non-covalent bonds.

multiple allelism An existence of a series of alleles (3 or more) for one gene. In human, e.g. there are three alleles (A, B and O) governing blood type.

multiplexing Using a laboratory approach that performs multiple sets of reactions in parallel (simultaneously), which greatly increases speed and throughput.

multipolar neuron or association neuron A neuron that has one axon and several dendrites extending from cell body. Stimulated by impulses which reach them from the sensory neurons.

muscarinic An acetylcholine receptor with no intrinsic ion channel. In humans, 5 subtypes of muscarinie receptors have been identified.

muscle A tissue that facilitates movement in animals. It is made up of sheets or bundle of cells called muscle fibres that are capable of contraction. They have been classified into three types as voluntary muscles, involuntary muscles and cardiac muscles.

muscle spindle A stretch receptor in tetrapod vertebrate muscle, running parallel to normal muscle fibres. Muscle spindles are responsible for the adjustment of muscle tone and play an important role in the subconscious maintenance of posture and movement.

muscular dystrophy The progressive degeneration of skeletal muscles mostly due to the genetic disorder, e.g. Duchenne muscular dystrophy.

muscularis mucosae A thin layer of smooth muscle that forms the outermost part of the mucosa of the stomach and intestine in mammals.

mushroom Fleshy, spore-bearing fruiting body of a fungus, typically produced above ground or on soil or on its food source.

mutagen An agent that causes mutation, e.g. X-rays, colchicine.

mutagenicity The capacity of a chemical or physical agent to cause mutation.

mutant Affected by or having a mutation.

mutation

1. A change in the structure of a nucleotide sequence, often, but not necessarily, in a gene.
2. The altered sequence or gene resulting from such a change.
3. A change in the karyotype.
4. An individual or type having such a change in its genetic make-up and any descendants of that individual in which it also occurs.

mutation frequency An average frequency with which a particular mutation occurs in a population. This frequency can be increased by radiation or by exposure to chemicals, such as mustard gas or hydrogen peroxide.

mutualism A form of symbiosis in which both participants benefit, e.g. the clown fish lives inside a sea anemone and is protected by it. In return, it brings scraps to the anemone, and lures larger fish into the anemone's tentacles.

myalgia The muscle pain.

Myasthenia gravis An autoimmune disorder of neuromuscular transmission in which acetylcholine receptors are blocked by auto antibodies.

mycelium A network of filaments (hyphae) that is a major component of the typical fungus.

Mycobacteria A genus of Gram-positive rod-shaped bacteria including the causative agents of leprosy and tuberculosis.

mycobiont The fungal component in association with algal component (phycobiont) in lichens.

mycology The scientific study of fungi.

mycoplasma Smallest, free-living microorganisms. These are prokaryotes lacking cell wall, variable in form and able to pass through bacteria- retaining filters. Grouped in genus–*Mycoplasma* and usually classified with bacteria. Some are saprophytes, others cause diseases of animals and man.

mycoprotein Any protein produced by a fungus.

mycorrhiza The symbiotic association between a fungus and the roots of a plant. It is a very common form of mutualism the fungus benefits by obtaining soluble organic nutrients and in turn enhances the absorption of mineral ions by roots. It is of two types:

1. Endotrophic in which fungus is within cortex cells.
2. Ectotrophic in which fungus is external.

mycosis Any infection caused by a fungus to animals, including humans.

myelin A phospholipid produced by the Schwann cells of the nervous system. It forms an insulating layer around the nerve fibres.

myelin sheath (medullary sheath) The layer of fatty material that surrounds and electrically insulates the axons of most vertebrates and some invertebrate neurons. It enables a more rapid transmission of nerve impulses (at speeds up to 120 ms^{-1}). Membrane layers derived from Schwann cells forms a part of it. The sheath is interrupted at intervals along the axon by nodes of Ranvier; myelinated sections of axon are called internodes.

myelocyte Cells in bone marrow that give rise to myeloid cells. They are derived from myoblasts.

myelodysplasia A group of disorders in which bone marrow malfunctions and fail to produce a normal number of blood cells.

myelogenic A originating or produced in the marrow.

myeloid tissue The tissue that manufactures white blood cells. It occurs in red bone marrow, surrounding the blood vessels.

myeloma A malignant cancer of myeloid tissue, causing anaemia, especially in the middle aged and elderly.

myoblast An embryological cell that gives rise to muscle cells.

myocardium The cardiac muscle that forms the walls of heart.

myocote A contractile cell, especially a muscle cell.

Myofibril Any of the small fibres that are arranged parallel to each other in skeletal muscle fibre. Appear as alternating light and dark bands due to the regular arrangement of proteins actin and myosin.

myogenic Originating within muscle tissue. Vertebrate heart is myogenic, since it has inbuilt mechanism of contraction even when its nerve supply is cut off.

myoglobin A pigmented protein storing oxygen in muscle cells.

myology Study of muscles.

myometrium The middle muscular wall of the uterus, made up of mainly smooth connective tissues and blood vessel. This layer stretches and expands during pregnancy to accommodate foetus.

myoneme A contractile fibril found in certain protoctists (e.g. *Vorticella*).

myopia Near-sightedness. It is due to the abnormally long eyeball, correction by concave lenses.

myosin A protein that forms thick filaments in the myofibrils. It is a polymerised protein with a globular head (Heavy Meromyosin or HMM) and tail (Light Meromyosin or LMM) in eukaryotic cells; interacts with actin to bring about cellular contraction; essential component of sarcomeres.

myotatic reflex *See* stretch reflex

myotome The part of somatic mesoderm which forms striped muscles.

myotube An embryonic precursor of a skeletal muscle fibre. During development in the early embryo, numerous individual precursor cells, called myoblasts, proliferate and migrate to sites where muscle formation will take place. There they fuse of form myotubes, each containing numerous nuclei and bounded by a single plasma membrane.

Myriapoda A subphylum of arthropods that comprises the classes-Chilopoda (centipedes), Diplopoda (millipedes), Pauropoda (pauropods) and Symphyla (symphilids). In other classifications, the Myriapoda is a class containing only the centipedes and millipedes.

myrmecochory The collection and dispersal of plant seeds by ants. A variety of plant species possess hard seeds that are inedible to ants but are nevertheless gathered by them and taken to the ants' nest. The ants perform this service because the seeds are equipped with special food bodies (elaiosomes). The ants detach elaisome and feed them to their larvae.

myxamoeba Naked cell characteristics of vegetative phase of slime fungi (Myxomycota) and some simple fungi; capable of amoeboid locomotion.

myxobacteria A rod-shaped bacteria that are distinguished by gliding movement in contact with solid surface and by delicate, flexible cell wall.

myxoedema The condition associated with malfunction of the thyroid, in which the subcutaneous tissue becomes filled with a mucin-like material.

Myxomycota *See* slime moulds

myxophycean starch Found in the Cyanophyta, myxophycean starch and has a similar structure to glycogen. This reserve product has a similar structure to glycogen.

myxovirus A one of a group of RNA containing viruses that cause diseases, such as influenza, mumps, measles, etc., of humans and other vertebrates.

N

NAA (Naphthalene Acetic Acid) A synthetic auxin which is useful in stimulating rooting and promoting flowering. It is more stable than IAA.

NAD (Nicotinamide Adenine Dinucleotide) A coenzyme, derived from the vitamin-B nicotinic acid, that participates in many biological dehydrogenation reactions undergo reduction. It often accepts one hydrogen and $2e^-$ to become NADH.

NADP (Nicotinamide Adenine Dinucleotide Phosphate) Possess an additional phosphate group. It functions in the same way as NAD although anabolic reactions (see anabolism) generally use NADPH (reduced NADP) as a hydrogen donor.

narcotic A drug that relieves pain and induces sleep, e.g. morphine and other opiates.

nares (nostrils) The paired openings of the nasal cavity in vertebrates. External nares which open to the exterior; in some species these are situated on a nose. Internal nares (or choanae) are present only in air-breathing vertebrates and open into the mouth cavity and beyond secondary palate in mammals.

Nasal cavity A cavity in the head of tetrapod vertebrates. Contains thin bones covered with a mucous membrane and lined with epithelium. In mammals, the large area of mucous membrane warms and moistens air as it passes to the lungs.

nasolacrimal duct A tube connecting the eye with the nasal cavity; functions in draining excess lacrimal fluid from the lacrimal sac into nasal cavity.

nasopalatine Pertaining to the nose and palate. Opens through glottis of the larynx region into the trachea.

nastic movements Movement of plant parts which is neither towards the stimulus nor away from the stimulus, i.e. independent of the direction of stimulus, e.g. movements of touch-me-not plant.

nasopharynx Portion of the pharynx above the soft palate and behind the nasal cavity. It communicates with oropharynx.

natality *See* birth rate.

national park Area protected by the state to conserve wildlife and forest resources.

native plant Plant species that is endemic (indigenous) in a geographic region or area; not introduced by humans intentionally or unintentionally.

natriuretic peptide One of the several peptide hormones that promote the excretion of sodium ions in urine (e.g. ANP produced by heart atrium natriuresis).

natural group A group of organisms of any taxonomic rank that are believed to be descended from a common ancestor. *See* cladistics.

natural history
1. Study of living organisms in their natural habitats.
2. Study of all natural phenomena.

natural killer cells (NK cells) A large lymphocyte with prominent lysosomes that attaches to the surface of virus-infected cells and secretes substances that puncture the target cell's membrane. This allows the entry of other substances that trigger the death of the cells.

natural selection A natural evolutionary process in which different heritable traits have a differential effect on the ability to produce offspring. According to Darwin, those traits aiding in the production of more offspring or that are well adapted for disease, competition in environment, become more common with time. They are naturally selected.

nauplius Free-swimming larva of many marine and freshwater crustaceans. It has an unsegmented oval body with a single eye at the front (the nauplius eye), mandibles, antennae and three pairs of appendages.

Neanderthal man A prehistoric race of humans that lived about 60,000 years ago. These were fairly short, strongly built and had heavy low brows but that the brain size was the same as, or larger than modern humans'. They were nomadic cave dwellers who buried their dead. Neanderthals became extinct abruptly.

near point The nearest point at which the human eye can focus an object. With advancing age, the near point recedes a condition known as presbyopia due to hardening of lens.

nearctic One of the six zoogeographical regions, including North America except for tropical Mexico. The fauna include mountain goat, pronghorn antelope, caribou and muskrat.

necrosis (also gangrene, mortification) Tissue death, due to infection or injury.

nectar A sweet, insect-attracting fluid produced by many flowering plants.

nectary A glandular structure in flowers that secretes a sugary fluid (nectar).

nectivorous Animals, birds or insects that rely on nectar as a source of food.

negative feedback The output of a reaction is used to reduce the input, e.g. regulation of blood glucose.

nekton Free-swimming animals of pelagic zone of a sea or lake, such as fishes, whales, etc.

nematoblast *See* thread cell.

nematocyst A specialised stinging cell in coelenterates, contains a hair-like structure that can be ejected.

Nematode Member of the Phylum–Nematoda, composed of the true roundworms, many of which are parasitic. They are characterised by a smooth narrow cylindrical unsegmented body tapered at both ends. They shed their tough outer cuticle four times during life to allow growth. e.g. *wuchereria.*

neo-Darwinism (modern synthesis) The current theory of evolution formulated between about 1920 and 1950, that combines evidence from classical genetics with the Darwinian theory of evolution by natural selection. It makes use of modern knowledge of genes and chromosomes to explain the source of the genetic variation upon which selection works.

neogene Second period of the Cenozoic era. Extending from 23.03 to 25.88 million years ago, it includes two epochs, i.e. the Miocene and Pliocene.

neo-Lamarckism One of the comparatively modern theories of evolution based on Lamarckism. Evolutionary change is the direct product of the interaction of organism and environment.

neolithic The phase of human prehistory when plants were cultivated and animals domesticated. Started about 10000 years ago.

neopallium (neocortex) Area of the cerebral cortex of the vertebrate brain. It is most highly developed in mammals, in which it forms a surface layer covering most of the forebrain.

neoplasm (tumour) Any new abnormal growth in cells, forming either a harmless (benign) tumour or a malignant one.

neoteny The retention of larval or other juvenile features beyond the normal stage in the mature state during development of an animal, e.g. Axolotl resemble its juvenile stage as adults.

neotropical Of or pertaining to the tropical regions of the new world. Consisting of South and central America, including tropical Mexico.

nephrideopore External opening of metanephridium (occurs in earthworm and some annelids).

nephridium A tube-like excretory organ of many invertebrates such as molluscs and earthworms. A funnel-like structure called nephrostome is present on one end of a nephridium. *See* metanephridium, protonephridium.

nephritis An inflammation of the kidney.

nephron Excretory unit within the vertebrate kidney.

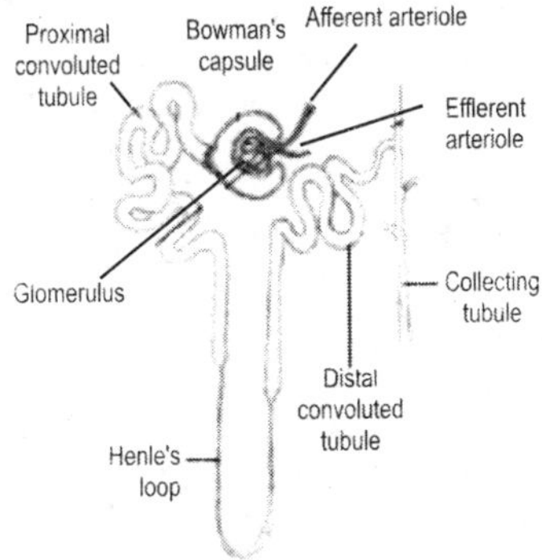

Structure of a nephron

nephrostome The internal opening of metanephridium in some annelids and earthworms.

nephrotoxin Any toxic substance that targets the kidneys. Example are mercury salts and certain herbicides, such as Paraquat.

neritic zone Shallow ocean region over the continental shelves, approx. 200 m in depth.

nerve A bundle of sensory and motor nerve fibres, enclosed in a connective tissue sheath connecting the central nervous system with other parts of the body.

nerve cell *See* neuron.

nerve cell adhesion molecules Proteins in the plasma membrane of animal cells that are important in ensuring proper cell-cell contacts during development of nervous system and muscle tissue.

nerve cord The solid strand of nervous tissue, forming the part of nervous system of invertebrates.

nerve fibre The axon of a neuron, together with tissues associated with it.

nerve impulse *See* impulse.

nerve net A network of nerve cells connected with each other by synapses or fusion. In certain invertebrates (e.g. coelenterates and echinoderms), a nerve net is present in the body wall. Makes up all or most of the nervous system.

nervous system The system of cells and tissues in multicellular animals by which information is conveyed between sensory cells and organs and effectors (such as muscles and glands). Its function is to receive, transmit and interpret information and then to formulate appropriate responses for the effector organs. It also serves to coordinate responses that require more than one physiological.

nervous tissue Tissue composed of neurons and glial cells. Nervous tissue is one of the four classes of tissue in animals.

Net Assimilation Rate (NAR) Increase in plant dry mass per unit leaf area per unit time. It enables comparisons of the efficiency of production between plants of different sizes.

Net Primary Productivity (NPP) Rate at which producer biomass is produced in a community.

Net Secondary Productivity (NSP) Rate at which consumer and decomposer biomass is produced in a community.

neural arch Arch of bone or cartilage that arises dorsally from the centrum of a vertebra and encloses the spinal cord. It may bear a number of projections, such as a neural spine for the attachment of muscles.

neural canal

1. Canal within the vertebral column through which runs the spinal cord.
2. Cavity or system of cavities in a vertebrate embryo that form the central canal of spinal cord and ventricles of the brain.

neural crest A crest of specialised cells at the edge of the neural plate in vertebrate embryos that comes to lie above the neural tube when it closes over. These cells form pigment cells, Schwann cells and gill arches.

neural plate A thickened plate of ectoderm, above the notochord, that lies along the central axis of the early embryos of chordates and develops into neural tube and crest by a process is neurulation. During the stage when the neural tube is forming, the embryo is known as a neurula.

neural tube A hollow tube of tissue in the early embryo of vertebrates that subsequently develop into brain and spinal cord.

neuralgia Shooting pain along the length of a nerve.

neurilemma cell *See* Schwann cell.

neurite Any projection from the cell body of a neuron, especially an immature neuron. It can become either dendrites or axons, as the neuron develops.

neuroblast The embryonic cell that develops into a nerve cell.

neurocranium The part of the cranium that surrounds and protects the brain.

neuroendocrine system System involving both nervous and endocrine factors to control functions of the body.

neurofibril Fibres in the cytoplasm of a nerve axon which include **neurofilaments.**

neurofilament A type of intermediate filament found in the axons of nerve cells. Neurofilaments serve as elements of the cytoskeleton supporting the axon cytoplasm.

neuroglia Cells found in nervous tissue that supports the neurons and electrically insulate them by producing myelin.

neurohaemal organ A discrete body formed by a cluster of neurosecretory-cell terminals, where neurohormones are released into an adjacent blood space or capillary bed. Found in both invertebrates and vertebrates.

neurohormone A hormone that is produced by specialised nervous tissue. e.g. nor-epinephrine.

neurohypophysis The posterior lobe of the pituitary gland in higher vertebrates. Releases oxytocin and vasopressin.

neurolemma (Neurilemma) White insulating sheath of Schwann cells covering most of the neurons.

neurology The study of nervous system and its related diseases.

neuromuscular junction A specialised region in which a nerve ending makes close contact with a muscle. Impulses arriving at the nerve ending cause it to rebase a chemical transmitter, which diffuses across the intervening gap and stimulates the muscle to contract.

neuron (nerve cell) Functional unit of nervous system. A neuron is composed of a cell body, an axon and one or more dendrites over which nerve impulses pass.`

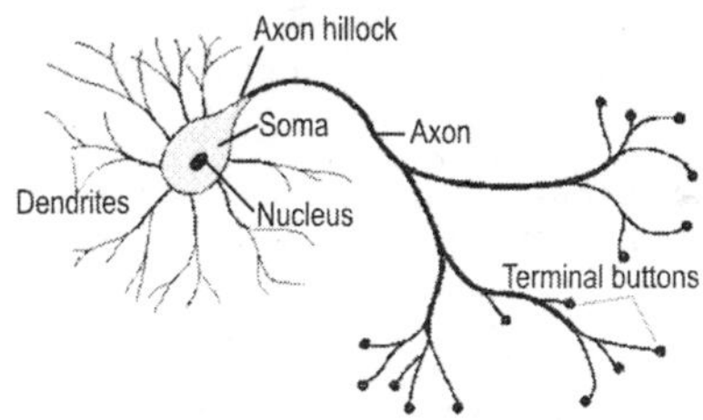

Structure of a neuron

neuronal network (neural network) A circuit of interconnected neurons. All types of behaviour depend on nerve impulses, being transmitted via synapses between individual neurons (nerve cells) through neuronal networks. Such network range from the relatively simple connections of a reflex arc, to the highly complex circuits involved in processing information in the brain.

neuropeptide Any of numerous peptides that influence activity of neurons. Example include the hypothalamic releasing hormones, ADH, etc.

neuropeptide Y (NPY) A neuropeptide that is found in the animal kingdom and in various body systems and has key roles in energy metabolism and in regulating the activity of the heart and blood vessels.

neurophysin Two cysteine-rich proteins that bind to oxytocin (neurophysin I) and antidiuretic hormone (neurophysin II) in the neurohypophysis of pituitary and are secreted with them.

neurosecretion A secretion of neurohormones by neurosecretory cells (possess characteristics of both nerve cells and endocrine cells), e.g. in hypothalamus, where they receive nerve impulses from other parts of the brain but transmit these signals to the pituitary gland by neurohormones that are released into the blood. Neurosecretory cells are similar in structure to neuron.

Neurospora The pink bread mould, an Ascomycetes fungus; used in many genetics and cell biology experiments.

neurotoxin Toxin that interferes with nerve functions, e.g. 6-hydroxydopamine.

neurotransmitter A chemical released from one of the synaptic knobs at the tip of the axon into the synaptic cleft. Neurotransmitters diffuse across the cleft and bind to the post synaptic cell, which results in its stimulation. It mediate the transmission of nerve impulse across a synapse or neuromuscular junction.

neurotubule Microtubules that play a role in the transport of proteins and other substances within the cytoplasm.

neuston An organism that colonise the surface of an aquatic habitat. They are most abundant in freshwater habitats and include pond skaters, certain types of water beetles and floating plants.

neuter An organism that does not possess either male or female reproductive organs, e.g. cultivated ornamental flowers that have neither pistils nor stamens.

neutral theory of molecular evolution Theory that proposes that most evolutionary changes at the molecular level are due to the random process of genetic drift acting on mutations, rather than natural selection. The great majority of the differences in macromolecular structures observed between individuals in a population are of no adaptive significance and have no impact on the reproductive success of the individual in which they arise.

neutrophils The most abundant type of leukocyte in mammals and the first to respond to inflammation; stains pink; targets bacteria and fungi.

niacin Nicotinic acid. Its deficiency causes pellagra. Vitamin of B-group.

niche Role played by a particular species in its environment.

nicotinamide adenine dinucleotide *See* NAD.

nicotinamide adenine dinucleotide phosphate *See* NADP.

nicotine A colourless poisonous alkaloid present in tobacco. It is used as an insecticide.

nicotinic One of the two main classes of acetylcholine receptors. The effect of acetylcholine on them can be mimicked by nicotine. Nicotinic receptors are found at neuromuscular junctions in skeletal muscle and at post-ganglionic cells in ganglia of both divisions of the vertebrate autonomic nervous system. They are fast-acting receptors containing their own intrinsic ion channel, is directly activated by the receptoragonists.

nicotinic acid (niacin) A vitamin of the vitamin-B complex, can be manufactured by plants and animals from the amino acid tryptophan. Its amide derivative, nicotinamide is a component of the coenzymes NAD and NADP. Deficiency causes the disease pellagra in humans. Good sources are liver and groundnut and sunflower meals apart from tryptophan rich protein.

nictitating membrane The third eyelid in amphibians formed by a clear membrane in birds and in some mammals (e.g. rabbit). Present in the inner corner of the eye can be flicked across the eye to wash it. It is a fold of conjunctiva.

nidation *See* implantation.

nidicolous Species that spent long time in the nest after hatching from the eggs.

nidifugous The phenomenon of leaving the nest within a few days of hatching. Nidifugous species of bird are capable of locomotion.

nidology The scientific study of nests of birds.

night blindness (nyctalopia) Disease in which a person becomes unable to see in the dim light especially in the evening or during night. May exist from birth or be caused by injury or lack of vitamin-A.

nipple The central conical projection in breast. 15-20 milk ducts open into the nipple. Each milk duct continue inward in a branching manner to join a cluster of 15-20 lobes of mammary glands.

Nissl granules (Missl bodies) Granular bodies in the cytoplasm of neurons, rich in RNA and stain strongly with basic dyes.

nitrification The conversion of ammonia and nitrites to nitrates by the action of nitrifying bacteria. One of the important parts of nitrogen cycle, because nitrogen cannot be taken up directly by plants except as nitrates.

nitrifying bacteria Bacteria that live in the soil and are capable of forming nitrate compounds. e.g. *Nitrosomonas.*

Nitrobacter A genus of bacteria playing an important role in the nitrogen cycle by oxidising nitrite into nitrate in soil.

nitrogen cycle One of the major biogeochemical cycle for circulation of nitrogen atoms, brought about mainly by living organisms. Inorganic nitrogenous compounds (chiefly nitrates) are absorbed by autotrophic plants 'from soil or water and synthesised into organic compounds. These autotrophs die and decay or are eaten by animals and the nitrogen, still in the form of organic compounds (e.g. proteins, nucleic acids), returns to the soil or water *via* excretion or by death and decay. Ammonifying and nitrifying bacteria then convert them to inorganic compounds (see ammonification, nitrification). Some nitrogen is lost to the atmosphere as nitrogen gas by denitrification.

nitrogen-fixation The conversion of elemental nitrogen, by certain prokaryotes, into compounds (such as ammonia, nitrate and nitrogen dioxide) that can be utilised by other organisms.

nitrogen oxides Oxides of nitrogen (NO_x), such as nitric oxide (NO) and dinitrogen oxide (N_2O), many of which are pollutants contributing to acid rain. Nitrogen oxides are expelled in the emissions from car exhausts, aircraft and factories. Nitric oxide is biologically important with various roles in the body.

nitrogenase An enzyme complex that is present in those microorganisms that are capable of fixing atmospheric nitrogen. It catalyses the conversion of atmospheric nitrogen into ammonia, which can then be used to synthesise nitrites, nitrates or amino acids. The two main enzymes within the nitrogenase complex are dinitrogenase reductase and dinitrogenase.

nitrogenous base A basic compound containing nitrogen often, used especially of organic ring compounds, such as adenine, guanine, cytosine and thymine, which are constituents of nucleic acids.

nitrogenous waste The waste product that contains nitrogen. The most common nitrogenous waste products in terrestrial animal are urea and uric acid. Freshwater fish excrete ammonia and marine fish excrete both urea and trimethylamine oxide.

nitrosamines A group of carcinogenic compounds which are a component of cigarette smoke, cause cancer in number of organs, particularly in the liver, kidneys and lungs. e.g. dimethylnitrosamine with two methyl side groups (CH_3^-).

NK cell *See* natural killer cells.

nival Living in or under snow; living in a snowy habitat.

nociceptor Sensory neurons that can detect potentially dangerous stimuli; such detection is perceived as pain.

nocturnal Occurring or active at night.

nodes

1. Points on a plant's stem where the leaves are attached (in botany).
2. A natural thickening or bulge in an organ or part of the body. Examples are the sinoatrial node that controls the heartbeat in anatomy.
3. A branch point in a phylogram in phylogenetics (vertex).

nodes of Ranvier One of several constrictions of neurolemma occurring at intervals along the peripheral myelinated nerve fibres.

nodule formation The sequence of multiple interactions between *Rhizobium* and roots of the host plant of Leguminosae. Nodule contains all the necessary biochemical components, such as nitrogenase, leghaemoglobin. *See* root nodules.

non-competitive inhibition The inhibition in which the inhibitor binds to a part of the enzyme or enzyme-substrate complex, on an allosteric site. This deforms the active site so that the enzyme cannot catalyse the reaction. It cannot be reversed by increasing the concentration of the substrate. The toxic effects of many substances are produced in this way. *See* inhibition.

non-cyclic photophosphorylation In this electrons derived from the photolysis of water are raised to higher energy levels in photosystems- I and II and pass along an electron transport chain of carrier molecules to NADP reductase. This enzyme transfers electrons to $NADP^+$ to make NADPH, which provides reducing power for the light-independent reactions of photosynthesis.
See photophosphorylation.

non-disjunction There are two types of non-disjunction:

1. Primary non-disjunction, failure of homologs to separate during meiosis-I.
2. Secondary non-disjunction, failure of sister chromatids to separate during meiosis-II. Non-disjunction results in some daughter cells receiving no copy and others receiving two copies, of the chromosome in question.

non-recombinant Refers to the offspring in mapping studies that has alleles arranged as the original parents.

non-reducing sugar Sugar that cannot donate electrons to other molecules and therefore, cannot act as a reducing agent, e.g. sucrose.

non-renewable energy sources Sources of energy that use up the earth's finite mineral resources; these include fossil fuels. Concern about the exhaustion of non-renewable energy sources, together with the fact that burning fossil fuels contributes to air pollution and the greenhouse effect, is leading to increased use or investigation of renewable energy resources, which are not exhaustible.

non-sense codons Codons which do not code for any amino acid but acts as a signal for termination of protein synthesis. These are three codons, UAA (ochre), UAG (amber) and UGA) (opal).

non-sense mutation A type of point mutation that converts a codon normally specifying an amino acid to one of the stop codons, thus signalling termination of translation and causing synthesis of the polypeptide chain in the cease prematurely.

norepinephrine (noradrenaline) A catecholamine secreted as a hormone by the adrenal medulla, that regulates heart muscles and glands. In the brain, levels of norepinephrine are related to mental functions; lowered levels lead to mental depression.

norm of reaction The range of phenotypes that a single genotype can produce under the influence of a range of environmental conditions.

normocytosis A normal state of the blood in which blood cell counts are normal.

Northern blotting It is the gel-based laboratory procedure that locates *m*RNA sequences on a gel that are complementary to a piece of DNA used as a probe.

nose It is the protuberance on the face of some vertebrates that contains the nostrils and part of the nasal cavity. It therefore forms part of the olfactory system and the external opening of the respiratory system.

nosography The systematic description of diseases.

notochord In chordate embryos, an elastic skeletal rod between the gut and nerve cord. It develops into the bones of the vertebral column in adult.

nucellus The central, main portion of the ovule, of seed plants, which contains the embryo sac and nutritive tissue. Cells of the nucellus have abundant reserve food materials. It may persists after fertilisation.

nuclear-cytoplasmic ratio Measure of the size of a cell nucleus in relation to the cytoplasm. It is often used as an index in the comparison of cells from normal and abnormal tissues. For example, cultured cancer cells show an increase in the nuclear-cytoplasmic ratio.

nuclear envelope (also nuclear membrane) The membrane separating nucleoplasm from cytoplasm eukaryotic cells. It is composed of an inner and outer membrane and is perforated with nuclear pores, the holes facilitating and governing the passage of macromolecules into and out of the nucleus. Both membranes are lipid bilayers and are separated by a perinuclear space. The rough endoplasmic reticulum is continuous with the outer membrane.

Nuclear Magnetic Resonance (NMR) A technique used for analysing the atomic composition of chemical compounds. NMR provides information about the structure of molecules because different types of atoms in a magnetic field resonate at different frequencies.

nuclear pore Perforation through the nuclear envelope facilitating and regulating movement of macromolecules into and out of the nucleus.

nuclear transfer The technique of removing of nucleus from a cell and inserting it into an oocyte that had its own nucleus removed. It is used in cloning animals as experimented in making Dolly (first cloned mammal).

nucleases Enzymes inducing hydrolysis of nucleic acids. Include restriction enzymes. Exonucleases remove nucleotides from the ends of the DNA, whereas endonucleases make cuts at specific positions within the DNA.

nucleic acid A nucleotide polymer; nucleic acids are of two types, i.e. RNA and DNA.

Nucleocytoplasmic Large DNA Virus (NCLDV; giant virus) Diverse group of DNA viruses that have unusually large capsids, comparable in size with small bacteria and contain large genomes, e.g. *Mimivirus*, which infects a type of *Amoeba* is the largest known virus, measuring 600 nm in diameter and has a genome size of roughly 1.2 Mb.

nucleoid (nuclear region) The concentrated mass of DNA in a prokaryotic cell.

nucleolus A round body within nucleus of non-dividing eukaryotic cell. It is associated with nucleolus organising region. It is composed of the primary products of the *r*RNA genes, as well as their proteins and enzymes.

Nucleolus Organising Region (NOR) A chromosomal region containing genes that encode ribosomal RNA.

nucleoplasm Fluid within the nucleus of a eukaryotic cell.

nucleoprotein A complex of nucleic acid and protein, e.g. chromosomes (DNA and histone proteins).

nucleoside A purine or pyrimidine base attached to a sugar (ribose or a deoxyribose). The four nucleosides found in RNA are adenosine, cytidine, guanosine and uridine. The four found in DNA (deoxyribosides) are adenine deoxyriboside, cytosine deoxyriboside, guanosine deoxyriboside and thymidine.

nucleosome The basic sub-unit of chromatin structure. Nucleosomes are visible as bead-like regions along the chromosomes during meiotic prophase. Each is composed of DNA wound around a histone protein core, which itself consists of two copies each of four different types of histones.

nucleotidase The enzyme that catalyses that breakdown of nucleotides. It is present in the epithelial cells of the small intestine and helps in the digestion of proteins.

nucleotide One of the monomeric units of which DNA and RNA polymers are composed. A nucleotide is made up of a phosphate group covalently bonded to a sugar molecule (deoxyribose in DNA and ribose in RNA and nitrogen bases purines and pyrimidines).

nucleus (pl. nuclei) An organelle in eukaryotic cell that contains the chromosomes. It functions as the control centre of the cell and is bounded by a double membrane (the nuclear envelope). When the cells are not dividing, a nucleolus is present in the nucleus and the chromosomal material (chromatin) is dispersed in the nucleoplasm. In dividing cells, the chromosomes become much shorter and thicker and the nucleolus disappears. The contents of the nucleus constitute the nucleoplasm. In certain protozoans, there are two nuclei per cell, a macronucleus and a micronuclues.

nudation Formation of an area devoid of plants; by natural or artificial means. Pre-requisite of plant succession.

null allele Allele whose effect is either an absence of normal gene product at the molecular level or the absence of normal function at the phenotypic level.

nullipara A woman or female animal who has not produced offsprings.

nullisomy Absence of a pair of chromosome from the somatic set.

numerical taxonomy Taxonomy that uses mathematical procedures to assess similarities and differences and establish taxonomic groups.

nurse cell Specialised cells of developing oocyte that provide material for further growth.

nut Dry single-seeded fruit that develops from more than one carpel and does not shed its seed when ripe. The fruit wall is woody or leathery. Many enclosed in a hard or membranous cup-shaped structure (cupule). The term 'nut' is often loosely used for any hard fruit, e.g., the walnut and coconut are in fact drupes and the Brazil nut is a seed.

nutation The autonomic growth movement in which the apex of a plant organ shows spiral growth. Also called circulation, e.g. coiling movements of the shoot tips of runner beans and of the tendrils of sweet peas.

nutrient Substance that is required for the nourishment of an organism, providing a source of energy or structural components. In animals, nutrients form part of the diet and include the major nutrients, i.e. carbohydrates, proteins and lipids as well as vitamins and certain minerals. Plant nutrients, derived from carbon dioxide in the atmosphere and water (containing minerals) absorbed from the soil by the roots, are divided into macronutrients and micronutrients.

nutrition Process of taking up of nutrients or inorganic substances and then utilising them in metabolism.

nyctinasty (sleep movements) The paratonic movement of variation of plant organs due to periodic alternation of day and night, e.g. opening and closing of flowers.

nycturia Excessive urination at night.

nymph Sexually immature stage of a hemimetabolous insect. An insect with incomplete metamorphosis, prior to the appearance of the first trace of wings. It resembles the adult except that the wings and reproductive organs are undeveloped. There is no pupal stage, the nymph developing directly into the adult, e.g. in grasshopper.

nyxis Piercing, puncturing.

O

obesity (cranial nerve III) The state of being immensely fat or overweight.

obligate aerobes Organisms that must need oxygen to live.

obligate anaerobes Organisms that cannot use free oxygen. Oxygen may inhibit their growth or kill them.

occipital Bone forming the posterior portion of the skull and surrounding the foramen magnum.

occipital condyle A single or paired bony knob that protrudes from the occipital bone of the skull and articulates with the first cervical vertebra (the atlas). Humans have a pair, one on each side of the foramen magnum.

occiput

1. The posterior portion of vertebrate skull where it joins the vertebral column.
2. A skeletal plate at the back of insect head.

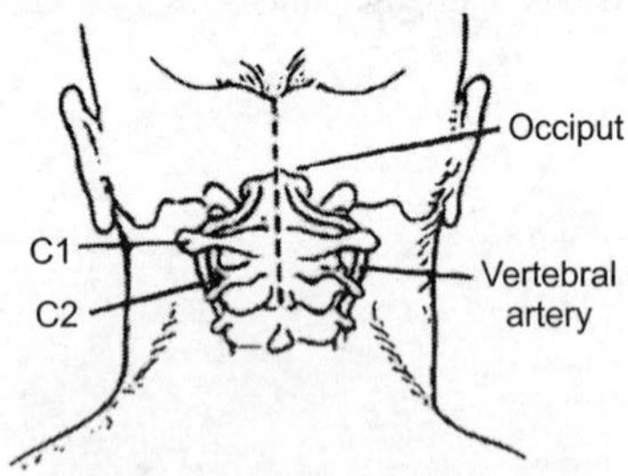

Location of occiput in human skull

occupational disease Disease or hazards related to a person's occupation, e.g. silicosis, asbestosis.

oceanic zone Region of the open sea, where the depth is greater than 200 m.

ocellus Simple eye of some invertebrates (e.g. insects). Ocelli are concerned with perception of the direction and intensity of light and do not form image.

ochre The UAA triplet of *m*RNA that forms the normal terminus of coding regions of many codons.

octoploidy A condition in which a cell has or the individual cells of an organism have eight complete sets of chromosomes.

Odonata An order of exopterygote insects with aquatic nymphs. Contains dragonflies and damselfies, large compound eyes, two pairs of similar wings, folded over the back in damselfies and horizontally in dragonflies.

odontoblast Cells that form the dentine of a vertebrate teeth.

odontogenesis Origin and growth of teeth.

odontology Study of teeth.

oedema Swelling of tissue due to the accumulation of tissue fluid.

oelic acid An unsaturated fatty acid. One of the most abundant constituent fatty acids of animal and plant fats, occurring in butterfat, lard, tallow, groundnut oil, soybean oil, etc.

oesophagus (gullet) Tube that connects the pharynx to the stomach. It is a muscular tube that functions to transfer food by peristalsis.

oestrogen Female sex hormones, produced mainly by the ovaries, that promote the development of secondary sexual characteristics and control the oestrus cycle. Oestradiol is most important, small amounts of oestrogen also produced by adrenal glands and testes.

oestrus cycle Cyclical changes during reproduction in non-primate mammals like cows, sheep, rats, dogs, tiger, etc. There are four phases:

1. **Pro-oestrus** (follicular phase) Graafian follicles develop in the ovary and secrete oestrogens.
2. **Oestrus** (heat) Ovulation normally occurs, the female is ready to mate.
3. **Metoestrus** (luteal phase) Corpus luteum develops from ruptured follicle.
4. **Dioestrus** Progesterone secreted by corpus luteum prepares uterus for implantation. Compare menstrual cycle.

offset A short runner (stem growing horizontally) that is used in vegetative propagation, e.g. those of houseleek.

offspring (progeny) New individuals that results from sexual or asexual reproduction. *See* F_1, F_2.

Okazaki fragment Series of short DNA fragments formed on the lagging strand during DNA replication. These vary in length, about 100-200 nucleotides in eukaryotes and 1000-2000 nucleotides in prokaryotes. See lagging strand, discontinuous replication.

olecranon process Bony process on ulna invertebrates extending beyond elbow joint, for the attachment of muscles which straighten forelimb.

oleosome *See* spherosome.

olfactology Study of the sense of smell.

olfacto-receptors Receptors or sense organs which receive smell.

olfactory lobe Either member of a pair of lobes in the forebrain, at the anterior end of the cerebrum. They are concerned with the sense of smell, as contain the endings of olfactory nerve.

olfactory nerve First cranial nerve of vertebrates, running from olfactory area of cerebral cortex to olfactory bulb.

Oligocene The third epoch of the Palaeogene period. It lasted from 33.9 – 23.03 million years ago. Characterised by the continued rise of mammals.

Oligochaeta A class of hermaphrodite annelid worms that bear only a few chaetae. They are very abundant in freshwater and terrestrial habitats, e.g. the earthworm (*Lumbricus*) and the freshwater bloodworm (*Tubifex*).

oligodendrocytes One of the class of glial cell. Form insulating sheaths of myelin round neurons, preventing impulses from travelling between adjacent neurons.

oligogenic trait Phenotypic trait produced by two or more genes working together.

oligonucleotide Short segment of RNA or DNA (short polymer of nucleotides). They are used as synthetic gene and DNA probes.

oligolecithal An egg containing very little yolk, e.g. of all higher mammals.

oligosaccharide Saccharide chain composed of several (usually 10 or fewer), simple sugars.

oligotrophic Water body with a poor nutrient supply and low rate of photosynthesis.

omasum The third chamber of a ruminant stomach .

ombrophile Plant that thrives in regions having high levels of rainfall e.g. *Selaginella*

omitis Inflammation of the shoulder.

ommatidium Basic unit of the arthropod compound eye. An arthropod eye can consist of thousands of ommatidia. It consists of outer cuticle covering lens, beneath which are 6-8 retinal cells.

omnivore An organism that eats both animal and vegetable matter, e.g. pigs.

onchium A hook found in or on the margin of the oral cavity of certain nematodes.

oncogene Mutated protooncogene that is known to cause cancer. Many oncogenes are normally involved, directly or indirectly, in controlling the rate of cell growth.

oncogenic A chemical, organism or environmental factor that causes the development of cancer, e.g. viruses such as human adenovirus.

oncogenic virus Cancer causing virus, having gene called cellular oncogenes.

oncology Medical study and treatment of tumours.

oncosphere An embryonic larva of a tapeworm, which are armed with six hooks serving as anchors.

one gene-one enzyme hypothesis Asserts that each gene codes for a single, specific enzyme (polypeptide).

ontogeny Course of an embryo's development, from fertilised egg through to maturity.

Onychophora (The velvet worms/Peripatus) Small invertebrate phylum, with caterpillar like invertebrates, inhabiting moist dark terrestrial habitats. They have some annelid and some arthropod features and are often referred to as missing links between the two groups.

oocyte Cell undergoing meiosis during oogenesis in females. Primary oocyte, undergo the first meiotic division. Scondary oocyte undergo the second meiotic division. *See* oogenesis.

oogamous Having a small mobile male gamete and a large immobile female gamete embryo.

oogamy Form of sexual reproduction in which the female egg is large and non-motile and is fertilised by a small, motile male gamete.

oogenesis (also ovogenesis) Process of producing female gametes (ova).

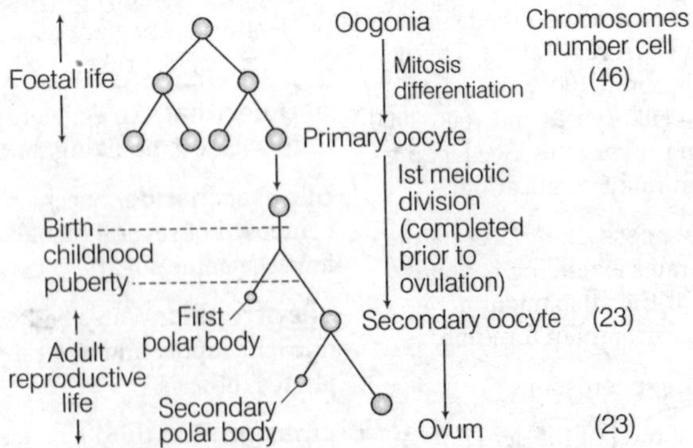

Oogenesis in human female showing formation of ovum

oogonium

1. Cell in the ovary of an animal that undergoes a period of multiplication and growth to give rise to an oocyte.
2. Female sex organ of certain algae and fungi.

Oomycota Phylum of eukaryotes that includes the water moulds, downy mildews and potato blight (*Phytophthora*), formerly classified as a class of fungi (Oomycetes). They are coenocytic and the cell wall is made of cellulose.

oophorectomy The surgical removal of one or both ovaries.

ooplasm Central portion of oogonium of some oomycetes, which is more or less the undifferentiated egg.

oosphere A large, naked, non-motile female gamete in plants and some algae.

oospore Thick-walled resting spore formed from fertilised egg in algae, fungi and other plants.

ootheca Sturdy protein case containing multiple, usually many, eggs; produced by certain insects and molluscs.

opal codon Stop codon UGA on *m*RNA.

open circulatory system Circulatory system where the blood is carried in large vessels but spends some of the time bathing the organs directly rather than flowing through capillaries as would be the case in a closed circulatory system. Present in arthropods and molluscs.

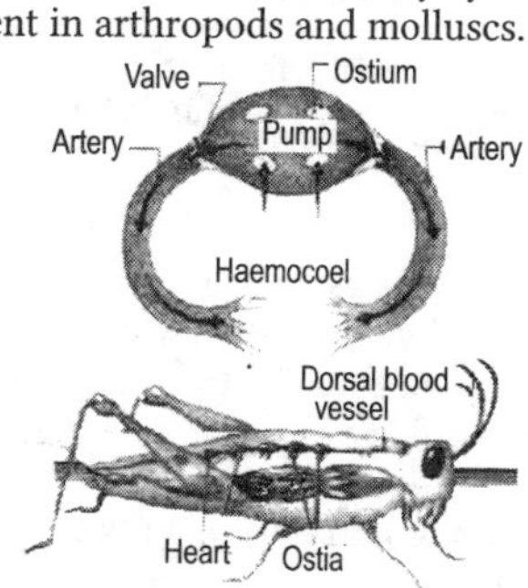

Open circulatory system

open reading frame Reading frame, consisting of a start codon and stop codon with a long sequence of codons that specify the constituent amino acids of the polypeptide. *See* reading frame.

operator gene An initial gene of an Operon which controls transcription of genes in it. The site of repressor binding.

operculum

1. A flap of skin covering the gill slit of fish and larval amphibians and the horny calcareous operculum secreted by many gastropod molluscs, which closes the opening of the shell when the animal is inside (in zoology).
2. The cone-shaped lid of the capsule of mosses which is detached forcibly to release the spores (in botany).

operon A set of adjacent genes with related functions. They are transcribed under the control of an operator gene (*See Lac* operon).

ophathakmic Relating to the eye and its diseases.

Opiates Group of drugs derived from opium (*Papaver somniferum*) that are CNS depressants.

opioids Drugs which bind to specific opioid receptors present in our central nervous system and gastrointestinal tractm, e.g. heroin.

opisthokonts An assemblage of eukaryotic organisms that includes all animals and fungi and certain flagellated protists. The grouping is based chiefly on molecular systematics, members are characterised by having flagellated cells propelled by a single posterior undulipodium (flagellum).

opisthosoma The posterior section of the body of arachnids and other arthropods of the phylum–Chelicerate. It consists of those body segments that do not bear legs. *Compare* prosoma.

opsin Protein part of the visual pigments of eye. Rods and three types of cones have different opsins.

opsonisation Process in which certain antibodies in the blood (opsonins) bind to the surface of an invading microorganism. This renders the antigen to be more susceptible to phagocytosis by macrophages and neutrophils, e.g. complement proteins.

optic chiasma Point where some of the fibres from both optic nerves cross forming an X-shaped structure. Sensory information from the right side of both the right and left eye is transmitted to the right visual cortex in the brain (*via* the right lateral geniculate nucleus), while information from the left side of each eye is carried to the left visual cortex.

optic lobe Either of the two lobes of the dorsal mesencephalon, containing primary visual centres.

optic nerve Second cranial nerve that runs from each eye to the brain. It is responsible for conveying visual stimuli received by the rods and cones in the retina to the brain for interpretation.

optic vesicle An outgrowth of the embryonic forebrain in vertebrates that develops into the retina of the eye.

optimal foraging theory A theory stating that natural selection favours animals whose behavioural strategies maximise their net energy intake per unit time spent foraging. Such time includes both searching for prey and handling (i.e. killing and eating) it.

oral cavity *See* buccal cavity.

oral pills (Oral contraceptive) A hormonal preparation in the form of a pill that prevents conception. Contain either progestogens or progesterone oestrogen combination. It alter/inhibit ovulation and implantation and also modify the quality of cervial mucus to prevent sperm entry, e.g. saheli.

oral groove A ciliated channel found in certain protozoa and aquatic invertebrates which directs food into the mouth.

orbit Bony cavity that holds the eyeball in skull of vertebrates.

order Refers to a taxonomic category between class and family. Usually, a class includes several orders and an order includes several families. Typically ends in-ales in botany (e.g. Rosales) and in zoology (e.g. Carnivora).

Ordovician period Geologic period between the Cambrian and the Silurian. It lasted from about 488-444 million years ago. Historically, the Ordovician was recognised later than the Cambrian and Silurian, with which it was at first confused. During the Ordovician, especially late in the period, sea levels were high.

oreximania An abnormal, extreme hunger resulting from an exaggerated fear of becoming thin.

organ A discrete portion of an organism serving a particular function or set of functions, e.g. lungs, kidneys (in animals).

organ culture The culture of complete living organs (explants) of animals and plants outside the body in a suitable culture medium. The organs must be small enough to allow the diffusion of nutrients to all the cells (in animals).

organ of Corti Structure located in inner ear containing hair. It is capable of transducing the energy of waves in a fluid medium into electrochemical energy of the action otential, in the auditory nerve.

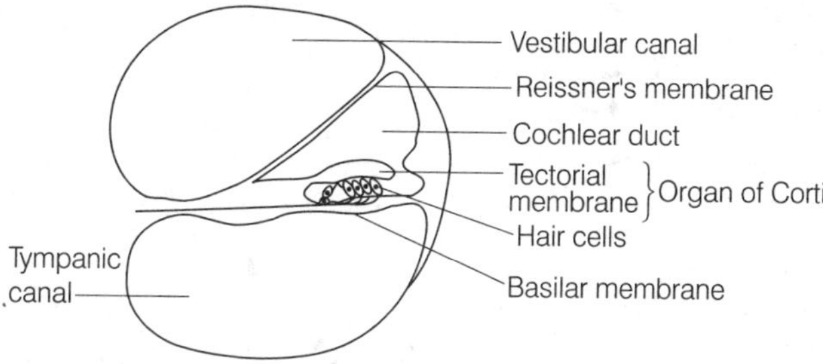

Organ of Corti

organelle One of the tiny, membrane-bounded structure with various specialised functions found in eukaryotic cell, e.g. mitochondria.

organic evolution Continuous change or transformation of living organisms in response to environmental changes.

organism A living system, that is capable of growth, maintenance and reproduction (i.e. animals, plants and microorganisms).

organiser An area of an animal embryo that causes adjacent areas of it to develop in a certain way. The primary organiser (blastopore lip or archenteron roof) directs the overall development of the gastrula. A secondary organiser directs the development of a subsidiary part or tissue.

organogenesis Formation of organs during development of embryo.

oriental region Faunal region covering tropical Asia and associated Continental Islands.

origin of life Process by which living organisms developed from inanimate matter. It thought to have occurred on earth between 3500 – 4000 million years ago.

ornithine (Orn) Amino acid, that is not a constituent of proteins but is important in living organisms as an intermediate in the reactions of the urea cycle and in arginine synthesis.

ornithine cycle (urea cycle) Cycle of chemical reactions in which the toxic nitrogenous products of protein catabolism are converted to urea which is nontoxic. In the cycle, (ammonia and carbon dioxide) first converted to citrulline, which is then converted to argino-succinic acid and then to arginine. Urea then splits off from arginine which produces ornithine. These reactions take place in the liver in the liver in mammals.

ornithology The scientific study of birds.

ornithophily Pollination by birds.

oropharynx Portion of the pharynx located below the soft palate.

orthogenesis An early theory of the nature of evolutionary change, proposing that organisms evolve along particular paths predetermined by some factor in their genetic make-up.

orthologous Describing homologous genes that are descendants of a single gene in a common ancestor. When a lineage splits to form two new species, any gene gives rise to two orthologues, which may subsequently diverge in their DNA sequence and function. Compare paralogous.

orthopnea An inability to breath easily when lying flat.

orthoptera An order of insects including the grasshoppers, locusts, and crickets. Characterised by large hindlegs and sound production.

orthotropismn The tendency for a tropism to be oriented directly in line with the stimulus concerned, e.g. vertical growth of main stems and roots in response to gravity (orthogeotropism).

orthotropus (atropus) Position of the ovule in the ovary, in which the ovule has developed vertically so that the micropylar end is directly over the funicle as in *Polygonum.*

osculum (pl. oscula)

1. The main opening through which water is discharged (in sponges).
2. Any of the suckers on the heat (scolex) of a tapeworm, by which it attaches itself to the gut wall of its host.

osmoconformer An animal whose body fluids are in osmotic balance with its environment. Such animals avoid the need to expend much energy on osmoregulation.

osmolyte An compound that protects cells from desiccation by maintaining a high intracellular osmolality. They are active in the process of osmoregulation and are of use in kidneys, e.g. urea.

osmometer An instrument that is used to measure osmotic pressure.

osmoreceptor Receptor situated in hypothalamus of brain that responds to an increase in the concentration of Extracellular Fluid (ECF). This results in release of ADH.

osmoregulation An ability of organisms to control the concentration of water and solutes in their body fluids. See osmoconformer and osmoregulator.

osmoregulator A animal that maintains a constant internal osmotic environment in spite of changes in its external environment, e.g. vertebrates and some aquatic invertebrates and some aquatic invertebrates (especially freshwater). They expend energy on osmoregulation to maintain cell volumes and achieve optimum conditions for metabolism.

osmosis Movement of a solvent, usually water, from a region where their concentration is high to a region where their concentration is low through a semipermeable membrane.

osmotic pressure Minimum pressure required to stop the flow of pure water into a solution across a partially permeable membrane. It characteristic of the solution. Thus, water will move from a region of low osmotic pressure to a region of high osmotic pressure.

osmotroph A heterotrophic organism that obtains its nutrients by absorbing organic matter in solution from its surroundings.

osphresiology Scientific study of the odours and the sense of smell.

ossicle A small bone or chitinous structure found in various skeletal parts of animals, e.g. ossicles that transmit vibrations through the middle ear, e.g. malleus, incus and stapes in mammals.

ossification The process of bone formation. *See* intramembranous and endochondral ossification.

Osteichthyes Class of vertebrates comprising the bony fishes. All have gills covered with a bony operculum and a layer of thin overlapping bony scales covers the entire body surface. Bony fish have a swim bladder and bony skeleton.

osteoblast Cells, found in bone, that secrete collagen and other substances that form the matrix of bone. These are derived from **osteoprogenitor cells** in the bone marrow and they eventually become osteocytes.

osteoclasts Multinucleate cells on bone surface, which breakdown the calcified intercellular substance of bone. It is stimulated by parathyroid hormone. Remodeling of bone shape by such breaking constantly accompanies bone growth.

osteocyte A bone cell, that perform the cellular activities, such as respiration and exchange of materials with the blood. They are derived from osteoblasts.

osteogenesis Formation of bone in embryo or during regeneration.

osteoid A soft material, consisting mainly of collagen secreted by osteoblasts. It constitutes the uncalcified matrix of bone and is converted into hard bone matrix when it combines with calcium phosphate (hydroxyapatite).

osteonectin A protein in bone that binds to collagen and is involved in the formation of growth sites for calcium phosphate crystals.

osteoporosis The age related disorder characterised by decreased bone mass and increase chances of fracture. Decreased levels of oestrogen is the common cause.

osteosarcoma A malignant tumor of the bone; usually affects the large bones of the arm or leg.

ostiole An opening in fruting body of certain algae and fungi by means of which gamete or spores escape out.

ostium (pl. ostia)

1. Any one of the pores in the body wall of a sponge through which water enters the body cavity.
2. Any one of the apertures in the arthropod heart through which blood enters from the haemocoel.

ostracum The calcified portion of an invertebrate's shell. While the organism is living, the ostracum is covered by layers of protein forming a periostracum.

ostreophagous Having a diet of oysters.

otalgia The ear pain.

otiatrics The treatment of ear diseases.

otic Pertaining to, near, or in the ear.

otitis Inflammation of the ear (otitis media-inflammation of the middle ear).

otology Scientific study of the ear, its diseases and treatments.

otolith Also called statoconium or otoconium, is a structure in the saccule or utricle of the inner ear, specifically in the vestibular labyrinth of vertebrates. The saccule and utricle together constitute the otolith organ. They detect linear acceleration and gravity orientation.

outbreeding (also crossbreeding) Mating of unrelated individuals. Increases the number of heterozygous individuals.

outer ear (external ear) Region of the vertebrate ear that is external to the ear drum (tympanum). It is present in birds, mammals and some reptiles. Consists of a passage (the external auditory meatus) and the pinna (in mammals only).

outgroup A species or higher taxon used in systematics for comparison with a group of closely related species or taxa in order to assess whether particular characters shared by the group members are derived or ancestral.

oval window (fenestra ovalis) A membrane that separates the middle and inner ear in tetrapods.

ovarian Pertaining to, in or like the ovary.

ovarian follicle *See* Graafian folicle

ovary

1. In animals, a female gonad.
2. In plants, the portion of a carpel containing the ovules.

overdominance Condition in which a heterozygour organism has greater fitness than the corresponding homozygotes.

overexploitation Exploitation to the point of diminishing returns.

overlapping inversion An inversion that involves a chromosomal region already affected by an inversion.

overwinding Positive super coiling of DNA, resulting in further tension in the direction of winding of two strands of a duplex about each other.

oviducts Egg carrying tubes from the ovary to uterus or outside in animals.

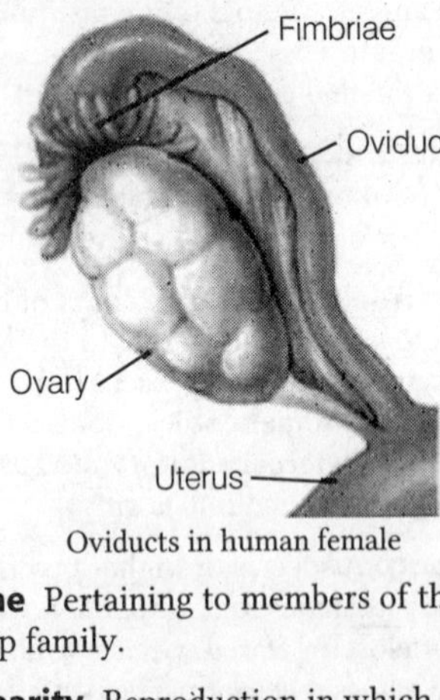

Oviducts in human female

ovine Pertaining to members of the sheep family.

oviparity Reproduction in which fertilised eggs are laid by the mother and are hatched outside her body. It occurs in most animals except marsupials and placental mammals.

oviparous Laying eggs in which embryo develops outside mother's body and eventually hatches.

oviposition Process of egg laying by insect female.

ovipositor An organ for laying eggs, at the hindend of the abdomen in female insects, through which eggs are laid.

ovotestis Gonads (sex glands) containing both ovarian and testicular tissue.

ovoviviparity Reproduction in which fertilised eggs develop and hatch in the oviduct of the mother. It occurs in many invertebrates and in some fish and reptiles (e.g. the viper), compare oviparity, viviparity.

ovulation Release of an egg cell from the ovary, which is mammals is stimulated by luteinising hormone. The developing egg cell within its follicle migrates to the ovary surface, when mature, it is released from the follicle (which breaks open) into the body cavity, from where it passes into the oviduct.

ovule Structure that develops into a seed after fertilisation, in seed plants. It consists of the nucellus, embryo sac and integuments.

ovuliferous scale One of the group of large woody specialised leaves that form the female cone of conifers and related trees. It bears the ovules, which develop into seeds.

ovum (egg cell) (pl. ova)

1. The mature reproductive cell of female animals, which is produced by the ovary. It is spherical, has a nucleus, is covered with an egg membrane (in zoology) and is not mobile.
2. The oosphere of plants (in botany).

oxaloacetate (oxaloacetic acid) An organic acid with four carbons that is the starting material for the Krebs' citric acid cycle. Its function is to pick up two energy rich carbons from other parts of metabolism feeding into the Krebs' cycle.

oxidase The respiratory enzyme which brings about the oxidation of a substrate.

oxidative burst Production of various highly reactive oxygen derivatives by certain cells notably macrophasges of the vertebrate immune system. These oxygen derivatives are used to kill bacteria engulfed by the macrophage.

oxidative deamination Reaction involved in the catabolism of amino acids that assists their excretion from the body, e.g. conversion of glutamate to α-ketoglutarate, a reaction catalysed by the enzyme glutamate dehydrogenase.

oxidative decarboxylation Reaction in the Krebs 'cycle in which oxygen, is used to oxidise two carbon atoms to two molecules of carbon dioxide. The two carbon atoms results from the decarboxylation reactions that occur during the Krebs' cycle as the six-carbon compound citrate is converted to the four carbon compound oxaloacetate.

oxidative phosphorylation Production of ATP from phosphate and ADP in aerobic respiration. It occurs in mitochondria, the energy being provided by steps in the electron transport chain.

oxidoreductases Group of enzymes that catalyses oxidation and reduction reactions. Includes the oxidases and dehydrogenases.

oxygen cycle The cycling of oxygen between the biotic and abiotic components of the environment. The oxygen cycle is closely linked to the carbon cycle and the water cycle. In the process of respiration oxygen is taken in by living organisms and released into the at-mosphere, combined with carbon, in the form of carbon dioxide. Carbon dioxide enters the carbon cycle or is taken up by plants for photosynthesis. During photosynthesis, oxygen is evolved by the chemical splitting of water and returned to the atmosphere. In the upper atmosphere, ozone is formed from oxygen and dissociates to release oxygen.

oxygen debt A physiological state that occurs when a normally aerobic animal is forced to respire anaerobically during insufficient supply of oxygen, e.g. due to the violent muscular exertion. Lactic acid is released which is toxic and needs oxygen to breakdown. The debt is repaid when oxygen is made available and allow oxidation of lactic acid.

Oxygen Evolving Complex (OEC) A small complex of proteins and maganese ions associated with the reaction centre of photosystem II (PS-II) that splits (oxidises) water molecules to yield oxygen during photosynthesis.

Oxygen Quotient (QO_2) Rate of oxygen consumption of an organism or tissue. Smaller organisms tend to have higher QO_2 than larger ones.

oxygen sag curve The curve obtained when the concentration of dissolved oxygen in a river (with sewage or some other pollutant) is plotted against the distance downstream from the sewage outlet. Samples of water are taken at areas upstream and downstream from the sewage outlet. The presence of sewage reduces the oxygen content of the water and increases the biochemical oxygen demand.

This is due to the action of saprotrophic organisms that decompose the organic matter in the sewage and in the process use up the available oxygen.

oxyhaemoglobin The oxygenated haemoglobin.

oxyntic cells (parietal cells) An acid (HCl) secreting cells in the stomach wall also produce intrinsic factor.

oxytocin Hormone secreted by the posterior lobe of pituitary whose principal action is on milk ejection in the lactating animal.

ozonation Formation of ozone (O_3) in the earth's atmosphere. In the stratosphere (about 20-50 km above the surface of the earth).

$$O_2 \xrightarrow{\text{UV rays}} O + O.$$

These atoms combine with oxygen molecules to form ozone. Ozone is also formed in the lower atmosphere (troposphere) from nitrogen oxides and other pollutants by photochemical reactions.

ozone (O_3) Form of oxygen, near the earth's surface. It is a pollutant produced by internal combustion engines, but it also forms a protective high altitude atmospheric layer that blocks ultraviolet radiation.

ozone hole In the 1980s, it was found that depletion of the ozone layer was occurring over both the poles, creating ozone holes. The cause being a series of complex photochemical reactions involving nitrogen oxides and Chlorofluorocarbons (CFCs) and halons. CFCs rise to the stratosphere, where they react with ultraviolet light to release chlorine atoms where atoms, which are highly reactive, catalyse the destruction of ozone.

ozone layer (ozonosphere) A layer of the earth's atmosphere in which most of the atmosphere's ozone is concentrated. It occurs 15-50 km above the earth's surface. In this layer, most of the Sun's ultraviolet radiation is absorbed by the ozone molecules.

ozostomia Bad breath.

P

P (Parental generation) Organisms selected to start breeding experiment. Cross between them form F_1-generation.

pacemaker

1. (or sinoatrial node) A small mass of specialised muscle cells in the mammalian heart, found in the wall of the right atrium, near the opening for the vena cava. By their rhythmic and spontaneous contractions, they stimulate contraction, of the atria. The cells themselves are controlled by the autonomic nervous system, which determines the heart rate. Similar pacemakers occur in the hearts of other vertebrates.
2. An electronic or nuclear battery-charged device that can be implanted surgically into the chest to produce and maintain the heartbeat. These devices are used when the heart's own pacemaker is defective or diseased.

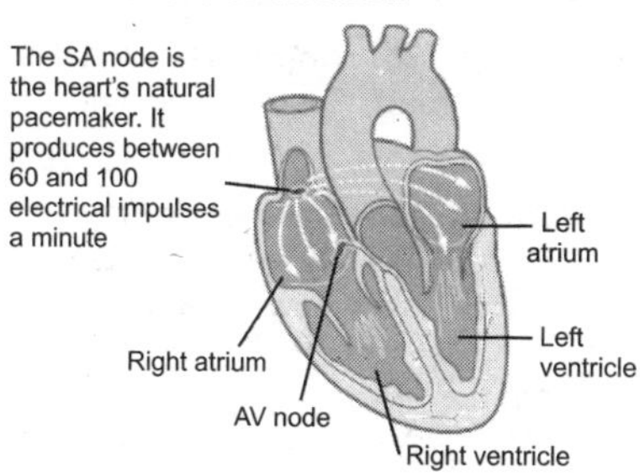

Position of natural pacemaker in human heart

pachytene A stage in prophase-I of meiosis that is characterised by contraction of paired homologous chromosomes. At this point, each chromosome consists of a pair of chromatids and the two associated chromosomes are termed as tetrad.

pacinian corpuscles The sensory receptors in the subcutaneous layers of skin that are sensitive to pressure. Each one consists of a nerve ending surrounded by oval capsule of connective tissue. Also called lamellated corpuscles.

paedogenesis The reproduction in larvae or other pre-adult forms. e.g. neotenic larvae like axolotl larva, can produce individual to self in thyroid deficiency.

paedomorphosis An evolutionary process in which larval or juvenile features of an ancestral organism are displaced to the adult forms of its descendants. It can arise by neoteny or progenesis.

pain Severe physical or mental discomfort or distress. Perception of and susceptibility to pain are determined by higher centres in the cortex and can be influenced by such factors as emotional state.

pairing (synapsis) The close association between homologous chromosomes that develops during the first prophase of meiosis.
The two chromosomes come together and a synaptonemal complex of proteins is formed between them. It makes sure that exact pairing of corresponding points along their lengths, as they lie side by side. The resulting structure is called a **bivalent.**

Palaearctic region The faunal region, comprising Europe and Asia North of the tropics.

palaeoanthropologist Scientist who studies the fossil remains, of human beings.

palaeobotany The study of extinct plants through fossils.

Palaeocene The earliest epoch of the Palaeogene. It lasted from 65.5-55.8. million years ago. By its end, primates and rodents had evolved.

palaeocontinent A formerly existing continent, now broken up or fused with other continents due to the effects of rifting and continental drift.

palaeoecology The branch of biology dealing with pre-historic ecology as revealed by studying fossils and their artifacts, pollen samples and the mineral deposits.

Palaeogene First period of the Cenozoic Era. Extending from 65.5 to 23.03 million years ago, it includes three epochs—the Paleocene, Eocene and Oligocene.

palaeolithic Old stone age. The palaeolithic is the first period of stone tool use. It extends from the time of the earliest known stone tools, about 2.5 million years ago, to the beginning of agriculture.

palaeontology The scientific study of the life of past geological periods. It deals primarily with the documentation of the chronology of the history of life on earth.

Palaeozoic era The first and most lengthy era of the Phanerozoic Era. It lasted from 542 to 251 million years ago and is composed of six geologic periods, which are sequentially, the Cambrian, Ordovician, Silurian, Devonian, Carboniferous and Permian. Nearly all of the plant and animal phyla, with the major exception of the angiosperms, first became abundant during the Palaeozoic.

palaeozoology The branch of palaeontology that deals with study of animals, as revealed by their fossil remains.

palate The roof of the mouth cavity of vertebrates. The bony portion is the hard palate, the fleshy rear portion is the soft palate in mammals.
It separates the buccal and nasal cavities.

palea One of two bracts that protects the floret of grasses. It is smaller above the lemma.

palindromic A section of double-stranded DNA in which the sequence of bases on one strand is inverted and repeated on the other.
e.g. –ACTTGCAAGT–
–TGAACGTTCA–
Common in DNA, these are the sites at which the DNA is cleaved by restriction enzymes.

palisade cells The mesophyll lying just beneath the upper epidermis. These are leaf cells containing many chloroplasts, where photosynthesis takes place.

pallium *See* cerebral cortex.

palmate Lobes of a leaf spread from the same point, like the fingers of a hand.

palmella stage A stage formed under certain conditions in various unicellular algae in which after division, the daughter cells remain within the envelope of the parent cell and are thus rendered immobile.

palmitic acid A saturated fatty acid occurring widely in fats and oils of animal and plants.

palp The elongated sensory organ, usually near the mouth, in many invertebrates. e.g. tactile head appendages of polychaete worms.

palpation The application of the hands to the external surface of the body for the purpose of determining the state of an underlying organ or tissue.

paludicolous Living in marshes.

palynology The science dealing with the study of the morphology of pollen grains preserved in deposits, by scanning electron microscope, etc. to determine the nature of the flora.

pancreas A gland in vertebrates near the duodenum that produces digestive enzymes, which are secreted into the small intestine. Also, an endocrine gland that produces hormone insulin and glycogen.

pancreatic islets *See* Islets of Langerhans.

pancreatin A mixture of digestive enzymes that have been extracted from the pancreas.

pancreozymin *See* cholecystokinin.

paneth cells The cells of crypts of Leiberkuhn that secrete intestinal juice.

pangea (also Pangaea) A great supercontinent that existed from the Permian to the Jurassic. It broke up to produce all of the modern continents.

panicle A type of inflorescence. A compound raceme formed by branching of peduncle, each branch bearing a raceme, e.g. oat.

panmictic population The population of organisms in which mating is random.

pantothenic acid A vitamin of the vitamin-B complex. It is a constituent of coenzyme-A, that performs a crucial role in the oxidation of fats, carbohydrates and certain amino acids. Occurs in many foods, especially cereal grains, peas, egg yolk, liver and yeast, so defeciency rarely occurs.

papain A proteolytic enzyme in the unripe fruit and green leaves of papaya. Used as a meat tenderiser and digestant.

papilla Cone-shaped protuberance projecting from the surface of an organ or organism, e.g. on the tongue, in the kidneys and in plants, on the surface of many petals.

papovavirus One of the group of small DNA containing viruses, about 50 nm in diameter, that cause tumours in their hosts, e.g. papillomaviruses produce malignant tumours.

pappus The ring of modified sepals that makes up the calyx in flowers of the Compositae. It persists on the fruit and serves to aid the wind dispersal of seeds.

papule A raised, red bump on the skin; the term is applied to any bump up to about the size of a pea.

parabiosis The surgical joining of two organisms, or parts of organisms, so that they share a common circulation and can exchange hormones or other internally secreted substances.

paracellular pathway The route between cells. e.g. substances can travel through epithelia by a paracellular pathway, if the tight junctions between constituent cells are not fully continuous.

paracentric inversion The inversion that does not include the centromere.

parallel evolution The development of related organisms along similar evolutionary paths due to strong selection pressure, acting on all of them in the same way.

parallelophyly Independent emergence of the same trait in two related lineages because of a genetic predisposition inherited from a common ancestor. Such a phenomenon could account for certain instances of apparent homoplasy.

paralogous Describing homologous genes that have arisen by duplication of an ancestral gene.

paramutation An example of epigenetic phenomenon. Certain genes behave as if they have undergone mutation after being in the same genome, as other (paramutagenic alleles). The change persists in subsequent generation. It is seen in plants and mice.

parapatric Occupying different, non-overlapping, but contiguous geographic regions.

parapatric speciation The speciation involving geographical isolation and divergence of two or more populations from a parent species. The speciation process is completed with the evolution of species isolating mechanisms which prevent mistaken interbreeding. Parapatric speciation is different than allopatric speciation in the completion of speciation involves the evolution of post zygotic species isolating mechanisms.

paraphyletic A group of organisms that excludes one or more descendants of a particular single common ancestor (in systematics) e.g. 'reptiles' used in evolutionary systematics is paraphyletic because it excludes birds and mammals, which share the same common ancestor as the reptiles.

paraphysis (pl. paraphyses) Sterile filaments of cells that are found around the sex organs of mosses, brown algae and certain fungi (the Ascomycota and Basidiomycota).

parapodium The lateral appendage present in members of phylum– Annelida which help in swimming.

parasexual cycle A type of reproduction in fungi that results in genetic recombination without the formation and fusion of gametes, e.g. in heterokaryotic mycelia, in which two haploid nuclei fuse and then divide by mitosis, during which a form of crossing over occurs.

parasite An organism that absorbs nutrients from a living organism. Generally, a parasite is smaller than its host.

parasitism A form of symbiosis in which one of the two participants benefits at the expense of the other.

parasitoid An animal that lives in or on another animal (the host), which it consumes and eventually kills, i.e. intermediate between a true parasite and a predator. Many wasps and other hymenopterans and also some flies, are parasitoids for part of their life cycle, laying their eggs commonly in or on the larvae of other insects.

parasitology The study of parasites.

parasympathetic nervous system The part of the autonomic nervous system. Its nerve endings release acetylcholine as a neurotransmitter and its actions tend to antagonise those of the sympathetic nervous system. e.g. increases salivary gland secretion, decreases heart rate and promotes digestion, while sympathetic nervous system has opposite effects.

parathyroid glands Four small oval-shaped structures embedded in or behind the thyroid gland. They produce parathyroid hormone and calcitonin.

parathyroid Hormone (PTH; parathormone; parathyrin) A peptide hormone secreted by the parathyroid gland in response to low levels of calcium in the blood. It acts to maintain normal blood levels of calcium by increasing the number of osteoclasts, to increasing the reabsorption of calcium and magnesium ions in the kidney tubules, converting vitamin-D to its active form, which increases calcium absorption in the intestine. It acts opposite to calcitonin.

Parazoa The animal sub-kingdom consisting of the sponges and placozoa.

parenchyma

1. In plants, relatively undifferentiated cells, where metabolism occur. Parenchyma stores metabolic products and develops into differentiated tissues.
2. A loose connective tissue formed of large cells, functions in packing the spaces between organs in some simple acoelomate animals, such as flatworms (Platyhelminthes)

parent

1. Either male or female partner that together produce offspring in the process of sexual reproduction.
2. Denoting an organism or cell that gives rise to new organisms or cells, as by asexual reproduction or cell division.

parental care The behaviour pattern in which a parent invests time or energy in feeding and protecting its offspring. It differs widely in species, e.g. most species of fish show little or no parental care, while humans and many other mammals care for their offspring until they reach adolescence.

parietal bones Two bones forming the roof and sides of the brain cavity.

parietal cell *See* oxyntic cell.

parietal placentation Ovules develop on the inner wall of the ovary or an peripheral part e.g. in mustard and *Argemone.*

Parkinson's disease A progressive disorder of the nervous system that affects movement.

parotid gland A paired, largest salivary gland situated below the ear. Saliva is released through an opening called parotid duct which enters the mouth on the inside of the cheek next to the upper molar teeth.

parthenocarpy The development of fruit from unfertilised flowers, resulting in seedless fruits. Naturally, it occurs in banana. It can be induced artificially by the application of auxin, as in commercial production of tomato.

parthenogenesis The development of the embryo from an unfertilised egg. e.g. in aphids.

partially permeable membrane A membrane that is permeable to the small molecules of water and certain solutes but does not allow the passage of large solute molecules.

particulate inheritance The transmission from parent to offspring of separate units that determine characteristics.

parturition Act of giving birth to young by females.

passage cells Endodermal cells that have casparian strips but remain otherwise thickened after deposits of ligin and cellulose have been laid down elsewhere in the endodermis. Particularly common in the endodermis of older monocotyledon roots.

passive immunity A type of apparent resistance to a foreign antigen obtained by injecting the patient with an appropriate set of antibodies produced by another organism. Passive immunity is temporary and used in situations, when the patient's own immune system could not mount a defence in time to save the individual.

passive transport *See* diffusion.

pasteurisation The method of partial sterilisation of drinks in which heating at temperature well below boiling point destroys bacteria. Best known for sterilisation of milk. Developed by Louis Pasteur.

patch clamp technique The method for studying the activation of voltage-dependent transmembrane channels in neurons. A small piece of membrane containing only a few channels is isolated and microelectrodes are sealed to both sides of the membrane.

patella (kneecap) A small rounded movable bone that is situated in a tendon in front of the knee joint in most mammals (including humans). Its function is to protect the knee.

pathogen A disease causing microorganism or virus. In addition to viruses, some bacteria, Protozoa and certain fungi are pathogens.

pathogenesis Steps leading to the development of a disease.

pathology

1. Study of diseases.
2. Changes caused by a disease.

patristic The similarity between organisms resulting from common ancestry.

pattern formation (patterning) The establishment of basic body symmetries and repetition of parts during development. In animal embryos, e.g. pattern formation defines the body axes, the number and polarity of body segments.

PCR *See* polymerase chain reaction

peat A mass of dark-brown or black fibrous plant debris produced by the partial disintegration of vegetation in wet places.

pecten The comb-like structure in animals. The pecten in the eyes of birds consists essentially of a network of blood vessels attached to the optic nerve and projecting into the vitreous humour. Its function is uncertain, but it may be involved in supplying the retina with nutrients and oxygen.

pectin The structural polysaccharide present in plant cell wall. It is composed of (1-4) linked galactouronic acid residues and sugars like thamnose may also be present.

pectoral girdle (shoulder girdle) The bony or cartilaginous structure in vertebrates to which the anterior limbs such as pectoral fins, forelegs or arms are attached. It consists of two dorsal shoulder blades attached to backbone and 2 collar bones attached to sternum in mammals.

pediatrics The study and medical treatment of the diseases of children.

pedicel Stalk attaching a flower to the main floral axis. Sessile flowers do not have a pedicel.

pedicle One of two vertebral processes, each of which connects the body of a vertebra with one of the lamina.

pedigree analysis Inheritance of a particular trait which is represented in the family tree over generations. Analysis of traits in several generations of a family.

pedigree chart Diagram showing the ancestral history of a group of related individuals.

pedipalp Either of the second pair of appendages in arachnids, posterior to the chelicerae. Variously adapted for defence, killing prey or digging. In species with large chelicerae, the pedipalps are typically adapted for walking or sensory functions.

pedology The science of the study of soils, including their origin and characteristics and their utilisation.

peduncle Stalk supporting an inflorescence.

pelagic Living in the open sea.

pelecypoda *See* bivalvia.

pellagra Disease resulting from a deficiency of nicotinic acid, which is characterised by dermatitis and mental disorder.

pellicle A thin flexible transparent outer protective covering of many unicellular organisms, especially undulipodiated protozoan protocists, e.g. *Euglena* and *Paramecium*.

pelvic girdle (pelvis) A rigid bony or cartilaginous skeletal structure in the posterior region of the vertebrate body, to which the hind limbs or fins are attached. It articulates dorsally with the backbone and is made up of two halves, each produced by the fusion of the illium, ischium and pubis.

pelvis

1. *See* pelvic girdle.
2. The lower part of abdomen in the region of the pelvic girdle.
3. A concial chamber in the kidney into which urine drains from the kidney tubules before passing to the ureter.

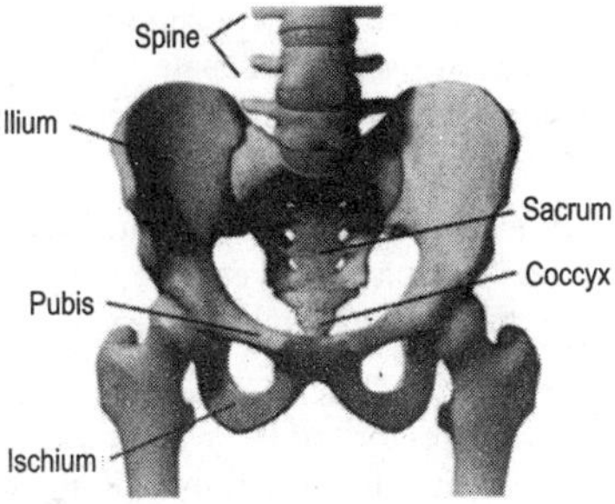

Pelvic bones

penicillin The antibiotic derived from the mould *Penicillium notatum;* other variants also available, all active against a wide variety of bacteria, producing their effects by disrupting synthesis of the bacterial cell wall and are used to treat a variety of infections caused by these bacteria.

penis The male reproductive organ used by many animals with internal fertilisation to introduce sperm into the female reproductive tract, in mammals, some birds and reptiles. In mammals, urine also leaves body through penis.

pentadactyl limb A limb with five digits, characteristic of tetrapod vertebrates (amphibians, reptiles, birds and mammals). It evolved from the paired fins of primitive fish as an adaptation to locomotion on land and is not found in modern fish.

pentose Sugar that has five carbon atoms per molecule.

pentose phosphate pathway (pentose shunt) A series of biochemical reactions that results in the conversion of glucose 6-phosphate to ribose 5-phosphate and generates NADPH. It provides reducing power for other metabolic reactions, such as synthesis of fatty acids. In plants, the pentose phosphate pathway also plays a role in the synthesis of sugars from carbon dioxide. In animals, the pathway occurs at various sites including the liver and adipose tissue.

pepo Berry that develops a hard outer rind, e.g. cucumber.

pepsin Enzyme in the stomach that digests proteins down to small polypeptides. Enzymes that breakdown peptide bonds are also called peptidases.

pepsinogen The inactive form of pepsin, secreted by stomach lining in vertebrates. It is converted to pepsin by HCl.

peptic ulcer The ulcer found on the inner membrane lining of stomach.

p

peptide bond A covalent bond between two amino acids; forms when the amino group of one amino acid bonds to the carboxyl group of another and water is eliminated.

peptidoglycan A polymer found only in prokaryotic cell wall. The cell wall of Gram positive bacteria have a thick layer of peptidoglycan, which stains purple in the Gram staining method. In Gram negative bacteria, the layer is thinner and stains pink.

peramorphosis The outcome arising from evolutionary changes in developmental rates. It involves the additon of new stages to the end of the ancestral development sequence.

perennation The survival of a plant for three or more seasons and normally flowering and fruiting atleast in the second and subsequent seasons.

perennial Plants having life cycle of more than two years.

perforin A cytotoxic protein that occurs in lytic granules of cytotoxic T-cells.

p

perianth The outer part of flower enclosing stamens and carpels, usually consisting of two whorls, the calyx and corolla or anyone of them. In monocots, no differentiation in calyx and corolla.

pericardial cavity The cavity in vertebrates that contains the heart and is bounded by a membrane.

pericardium The membrane enclosing the heart and the adjacent portions of the great blood vessels.

pericarp The outer wall of a plant ovary, made up of 3 layers. *See* epicarp, mesocarp and endocarp.

pericentric inversion The inversion that includes the centromere.

perichondrium A dense layer of fibrous connective tissue that covers the surface of cartilage.

periclinal Parallel to the surface of an organ or part in plants. In periclinal cell division, the plane of division is parallel to the surface of the plant body.

pericycle A layer of potentially meristematic cells immediately beneath a root's endodermis.

periderm In plants, the permanent outer layer that replaces the epidermis during secondary growth.

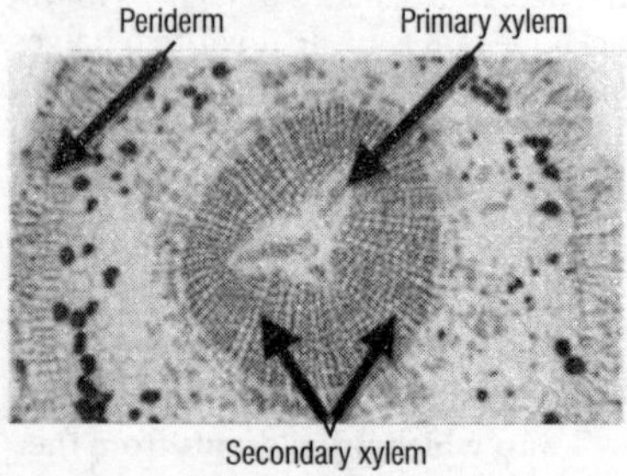

Periderm layer in plants

perigynous A flower in which the receptacle develops into a concave structure on which the sepals, petals and stamens are borne. e.g. cherry and plum flowers.

perilymph The fluid of the inner ear that fills the space between the bony labyrinth and the membranous labyrinth.

perimetrium The external, thin membrane present in the wall of uterus. Middle layer is myometrium and inner layer is endometrium.

perineum The area between the anus and the scrotum or valva.

periodontal membrane Membrane of connective tissue that surrounds the root of a tooth and anchors it to its socket in the jaw bone.

periosteum A fibrous coating surrounding all surfaces of a bone other than the articular surfaces.

peripatric speciation A form of allopatric speciation in which a founder population becomes established beyond the existing range of the main population and remains isolated because of difficult terrain or other factors.

Peripheral Nervous System (PNS) A type of nervous system which includes cranial nerves, spinal nerves and visceral nerves.

periphyton Organisms, collectively, that live attached to the stems and leaves of freshwater plants.

periplasm (periplasmic space) The zone between the cytoplasmic membrane and the outer membrane in Gram-negative bacteria. It contains a thin layer of peptidoglycan, with various type of proteins.

perisperm A nutritive tissue in the seeds of many Caryophyllaceae. It is derived from the nucellus or integuments rather than the embryo sac. Such seeds differ from other angiosperm seeds that have either cotyledons or endosperm tissue acting as the food store.

perissodactyla Order of mammals having hoofed feet with an odd number of toes, including all herbivores-the tapirs, rhinoceros and horse. The teeth are large and specialised for grinding. Cellulose digestion occurs in the caecum and large intestine.

peristaltic movement Movements involving involuntary contraction and expansion in the food pipe when food enters into it.

peristome

1. A ring of tooth-like structures around the opening of a moss capsule opening of the capsule.
2. Area around the mouth in many invertebrates and some protists, often assists in food collecting. e.g. spirally ciliated groove around the mouth of some ciliate Protozoans.

peritoneum In human anatomy, the serous membrane lining the abdominal cavity.

peritrichous The bacteria that possess flagella all over the cell surface, e.g *Proteus.*

permanent teeth The second set of teeth of most mammals, replacing the deciduous teeth (milk teeth). Usually appear between 6 to 21 years in humans, 32 in number.

permeable Able to let other substances pass through. Used to describe a membrane.

permease A membrane transport protein that serves as a gateway or channel to allow diffusion of substances through a plasma membrane. May refer to a gated or ungated channel.

permian The last period of Palaeozoic era, 270-225 Myr BP in duration. Among vertebrates, **palaeoniscoid** fish dominated and holosteans appeared; the labrythodontia survived among amphibians, but amphibians were much reduced in the later Permian; reptiles included pelycosaurs and their replacement the thrapsida, but there are few representatives of the dominant Mesozoic reptiles.

peroxisome In a eukaryotic cell, small organelles containing enzymes that transfer hydrogen to oxygen from various substrates. This process produces hydrogen peroxide as a byproduct, which is toxic. But, peroxisomes also contain an enzyme that prevent any build-up of hydrogen peroxide by breaking it down into water.

pest A destructive insect or other animal that attacks crops, food, livestock, etc.

petal Brightly coloured parts of the flower, which together make up the corolla.

petiole A stalk of a leaf, which joins it to a node on the stem.

Peyer's patches Numerous lymphoid organs found in the wall of the small intestine. They have a crucial role in protecting against infection. Each consists of a dome-shaped follicle containing chiefly B-cells surrounded by smaller numbers of T-cells.

Pfitzer's law According to the law, cell become progressively smaller during successive generations.

Phaeophyta (brown algae) A phylum of algae in which the green chlorophyll pigments are usually masked by the brown pigment fucoxanthin which vary in size from small branched filaments to ribbon-like bodies (known as kelps) many metres long. Brown algae are now considered to be stramenopiles.

phage *See* bacteriophage.

phagocyte A cell that is able to engulf and break down foreign particles, cell debris and disease producing microorganisms. e.g. macrophages and monocytes.

phagocytosis The intracellular digestion of foreign particles by phagocytes.

phagotroph (macroconsumer) A heterotrophic organism that feeds by ingesting organisms or organic particles, which are digested within its body.

phalanges Bones that make up the digits of the hand or foot in vertebrates. They articulate with the metacarpals of the hand or with the metatarsals of the foot.

phanerogam In early classification, any plant that reproduces by seed phanerogams are thus, equivalent to the spermatophytes.

phanerophyte A plant life form in Raunkiaer's system of classification. These are large shrubs and trees in which the perennating buds are located high above the ground. Thus, being at risk of exposure to drought, stress or frost and such plants occur mainly in regions where frost and drought are uncommon, such as the tropics.

Phanerozoic The most recent eon of geological time, represented by rock strata containing clearly recognisable fossils. It comprises the Palaeozoic, Mesozoic and Cenozoic eras and has extended for about 542 million years from the beginning of the Cambrian period.

pharate A larva or adult when inside the cuticle of the previous developmental stage. e.g. a newly metamorphosed adult insect may remain 'cloaked' for some hours or even days within the pupal case until it receives the appropriate environmental cue that triggers emergence.

pharmacodynamics The study of the action of drugs on the body.

pharmacogenomics (pharmacogenetics) The study of how genes affect the actions of drugs.

pharmacokinetics The movement of foreign substances, particularly drugs, throughout the body of an animal.

pharmacology The study of the properties of drugs and their effects on living organisms.

pharynx The passage connecting the mouth and nasal cavity with the oesophagus and windpipe.

phasic receptor A sensory receptor that adapts rapidly to a stimulus. These are also called fast-adapting receptors.

phellem *See* cork.

phelloderm Secondary inner cortex formed by the division of the cells of cork cambium.

phellogen *See* cork cambium.

phenetics Classification of organisms on the basis of overall similarity, regardless of the nature of their evolutionary relationship and genetic makeup.

phenocopy A phenotypic trait that seems the same as a genetic trait, but that is caused by environmental influences e.g. rickets.

phenology The study of periodicity phenomena in plants e.g. time of flowering in relation to climate and other environmental factors.

phenotype Observable characteristics of an organism. Physical appearance of an organism with respect to both the expression of its genetic material and environmental influences on its appearance.

phenotypic plasticity The ability of an organism to adapt to environmental changes by modifying its own development, form, behaviour or other trait. It depends on the capacity for a given genotype to respond to environmental factors and to adjust the phenotype accordingly.

phenylalanine One of the essential amino acids.

phenylketonuria A genetic disorder in which there is disordered metabolism of the amino acid phenylalanine. It lead to severe mental retardation of affected children. It is caused by the absence or deficiency of the enzyme phenylalanine hydroxylase, which results in the accumulation of phenylalanine in all body fluids. Both parents of such individuals are heterozygous carriers of the allele.

pheophytin First electron acceptor in the light-dependent reactions of photosynthesis. It accepts an electron from the excited form of photosystem II and passes it on *via* another acceptor. Q_A, to plastoquinone. It is a form of chlorophyll-*a* in which the magnesium ion is replaced by two hydrogen ions.

pheromone (ectohormone) The chemical produced by an organism that influences the behaviour or physiology of another member of the same species. Thus, the chemicals ants use for their ant trails would be pheromones. Also, humans do have sex pheromones but because of our relative intelligence, attraction involves a few more characteristics than the right smell.

phloem (bast) A vascular tissue in plants that carries sugars, especially sucrose, made in the photosynthetic regions of the plant, to other parts of the plant. It consists of hollow tubes (sieve tubes) that run parallel to the long axis of the plant organ and are formed from elongated cells (sieve elements) joined end to end and closely associated with companion cells.

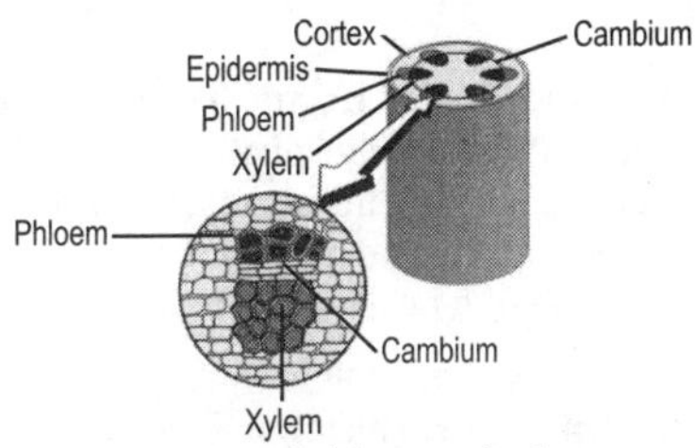

Location of phloem and other tissues in plant stem

phloroglucinol Red dye (usually acidified with hydrochloric acid) that stains lignin in plant cells red.

phonotaxis The movement of an organism in relation to a sound source. e.g. females are often attracted by the courtship song of a potential mate.

phoresy A method of dispersal in which an animal attaches itself to an animal of another species in order to be transported to a new site, where it releases itself, with little or no harm to the carrier. e.g. when their host dies, certain bird lice may attach themselves to blood-sucking flies in their effort to find a new avian host and some insect parasitoids.

phoronids (or phoronid worms) Marine worms comprising the phylum-Phoronida. They are commonly called horseshoe worms in reference to the configuration of their gut which is U-shaped, with the anus next to the mouth. They live in chitinous tubes, which they secrete. Another characteristic feature of phoronids is the presence of an umbrella-like feeding apparatus called a lophophore, which is also present in bryozoans and brachiopods, with which phoronids are sometimes grouped as lophophorates.

phosphagen A compound that provides a reserve of chemical energy in form of high energy phosphate bonds in animal tissue. e.g. creatine phosphate found in vertebrate muscles and nerves.

phosphodiester bond The covalent bond that links a phosphate group and a sugar group, by means of an oxygen bridge, in the sugar-phosphate backbone of a nucleic acid molecule. *See* DNA.

PhosphoGlyceraldehyde (PGAL) Three carbon compound found as an intermediate both in glycolysis and in the Calvin-Benson cycle of photosynthesis. Surplus PGAL harvested by the chloroplast is the true end product of the Calvin-Benson cycle and in turn is used by the plant cell to produce fructose, then glucose and from there do all the other things that plants and the rest of the living world needs to do.

PhosphoGlycerate Kinase (PGK) Transferase enzyme used in the seventh step of glycolysis.

phospholipid Lipid that contains phosphate esters of sphingosine or glycerol. Phospholipids have a head group, which is attracted to water and a tail group, which is made up of a long hydrocarbon chain repelled by water. Phospholipids are the primary constituent of the lipid bilayers of cells.

phosphoprotein A protein that contains phosphorus other than in a nucleic acid or a phospholipid.

phosphorescent Glowing in the dark.

phosphorus cycle A sedimentary nutrient cycle.

(i) Phosphorus is a major constituent of biological membranes, nucleic acids and cellular energy transfer systems.
(ii) Many animals also need phosphorus to make shells, bones and teeth.
(iii) The natural reservoir is rock, which contains phosphorus in the form of phosphates.
(iv) When rocks are weathered, minute amounts of these phosphates dissolve in soil solution and are absorbed by the roots of the plants.
(v) Herbivores and other animals obtain this element from plants.
(vi) Organic waste products and the dead organisms are decomposed by the phosphate solubilising bacteria releasing phosphorus.

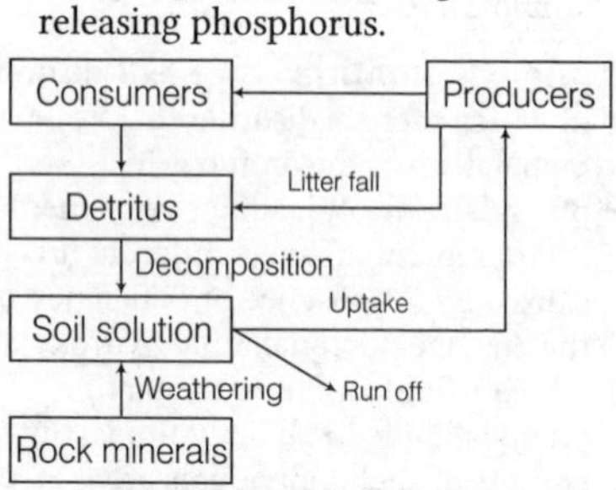

A simplified model of phosphorus cycling in a terrestrial ecosystem

phosphorylation Production of a phosphate derivative of an organic molecule, usually by adding a phosphate group from ATP. This process occurs, in respiration and photosynthesis.

photoautotroph Autotrophic organism that synthesises its organic materials from inorganic components using energy derived from the solar energy in the process of photosynthesis. e.g. green plant or a photosynthetic bacterium

photoblastic Describing a seed whose germination is influenced by light.

photochemical smog Injurious smog produced by the reaction of nitrogen oxides with hydrocarbons in the presence of ultraviolet light from the sun. The reaction is very complex and one of the products is ozone.

photoheterotroph The organism that uses energy derived from the sun to manufacture organic compounds from organic precursors in photosynthesis. e.g. under certain conditions, some purple sulphur bacteria use organic acids (rather than hydrogen sulphide) as a source of hydrgoen.

photoinhibition

1. Reduction in photosynthesis caused by exposure to abnormally high intensities of visible or ultraviolet light.
2. The (usually reversible) suppression of any biological process by exposure to light.

photolysis Chemical reaction produced by exposure to light or ultraviolet radiation. e.g. photolysis of water, using energy from sunlight absorbed by chlorophyll, produces gaseous oxygen, electrons and hydrogen ions in photosynthesis.

photomicrography Use of photography to obtain a permanent record (a photomicrograph) of the image of an object as viewed through a microscope.

photomorphogenesis Development of plants under the influence of light. All the processes crucial to the growth and development of plants are triggered by light.

photonasty (photonastic movement) Nastic movement in response to change in light intensity. e.g. opening of evening primrose at night.

photoperiodism The response of an organism to changes in day length. In plants, leaf fall and flowering are common responses to seasonal changes in day length, as are migration, reproduction and molting etc.

photophore A gland or organ that is specialised for the production of light. They contain the light-producing chemicals or symbiotic bacteria responsible for the bioluminescence.

photophosphorylation The formation of ATP from ADP and inorganic phosphate using light energy in photosynthesis. There are two pathways, non-cyclic and cyclic photophosphorylation, which occur in the thylakoid membranes of the chloroplasts.

photopic vision The type of vision that occurs when the cones in the eye are the principal receptors, i.e. when the level of illumination is high, colours can be identified.

photopigment A pigment whose chemical state depends on its degree of illumination, such as those in the retina of the eye.

photoprotection The protection of a plant's photosynthetic apparatus from the harmful effects of light, when there is high light intensity. Much of the excess energy is trapped and dissipated as heat by carotenoids of the xanthophyll cycle. Also, chloroplasts contain the enzyme superoxide dismutase, which scavenges superoxide radicals.

photoreceptor A sensory cell or group of cells that react to the presence of light. It usually contains a pigment that undergoes a chemical change when light is observed, thus stimulating a nerve.

photorespiration Metabolic pathway that occurs in plants in the presence of light, in which it accepts oxygen in place of carbon dioxide, resulting in the formation of a two-carbon compound, glycolate. There is no ATP production.

photosynthesis The process carried out in plants, algae and bacteria which uses energy from sunlight to convert carbon dioxide and water into glucose and oxygen. Photosynthesis is the source of atmospheric free oxygen and is the essential starting point for the construction of all organic molecules present in living things.

photosynthetic carbon reduction cycle *See* Calvin Cycle.

photosynthetic nitrogen use efficiency The ratio of a plant's rate of carbon assimilation to the concentration of ogranic nitrogen in its leaves. It is a measure of the efficiency with which the plant allocates nitrogen to its photosynthetic apparatus.

photosynthetic pigments The pigments responsible for the capture of light energy during the light dependent reactions of photosynthesis. In plants, algae and cyanobacteria are the green pigment chlorophyll-*a* is the **primary pigment**, absorbing blue and red light. The carotenoids and various other pigments are accessory pigments, absorbing light energy and passing this onto the chlorophyll-*a* molecules.

photosystems Functional and structural units of protein complexes involved in photosynthesis that together carryout the primary photochemistry of photosynthesis. The absorption of light and the transfer of energy and electrons. They are found in thylakoid membranes of plants, algae and cyanobacteria or in the cytoplasmic membrane of photosynthetic bacteria. Each photosystem contains about 300 chlorophyll molecules that trap light energy, which is then passed to a **reaction centre**, comprising a chlorophyll-*a* molecule, in each photosystem.

phototaxis (phototactic movement) Taxis in response to light. Many motile algae are positively phototactic, *e.g. Volvox*, while cockroaches are examples of negatively phototactic organisms.

phototroph An organism that uses solar energy to manufacture its food by the process of photosynthesis.

phototropin Class of photosensitive flavoproteins that are sensitive to blue light and trigger reorientating growth of a plant towards a light source, (i.e. phototropism). They also cause opening of stomata and movement of chloroplasts inside leaves.

phototropism (heliotropism) Growth of plant organs in response to light. Aerial shoots usually grow towards light, while some aerial roots grow away from light.

phragmoplast A barrel-shaped array of short microtubules that organises formation of the cell plate in dividing plant cells. It is orientated parallel to the mitotic spindle and accumulates vesicles in the plane of the metaphase equator, midway between the daughter nuclei. The vesicles fuse and their components form the new plasma membranes and cell walls of the daughter cells.

phrenic nerve The nerve that forms a network in the diaphragm and is involved in the control of ventilation.

phycobiliprotein Pigment in the blue-green bacteria, red algae and photosynthetic cryptophytes that act as accessory pigments in photosynthesis, it is attached to the thylakoid membranes. Each comprises a protein bound to a phycobilina, a prosthetic group, e.g. phycocyanin and phycoerythrin.

phycobiont Algal or bacterial component of a lichen.

phycocyanin Accessory photosynthetic pigment occurring mainly in cyanobacteria and red algae. It is a phycobiliprotein, in which the pigmented prosthetic group is phycocyanobilin, which gives it blue colour.

phycoerythrin Accessory photosynthetic pigment occurring mainly in the red algae and cyanobacteria. It is a phycobiliprotein, in which the pigmented prosthetic group is phycoerythrobilin, which gives it red colour.

phycomycetes Class of kingdom–Fungi. Members of Phycomycetes are *Mucor* and *Rhizopus.* Many are unicellular but those that form mycelia generally have hyphae lacking cross walls, which distinguishes them from the Ascomycota and Basidiomycota. They include the Zygomycota.

phycoplast The array of microtubules that organises cell division after mitosis in certain algae.

phyletic series Sequence of fossil forms that represent the succession of species in the evolution of a particular lineage.

phylloclade The modified stem which is flattened like a leaf and turns green in colour to carryout photosynthesis. Leaves either fall down or are modified into spines to check transpiration. e.g. cactus.

phyllode A flattened and widened petiole (leaf stalk) that looks like and functions as a leaf. e.g. *Acacia*

phyllomorphous Having an appearance or shape of a leaf.

phyllotaxy Arrangement of leaves on the stem (whorled, opposite or alternate.)

phylogenetic Describing a system of classification of organisms that aims to show their evolutionary history.

Phylogenetic Species Concept (PSC) The concept of a species as an irreducible group whose members are descended from a common ancestor and who all possess a combination of certain defining or derived traits.

phylogenetic tree

1. *See* evolutionary tree.
2. *See* phylogram.

phylogenomics The field of bioinformatics that integrates knowledge of phylogeny with structural and functional analysis of their genome (i.e. genomics) and proteins (proteomics). A basic supposition is that well-characterised genes (or proteins) in one organism provide valuable insights into the likely sequence and function of homologous genes or proteins in related organism.

phylogeny Evolutionary relationships of a group of organisms or an organism.

phylogram (phylogenetic tree) Diagram representing the evolutionary relationships and distances between different groups or species of organisms. The relationships are usually depicted by branching lines to form a tree-like pattern, with the lengths of the branches proportional to the distance of the relationship in evolutionary terms.

phylum (pl. phyla) In taxonomy, a division of a kingdom that c.ontains one or more similar or closely related classes.

physiognomy Characteristic appearance of an organism on the basis of which it can be recognised.

physiology Study of the vital functions of living organisms.

physiome Complete description of all processes that sustain the life of an organism or species and how these change in time and space. Such a description integrates all aspects of an organism's biology, from its genes and proteins through cellular metabolism and biochemistry to anatomy.

Physometra Distention of the uterus by gas.

phytoaccumulation (or phytoextraction) Uptake of contaminants by a plant and storage in its tissues. The plant can then be harvested, thus removing the toxin with the plant. e.g. alpine pennycress (*Thlaspi caerulescens*), which is an effective accumulator of cadmium and zinc.

phytoalexin An antipathogenic substance that is produced by a plant in response to infection and helps to defend the plant by inhibiting the growth of invading pathogen. They are isoflavonoides in legumes, terpenes in potato family.

phytochelatin A group of sulphur-rich peptides found in plants exposed to high concentrations of heavy metals (e.g. cadmium, zinc, lead, copper and mercury). These are thought to play a role in detoxification.

phytochrome A protein pigment present in very small quantity in plants. It controls photoperiodism and many other physiological responses of plants to light.

phytoecdysone Chemicals, structurally similar to insect ecdysones, found in plants. They may help to protect the plant by interfering with the moulting cycle of insect pests that consume plant tissues.

phytogeography *See* plant geography.

phytohaemagglutinin (PHA) Plant-derived compounds that induce changes in lymphocytes normally associated with antigen challenge. These changes include cell enlargement, increased RNA and DNA synthesis and finally cell division.

phytohormone *See* plant hormone.

phytopathology Plant diseases.

phytoplankton The plankton composed of photosynthetic algae.

phytoremediation Use of plants to decontaminate polluted land, water or air. Different plant species that are able to grow on contaminated sites have evolved various ways of countering high concentrations of heavy metals, oil, solvents or other toxic substances. *See* phytoaccumulation, phytotransformation and phytostabilisation.

phytostabilisation A phytoremediation measure in which the plant's root exudates or associated microorganisms precipitate the toxins (e.g. heavy metals) and make them harmless to other organisms.

phytotelm A small body of water found on a plant. e.g. the contents of a pitcher plant may contain abundant life and represent well-demarcated animal communities that are very amenable to ecological study.

phytotransformation A phytoremediation measure in which the plant takes up the toxin and degrades it by enzyme activity, thereby eliminating its toxic nature altogether.

piamater The innermost highly vascular membrane of meninges richly supplied with blood.

piebald Refers to a mammalian phenotype in which patches of skin are black and white because of absence of melanocytes.

pileus The umbrella-shaped cap of certain fungi, such as mushrooms. Spores are produced from gills or pores on the lower surface.

piliferous layer Region of the root epidermis that bears root hairs. It is located just behind the zone of elongation and is the main absorptive area of the root.

pilomotor Describing nerve fibres or the muscles innervated by them that can cause the position of the body hair to be altered. The arrector pili is a pilomotor muscle.

pilus (pl. pili) Hair-like appendage in bacteria as *Escherichia coli* that are involved in conjugation and the attachement of bacteria to other cells.

pineal eye *See* median eye.

pineal gland (epiphysis) An endocrine gland in the vertebrate forebrain that secretes melatonin and is involved with regulation of daily rhythms.

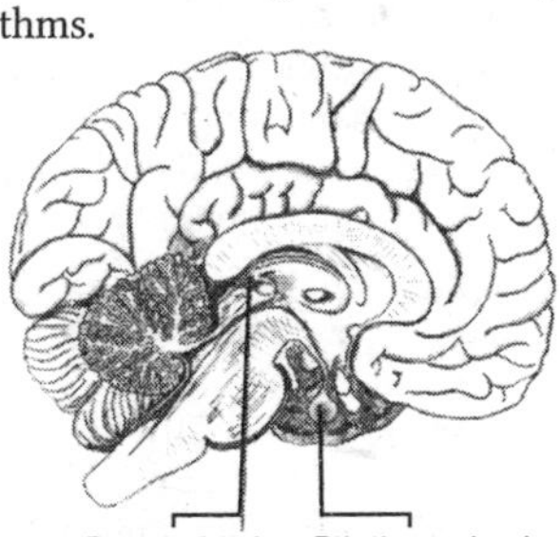

Location of pineal gland in human brain

pinna (auricle) The external and cartilaginous projecting portion of the ear which surrounds the auditory meatus.

pinocytosis The active process of taking in of bulk fluids by a cell using endocytosis. The term literally means cell drinking.

pioneer species Species that invade a bare area. e.g. lichens which later pave way to some small plants like bryophytes on rocks.

Pisces Superclass of the Gnathostomata (jawed chordates) comprising the fishes. The whole body of members is covered with. scales. The circulatory system is a single circuit with blood passing through two sets of capillaries, one at the gills and the other in the body tissues. The sense of smell is particularly well developed and pressure waves are detected by the lateral-line system. These are two classes of modern fish : Chondrichthyes (cartilaginous fishes) and Osteichthyes (bony fishes).

pistil Ovule-bearing female part of a seed plant. It consists either of a single carpel or a group of carpel.

pith (or medulla)

1. Cylinder of parenchyma tissue found in the centre of plant stems to the inside of the vascular tissue.
2. The white tissue below the rind of many citrus fruits.
3. To destroy the central nervous system of an animal, especially a laboratory animal such as a frog, by severing its spinal cord.

pitressin Extract of posterior lobe or pituitary, causing constriction of capillaries and arterioles when injected and diminished urine formation.

pituitary gland (hypophysis) An important endocrine gland associated with and largely controlled by the hypothalamus that secretes a wide range of hormones largely for the control of other endocrine glands. It consists of two main lobes, the anterior and the posterior, separated by a much smaller intermediate lobe.

p

The **anterior pituitary** (or adenohypophysis) secretes hormones as growth hormone, the gonadotrophins, prolactin, thyroid-stimulating hormone, and ACTH. These hormones regulate the growth and activity of several other endocrine glands, so the anterior pituitary is often referred to as the **master endocrine gland**. Activity of the anterior pituitary is regulated by specific releasing hormones produced by the hypothalamus. The **posterior pituitary** (or neurohypophysis) secretes the hormones oxytocin and antidiuretic hormone, and the intermediate lobe secretes melanocyte stimulating hormone.

placenta Oval, vascular structure in the uterus that nourishes the foetus, in mammals and other viviparous animals. It is composed of embyronic and maternal tissues: extensions of the chorion and allantois grow into the uterine wall so that oxygen, nutrients, etc., can pass between the blood of the embryo and its mother.
It is eventually expelled as part of the afterbirth. 2. A ridge of tissue on the ovary wall of flowering plants to which the ovules are attached.

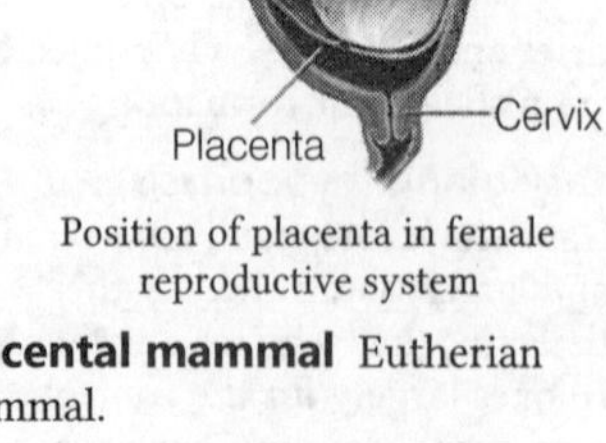

Position of placenta in female reproductive system

placental mammal Eutherian mammal.

placentation Arrangement of ovules within the ovary.

placode Area of thickening in the ectoderm of an early embryo that is the forerunner of a particular organ or structure. e.g. the inner ear develops from the otic placodes, located on either side of the hindbrain. Likewise, hair follicles arise from individual epithelial placodes.

placoid scale (denticle) Scales, characteristic of cartilaginous fish. These are small and tooth-like, with a projecting spine and a flattened base embedded in the skin. Made of dentine, have a pulp cavity and spine covered with enamel layer.

Placozoa Phylum of simple aquatic animals containing just a single known species, *Trichoplax adhaerens.*

plagiotropism (diatropism) Tendency for a growth response of a plant to be orientated at an angle to the line of action of the stimulus concerned. e.g. the growth of lateral branches and lateral roots is at an oblique angle to the stimulus of gravity (plagiogeotropism).

plankton Small sea organisms that move primarily, or exclusively, by water currents includes many microscopic organisms, such as algae, protozoans, various animal larvae and some worms. It forms an important food source for many other members of the aquatic community and is divided into zooplankton and phytoplankton.

planogamete A motile gamete, especially a ciliated one.

plant geography (phytogeography) Study of the distribution of world vegetation, with particular emphasis on the influence of the environmental factors that determine this distribution.

plant hormone (growth substance; phytohormone) Number of organic chemicals that are synthesised by plants and regulate growth and development. They are usually made in a particular region, such as the shoot tip and transported to other regions, where they take effect. See abscisic acid; auxin; brassin; cytokinin; ethylene; gibberellin.

plantae The kingdom–Plantae is also called as kingdom–Metaphyta. It includes all types of eukaryotic, multicellular, photosynthetic plants found in the biosphere.

plantigrade A gait in which entire palm or sole rests on the ground. The wrist and ankle are not raised from the ground, e.g. bear, baboon (primates).

plaque A film covering the teeth. It is made up of mucous from saliva with dissolved sugar and other food and it provides a breeding ground for bacteria.

plasma *See* blood plasma.

plasma cells Antibody-secreting cells found in lung and gut epithelium and also in bone forming tissue.

plasma membrane The limiting surface of the cytoplasm of a cell. It consists of a phospholipid bilayer with a variety of embedded proteins that act as channels and pumps, selectively moving, particular molecules into and out of the cell. The plasma membrane also plays roles in cell adhesion and in maintaining the shape of the cell *via* cytoskeleton attachment. Surface molecules on the plasma membrane allow specific recognition of each particular cell type. Also, called the cell membrane or plasmalemma. *See* fluid mosaic model.

plasma protein Protein constituents of blood plasma. They comprise albumins, globulins and fibrinogen.

plasmagel Specialised outer gel-like cytoplasm of living cells (such as *Amoeba*) that move by extruding part of the cell (pseudopodium) in the direction of motion.

plasmalemma A plasma membrane which bounds a cell, especially one immediately within the wall of a plant cell.

plasmasol The specialised inner sol-like cytoplasm of living cells that move by producing pseudopodia.

plasmids Autonomously replicating extra-chromosomal circular DNA molecules, distinct from the normal bacterial chromosome and non-essential for cell survival under non-selective conditions. Some plasmids are capable of integrating into the host chromosome. Often used as cloning vectors, in recombinant DNA technology.

plasmin (fibrinase; fibrinolysin) Enzyme in blood plasma, that breaks down a blood clot by destroying the fibrin threads of the clot and inactivating factors involved in blood clotting, such as prothrombin and the clotting factors. This occurs during fibrinolysis. It is derived from an inactive precursor, plasminogen.

plasminogen Inactive precursor of the enzyme plasmin. It minogen is incorprated into blood clots and is converted to plasmin during fibrinolysis.

plasmodesmata The cytoplasmic thread piercing the cell walls of plants, connecting adjacent cells.

plasmodium A mass of cytoplasm that contains nuclei. It is capable of slow movement and feeds by engulfing of organic matter.

plasmogamy The fusion of cytoplasm of two or more cells.

p

plasmolysis The separation of the plasma membrane from the cell wall resulting from shrinkage of the cytoplasm in a hypertonic environment.

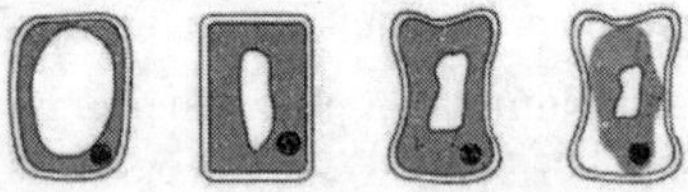

plastids Self-replicating organelles present in the cytoplasm of plant cells. Different types of plastids (*e.g.* chloroplasts, leucoplasts, chromoplasts) serve different functions.

plastocyanin The important blue copper-containing protein involved in electron-transfer during photosynthesis. Found in chloroplasts.

plastogene Gene present in a self-replicating plastid.

plastoglobulus (pl. plastoglobuli) A densely staining droplet found, often in large numbers, in plastids of plant cells. It consist of lipid pigment and are especially prominent in chromoplasts.

p

plastome In this, the genetic material is contained in a cell's cytoplasmic organelles, such as mitochondria or chloroplasts.

plastoquinone Quinone, found in chloroplasts, that functions as one of the carrier molecules of the electron transport chain in the light-dependent reactions of photosynthesis.

plastron The underside of a tortoise's or turtle's shell. *See* carapace.

plate count Estimate of the number of viable cells in a culture. A plate count of a bacterial culture is made by inoculating the culture plate with a dilute solution of the microorganisms and counting the number of cells that appear in the resultant culture.

platelet (thrombocyte) Minute, non-nucleated disc-shaped cell fragment in mammalian blood. Play an important role in blood clotting. These are about 250000 platelets per cubic millimetre of blood.

platycnemia (also platycnemism) Condition of having an abnormally wide tibia (laterally fattened).

Platyhelminthes (Plathelminthes) A phylum that includes flatworms. These are acoelomate invertebrates, characterised by flat unsegmented body. The simple nervous system shows some concentration of cells at the head end. The mouth leads to a simple branched gut without an anus. They are hermaphrodite but self-fertilisation is unusual. Many species are parasitic.

platyzoa A group of protostome animals based on molecular systematics. It includes the flatworms, rotifers, and certain other phyla.

pleiomorphism The existence of distinctly different forms during the life cycle of an individual, e.g. the caterpillar, pupa and winged adult of a butterfly.

pleiotropy Phenomenon in which a single gene exhibits multiple phenotypic expression. A single pleiotropic gene may produce more than one effect. e.g. phenylketonuria, a disorder caused by mutation in the gene coding the enzyme phenylalanine hydroxylase. The affected individuals show hair and skin pigmentation and mental problems.

Pleistocene Third epoch of the quaternary period of the Cenozoic era. Often called the Ice Age, extensive continental glaciers existed through much of this epoch, which began at 2.588 million years ago, and ended some 12000 years ago.

plesiomorphy (ancestral trait) The evolutionary trait that is homologous within a particular group of organisms but is not unique to members of that group. Thus, cannot be used as a defining character for the group. e.g. vertebrae are found in zebras, cheetahs and orangutans, but the common ancestor in which this trait first evolved is so distant that the trait is shared by many other animals. Therefore, possession of vertebrae sheds no light on the phylogenetic relations of these species.

pleura (pleural membrane) One of the two serous membrane that encloses the lungs and lines the thoracic cavity.

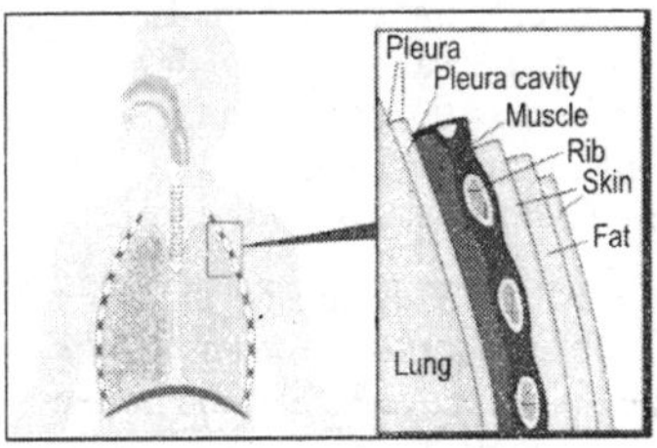

Pleura and pleural cavity in human respiratory system

pleural cavity In human anatomy, It is the space between the pleural membrane of the lungs.

Pliocene Second epoch of the Neogene period. It lasted from 5.332 to 2.588 million years ago.

ploidy Number of repetitions of the basic chromosome number occurring in a polyploid.

plumule

1. The shoot apex and first rudimentary leaves in the mature embryo and the seedling.
2. The down feather, a small feather in which the barbs are not held firmly together by their barbules.

pluripotency The capacity to differentiate into any cell or tissue type.

pneograph A machine for recording respiratory movements.

pneumatic bones Bones which are hollow and air filled with many air-cavities to add buoyancy during flight of arboreal animals.

pneumatophore

1. A specialised negatively geotropic root produced by certain aquatic vascular plants (e.g. mangrove). The aerial parts is converted with pores through which gases can diffuse to and from the highly developed system of intercellular air spaces.
2. The gas filled float of certain colonial cnidarians of the class-Hydrozoa, such as *Physalis* (Portuguese man-of-war).

pod *See* legume.

podocyte A type of epithelial cell found in Bowman's capsule. These possess major foot-like processes. These minor processes are interwoven with those from other podocytes to form a number of slits, through which filtration can take place.

pododynia Pain in the foot.

podsol The type of soil found in coniferous forests in temperate climates. It is strongly acidic and often lacks nutrients.

poikilothermy The passive variation in the internal body temperature of an animal, which depends on the temperature of the environment. All animals except birds and mammals exhibit poikilothermy.

point mutation (gene mutation) Change in the nucleotide sequence of the DNA within a gene forming a mutant gene or allele. The DNA sequence can be altered in several ways, e.g. by insertion, substitution, deletion, and inversion, such mutations can result in misreading of the genetic code during the translation phase of protein synthesis and can change the order of amino acids making up a protein, which may sometimes affect the function of that protein.

poison A substance that is capable of causing the illness or death of a living organism when introduced or absorbed.

polar body During oogenesis, one of upto three small haploid nuclei produced by each meiosis, in addition to the much larger egg cell. A polar body serves no function in reproduction.

polar nuclei Haploid nuclei in the centre of the embryo sac of flowering plants. These nuclei fuse with a male gamete nucleus to form a triploid endosperm nucleus, which subsequently divides to form the endosperm.

pollen Mass of pollen grains produced by a seed plant, containing the male gamete.

pollen grain In flowering plants, a granular microspore that germinates to produce the male gametophyte. Pollen grains are very durable and of great interest as fossils. The shape and size of the grains produced by each type of plant are so distinctive that they can be used to determine whether any given plant was present during past geologic eras. It contains two male nuclei a generative nucleus and a tube nucleus. The wall of the mature pollen grain consists of the tough outer wall (exine) and the more delicate narrower **intine**. The latter gives rise to the pollen tube.

pollen sacs Pouches in the anther where pollen is formed. In angiosperms there are usually four pollen sacs in each anther containing the microspore mother cells. In gymnosperms, variable numbers of pollen sacs are borne on the microsporophylls that make up the male cone.

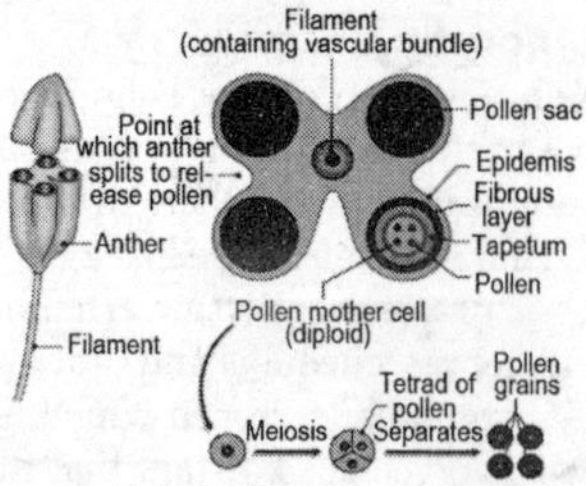

Anther depicting pollen sac

pollen tube Filamentous outgrowth of the pollen grain that in most seed plants, transports the male gamete to the ovule.

pollex The innermost digit on the forelimb of a tetrapod vertebrate. It contains two phalanges and in humans and higher primates it is the thumb, which is opposable (i.e. capable of facing and touching the other digits) and gives the hand greater manipulating ability. In some mammals a pollex is absent.

pollination The transfer of pollen from the anther to the stigma. If the pollen is compatible then the pollen grains germinate, producing a pollen tube that grows down the style carrying the pollen nuclei to the ovule. It can be self-pollination or cross-pollination.

pollinium (pl. pollinia) A clump of pollen grains produced by a single anther and transferred as a unit during pollination. e.g. orchids.

pollutant Substance produced and released into the environment as a result of human activities or natural calamities, that has damaging effects on living organisms. e.g. pesticides, CO_2 etc. There are two main classes of pollutants; i.e. **biodegradable** (e.g. sewage) that can be rendered harmless by natural processes and need therefore cause no permanent harm if adequately dispersed or treated; and **non-biodegradable** (e.g. heavy metals (such as lead) in industrial effluents and DDT and other chlorinated hydrocarbons used as pesticides), which eventually accumulate in the environment and may be concentrated in food chains.

pollution The undesirable change in the environment that results from the physical, chemical or biological side-effects of human industrial or social activities. Pollution can affect atmosphere, rivers, sea or soil. Other forms of pollution in the environment include noise (e.g. from jet aircraft, traffic, and industrial processes), thermal pollution (e.g. the release of excessive waste heat into lakes or rivers causing harm to wildlife), and light pollution (from street lights, buildings, etc., which can disorientate wildlife). Recent pollution problems include the disposal of radioactive waste; acid rain; photochemical smog; increasing levels of human waste; high levels of carbon dioxide and other greenhouse gases in the atmosphee (*see* greenhouse effect); damage to the ozone layer by nitrogen oxides, chlorofluorocarbons (CFCs), and halons; and pollution of inland waters by agricultural fertilisers and sewage effluent, causing eutrophication.

polyandry

1. The mating of one female with many males.
2. The condition of having numerous stamens.

Polychaeta A class of annelid worms in which each body segment has a pair of **parapodia** bearing numerous chaetae. All are aquatic and most of them are marine. They include the fanworms (*Sabella*), lugworms (*Arenicola*) and ragworms (*Nereis*).

polycistronic Describing a type of messenger RNA that can encode more than one polypeptide separately within the same RNA molecule, e.g. Bacterial messenger.

polydipsia Consumption of increased amount of water or Excessive thirst.

polyembryony

1. Occurrence of more than one embryos in a seed. Only one of these is a true sexual embryo, the others arising vegetatively. It is a common phenomenon in citrus fruits.
2. Formation of more than one embryo from a single animal zygote. Identical twins are produced in this way.

polygene Group of genes influencing a quantitative trait, such as height in humans.

polygenic inheritance
See quantitative inheritance.

polygenic trait Trait resulting from the combined action of multiple genes. Polygenic characters display continuous variation, each polygene having a small influence on the particular character in question.

polymerase Enzyme that catalyses the synthesis of a polymeric molecule, e.g. nucleic acid.

Polymerase Chain Reaction (PCR) Reaction in which multiple copies of specific DNA sequences are made (amplification) *in vitro*. The technique was developed by *Kary Mullis* in 1985 who received Nobel Prize for chemistry in 1993

(a) *PCR technique requires*
- **A DNA template** which is a double- stranded DNA that needs to be amplified. naughty
- **Primers** are small chemically made oligonucleotides of about 10-18 nucleotides that are complementary to a region of template DNA.
- **Enzymes** used are DNA Taq polymerase (from *Thermus aquaticus*).

(b) *Step in PCR*
- **Denaturation** of double- stranded DNA is carried out by applying high temperature of 95°C for 15 seconds.

Each separated single strand acts as a template for DNA synthesis.

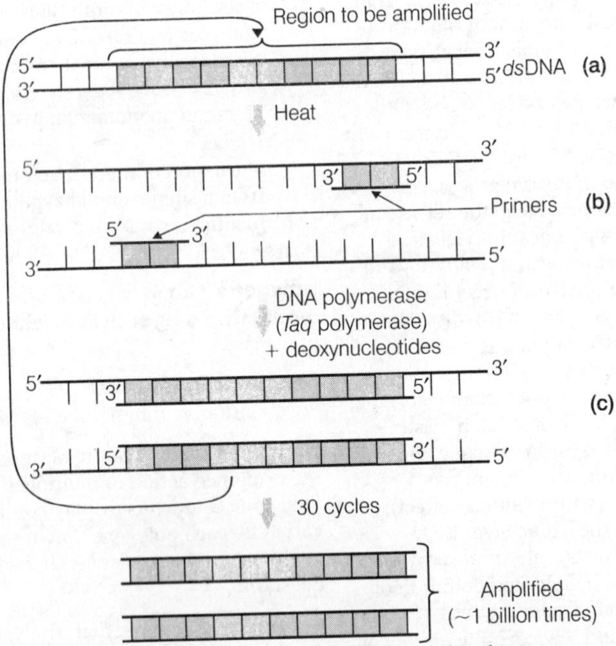

Polymerase Chain Reaction (PCR) Each cycle has three steps (a) Denaturation (b) Primer annealing and (c) Extension of

- **Annealing** is carried out by two sets of primers, which are added in the reaction. They anneal to the 3' end of each separated strand. Primers act as initiator of replication.
- **Extension** is done by DNA polymerase by adding nucleotides complementary to the template in the reaction.
- **A thermostable** DNA polymerase (Taq polymerase) is used in the reaction, which can tolerate the high temperature of the reaction.

(e) These steps are repeated many times to obtain several copies of the desired DNA.

polymorphism The existence of two or more distinct variants within a population; usually refers to the variation at a particular locus.

polymorphonuclear Having a nucleus with two or more lobes.

polynucleotide A long chain of nucleotides, e.g. DNA and RNA.

polyp (hydranth) The sedentary stage in the life cycle of the *Cnidaria*. It consist of a cylindrical body fixed at one end to a firm base and having a mouth surrounded by a ring of tentacles at the other. They reproduce asexually by budding to form either new polyps or medusae.

polypeptide Chain of amino acids covalently linked by peptide bonds.

polypetalous Flower in which the corolla consists of separate petals. *Compare* Gamopetalous.

polyphagia Excessive eating.

polyphenism (environmental polymorphism) Occurrence of distinct morphological types within species, in response to environmental cues.

polyphyletic Taxon in which members are thought to have distinct evolutionary histories.

polyphyodont Type of dentition in which the teeth are continuously shed and replaced during the lifetime of the animal e.g. in sharks and frogs. *Compare* Diphyodont, Monophyodont.

polyploidisation Any genetic process producing a polyploid.

polyploidy Condition in which the individual cells of an organism have more than two complete sets of chromosomes.

polyribosome (polysome) Aggregate of ribosomes in association with a single messenger RNA molecule during the translation process of protein synthesis. These are attached to the surface of the rough endoplasmic reticulum and the outer membrane of the nucleus in eukaryotes while free in the cytoplasm in bacteria.

polysaccharides (glycan polysaccharoses) A group name for those carbohydrates that can be broken down into three or more simple sugars; usually insoluble in water. When soluble, they form colloidal solutions. Common examples are starch and cellulose.

polysepalours Flower that possesses a calyx consisting of separate sepals. *Compare* Gamosepalous.

polysome *See* polyribosome.

polyspermy Penetration of several sperm into one ovum at fertilisation; only one sperm actually fuses with the ovum nucleus. It occurs in a few animals with yolky eggs, e.g. birds.

polystely Having many steles, as in the stems of some *Selaginella* species.

polysynaptic reflex A reflex action that involves an electrical impulse being transferred from a sensory neuron to a motor neuron via at least one connecting neuron e.g. in the spinal cord. Stimulation of pain receptors in the skin initiates a **with drawal reflex**, which involves several synapses with several motor neurons and results in the removal of the organism or part from the stimulus.

Polytene Chromosome condition caused by chromatids not separating after duplication. It leads information of giant chromosomes.

polytene chromosome A type of giant chromosome in which many identical chromatids lie alongside one another. A pattern of bands is visible because of differential degrees of hypercoiling. Polytene chromosomes are found in a rather limited variety of organisms, but have been extensively studied in the salivary glands of *Drosophila.*

polythetic A type of classification in which members of a taxon are based on its constitutent organisms sharing a large number of characteristics.

polytopic Describing a taxon that has arisen and exists in more than one geographical region.

polytypic Composed of a variety of types, sorts or forms.

polyuria Excessive urination.

pome A fleshy pseudocarpic fruit, characteristic of the family *Rosaceae.* (e.g. apple and pear) It is the flesh of the fruit develops from the receptacle of the flower, which completely encloses the fused carpels. After fertilisation the carpels form the 'core' of the fruit, which contains the seeds.

pons (pons cerebellipons varolii) A thick band of nerve fibres in the mammalian brain that passes across the medulla oblongata to link the two hemispheres of the cerebellum.

population A group of individuals of the same species (or other groups within which, genetic information can be exchanged) occupying a particular space. The nature of a population is determined by such factors as density, sex ratio, birth and death rates, emigration and immigration.

population density Number of individuals per square kilometre at any given time.

population dynamics Study of the fluctuations that occur in the number of individuals in animal and plant populations and the factors controlling these fluctuations.

population ecology The ecological study of the dynamics of populations of a single species.

population genetics The study of distribution of inherited variations in a population of individuals.

p

population growth Increase in a population that occurs when the birth rate is higher than the death rate, or when immigration exceeds emigration, or when a combination of these factors is present. A growth curve, obtained by plotting population size against time, is typically S-shaped (sigmoid) or J-shaped (see graphs).

Porifera The phylum of marine and freshwater invertebrates that comprise the sponges. The body is hollow and consists of an aggregation of cells between which there is little nervous coordination, supported by an internal skeleton of spicules of chalk, silica, or fibrous protein (bath sponges have protein skeletons). Undulipodium bearing (flagellated) cells (choanocytes) cause water to flow in through **ostia** in the body wall and out through **oscula** at the top; food particles are filtered from the water by the choanocytes.

porins Proteins that form water-filled channels in the outer membrane of the cell membranes. Allowing molecules up to the size of small proteins to pass through.

porogamy Fertilisation method, usually seen in angiosperms in which, pollen tube enters the ovule through micropyle.

porphyrin A group of organic pigments characterised by the cyclic group of four linked nitrogen containing rings the nitrogen atoms of which are often coordinated to metal ions. They include the chlorophylls, which contain magnesium and haem, which contains iron and forms the prosthetic groups of haemoglobin, myoglobin and the cytochromes.

portal vein A vein connecting capillary network of two particular regions that allows blood from one to be regulated by the other.

position effect Phenomenon in which phenotype of a gene is altered by change in position of the gene within the genome.

positional cloning Technique used to identify genes, usually those that are associated with diseases, based on their location on a chromosome.

positive feedback The biochemical control where the accumulation of the product stimulates production of an enzyme responsible for that product's production.

positive interference Phenomenon where the occurrence of one crossover event reduces. The likelihood of a second crossover event occurring in the vicinity.

Positron Emission Tomography (PET) A non-invasive imaging technique that can produce three-dimensional pictures of certain biochemical changes within the body, such as areas of increased tissue metabolism. It is used for research in several areas, including neurology and pharmacology and as a diagnostic technique in medicine, e.g. in detecting brain tumours.

postcaval vein *See* vena cava.

posterior

1. Designating the part of an animal that is to the rear, i.e. that follows when the animal is moving. In humans and bipedal animals (e.g. kangaroos) the posterior surface is equivalent to the dorsal surface.
2. Designating the side of a flower or axillary bud that faces towards the flower stalk or main stem, respectively.

postreplicative repair (recombinational repair) A form of DNA repair that fills large gaps in a new DNA strand resulting from failure of replication.

postsynaptic membrane The membrane at the end of a neuron that receives an impulse at a synapse.

potometer The device that measures the rate of water loss by a plant.

poxvirus One of a group of DNA containing viruses, often enclosed in membrane, that produce skin lesions in vertebrates. They include the viruses causing smallpox (variola) and cowpox (vaccinia) and also the myxoma virus responsible for myxomatosis.

P-protein (phloem protein) The protein found in large amounts in the sap-conducting sieve elements of phloem tissue in plants. It takes various forms in the mature sieve element, depending on plant species.

preadaptation The anatomical structure, physiological process, or behaviour pattern in an organism that is by chance highly suited to a new habitat to which the organism migrates or that improves the chance of the organism surviving a change in environmental conditions. e.g. the lungs that developed in certain fish were probably originally a buoyancy aid before these fish began to adapt to a new environment on dry land.

Precambrian The informal name for the time period from the formation of Earth to the first abundant appearance of macroscopic hard-shelled animals that defines the beginning of the Cambrian.

precaval vein *See* vena cava.

precocial species Capable of significant independent action soon after hatching or birth. Precocial birds or mammals are those with newly hatched or born young being less dependent on their parents (i.e. ones that can move about, open their eyes and that have down or hair).

pregnancy The condition of actually bearing an embryo or foetus within the uterus of Fallopian tubes in humans.

premolar Crushing teeth of mammals, having usually more than one root and a pattern of ridges and projections of biting surface. They have no temporary variants.

premotor cortex Region of cerebral cortex in the vertebrate brain situated immediately anterior to the primary motor cortex. It is thought to play a crucial role, in performing complex movements and it stores learned motor patterns, such as that employed in writing one's signature.

premutation Gene variant (allele) that produces a normal individual but is predisposed to become a full mutation in subsequent generations.

prepuce Loose skin or flab which protects the glans penis in mammal.

preputial gland Modified sebaceous glands on the inner surface of prepure. Secrete smegma.

presbyopia Stiffening of the eye lens which makes accommodation to nearby objects difficult. It normally develops over the age of 45-50 years.

pressure potential (ψ_P) The component of water potential due to the hydrostatic pressure that is exerted on water in a cell. In turgid plant cells, it has a positive value as the entry of water causes the protoplast to push against the cell wall. In xylem cells, there is a negative pressure potential, or tension, as a result of transpiration. Water at atmospheric pressure has a pressure potential of zero.

presumptive Describing embryonic tissue that is not yet determined but which will eventually develop into a certain kind of tissue by virtue of its position in the embryo.

presynaptic membrane The membrane of a neuron that releases neurotransmitter into the synaptic cleft between nerve cells.

prey An animal that is a source of food for a predator.

Pribnow box (minus 10 sequence) A consensus sequence of nucleotides-TATAAT- occurring in the promoter region of prokaryote genes about 10 nucleotides before the start of transcription.

prickle A hard sharp protective outgrowth, many of which may cover the surface of a plant. It contains cortical and vascular tissue and is not regarded as an epidermal outgrowth.

primary growth Increase in size of shoots and roots of plants that results from the activity of the apical (tip) meristems and subsequent expansion of the cells produced.

primary meristem A meristem that has persisted from embryo, e.g. apical meristem, cambium.

primary oocyte Germ cell of the female. It undergoes first meiotic division to form a secondary oocyte and a polar body.

primary productivity Total amount of organic matter synthesised by the producers (e.g. green plants) of an ecosystem.

primary structure Protein molecules consist of one or several long chains (polypeptides) of amino acids linked in a characteristic sequence. This sequence is called the **primary structure** of the protein.

primary spermatocyte Male germ cell. It undergoes meiosis to form two secondary spermatocytes.

primase Enzyme that creates the short primer sequences needed for the discontinuous replication of DNA. It is an RNA polymerase that synthesises a short segment of RNA, consisting of 10 to 12 nucleotides, attached to the lagging strand of the unwound DNA molecule.

Primates The order of mammals that includes the monkeys, apes and humans.

primatology The study of primates. Monkeys, apes, humans.

primer A short nucleotide sequence used by a DNA or RNA polymerase to initiate transcription.

Primer Binding Site (PBS) The site on a retrovirus or retrotranspons, where the *t*RNA primer binds to initiate reverse transcription.

primitive node (primitive knot) *See* Hensen's Node.

primitive streak Longitudinal groove that develops in the gastrula during the development of bird and mammal embryos. The cells in the primitive streak proliferate rapidly to form mesoderm cells, which migrate to the interior of the embryo.

primordium Progenitor cells of an organ or tissue at its earliest stage of development in plants.

primosome

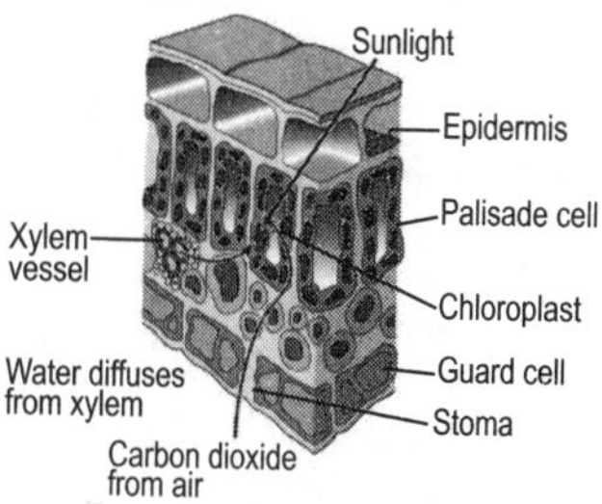

Palisade cells in leaves

A complex of enzymes and additional proteins that associates with the DNA molecule to initiate DNA replication in prokaryotes, such as *E.coli.* Its chief component is a primase enzyme but it also contains helicase.

prion An infectious proteinaceous agent causing a number of neurological disease in humans and animals. This infectious protein is formed due the misfolding of protein, e.g. Creutzfeldt Jakob disease in humans.

probe A single-stranded DNA or RNA molecule, or some other biochemical, labelled either radioactively or immunologically and used to detect the presence of a gene, a gene product, or a protein.

Proboscidea The order of mammals that comprises the elephants. (Only two species of modern elephants the African and Indian). They are herbivorous, with a muscular trunk used for drinking, bathing and collecting food. The tusks are continuously growing upper incisors and the enormous ridged molar teeth are produced in sequence to replace worn teeth throughout life.

proboscis

1. The trunk of an elephant; a muscular and very flexible elongation of the nose.
2. The elongated mouthparts of certain invertebrates, such as the two winged flies (*Diptera*).

procambium The part of apical meristem which gives rise to the primary vascular tissues.

Procaryote *See* prokaryotes.

Proconsul A genus of extinct apes from fossils, 23-14 million years old. Found in East Africa.

proctodaeum The ectodermal depression of the caudal end of embryo with endoderm at the posterior end of gut, forming the anus or cloacal opening.

producer An organism which is considered as a source of energy for those above it in a food chain. Green plants, which convert energy from sunlight into chemical energy, are **primary producers**; herbivores are **secondary producers**, as they utilise energy from green plants and supply energy for carnivores.

productivity (production) Rate at which an organism, population, or community assimilates energy (gross productivity) or makes energy potentially available (as body tissue) to an animal that feeds on it (net productivity). The difference between these two rates is due to the rate at which energy is lost through excretion and respiration.

profundal Occuring in or the deep-water zone of an inland lake. Light intensity, oxygen concentrations and temperature are markedly lower than as compared to that of surface layer.

progenesis Maturation of gametes in an organism that is still otherwise at the juvenile stage of development.

progeny *See* offspring.

progesterone A steroid hormone secreted by the corpus luteum in the ovary after ovulation. It initiates preparation of the uterus for implantation of the ovum, development of the placenta and the development of mammary glands in preparation for lactation.

progestogen Hormone occuring naturally or synthetic, whose effects resemble those of progesterone. Used in therapy and oral contraceptives.

proglottid (also proglottis) One of the segments of a tapeworm (cestode).

prokaryotes (procaryote) Simple organisms lacking a membrane-bound nucleus; commonly known as bacteria. Their genetic material is not enlosed in a cell nucleus. These were the first cells to evolue.

p

prolactin A lactogenic hormone produced by the anterior pituitary gland. In mammals, it stimulates the mammary gland for milk production after oestrogen and progesterone priming.

proline One of the non-essential amino acids.

promoter Specific DNA site where RNA polymerase binds and initiates transcription.

pronation Rotation of the lower forearm so that the hand faces backwards or downwards with the radius and ulna crossed.

pronephros First type of vertebrate kidney to develop. It occurs in larval fish and amphibians composed of variable number of open ended ducts.

pronucleus The nucleus of a sperm, egg, or pollen grain.

prop root The modified root that arise from the stem of certain plants and provide extra support.

propagation

1. *See* vegetative propagation.
2. (in neurophysiology) A process whereby a nerve impulse travels along the axon of a neuron.

propagule A cellular structure produced by an organism that is capable of dispersing and surviving in the environment before developing into a new individual, e.g. seeds, spores and cysts and plant cuttings (artificial propagules).

prophage The phage genome integrated as a linear part of a host bacterium chromosome.

prophase The first stage of cell division. The chromosomes contract and divide along their length (except for the centromeres) into chromatids. In mitosis, the chromosomes remain separate from each other. In the first division of meiosis, homologous chromosomes become paired. By the end of first prophase, the two chromosomes begin to move apart.

proplastid A self duplicating meristematic plastid, found in meristematic regions of plants. They grow to develop into different type of plastids.

proprioceptor Receptor that is sensitive to movement, pressure, or stretching within the body. Occur in muscles, tendons and ligaments and are important for the coordination of muscular activity and the maintenance of balance and posture.

prosencephalon *See* forebrain.

prosenchyma The plant tissue consisting of elongated cells with tapering ends, occuring chiefly in vascular tissues.

prosoma The anterior section of the body of arachnids and other arthropods of the phylum–Chelicerata, which consists of the fused head and thorax (cephalothorax) and bears the chelicerae and other appendages.

prostacyclin (prostaglandin I_2) The eicosanoid normally produced from endothelial cells, lining the walls of blood vessels. It inhibits blood clotting by preventing the aggregation of platelets and causes vasodilation. Hence, it is an antagonist of thromboxane A_2.

prostaglandin Any of lipid soluble organic compound that are synthesised in the body from unsaturated fatty acids and are responsible for a variety of metabolic activities. They act as chemical messengers.

prostate gland A gland in male mammals that surrounds and opens into the urethra. During ejaculation, it secretes a slightly alkaline fluid into the semen that activates the sperms and prevents them from sticking together.

prosthetic group The non-protein part of an enzyme.

protamine The group of proteins of relatively low molecular weight found in association with the chromosomal DNA of vertebrate sperm cells. They contain a single polypeptide chain comprising about 67% arginine. Protamines serve in packaging the highly condensed DNA of the germ-cell chromosomes. Protamine sulphate is used therapeutically to treat heparin overdosage.

protandrous

1. Male gonads developing before the female gonads in hermaphroditcs.
2. The anthers ripening before the stigma of the same flower, ensuring that self-fertilisation does not occur, e.g. ivy.

protanopia Colour blindness in which individual lack cones sensitive to red light.

protease (peptidase; proteinase; proteolytic enzyme) An enzyme that catalyses the hydrolysis of proteins into smaller peptide fractions and amino acids, proteolysis, e.g. pepsin and trypsin.

proteasome A large complex of proteins found in the cytoplasm of cells that degrades cell proteins into small peptide units.

protein A molecule composed of one or more polypeptide chains.

protein kinase An enzyme that catalyses the transfer of a phosphate group from ATP to an intracellular protein, thereby affecting the biological activity of the protein.

protein sequencing The process of determining the amino acid sequence of a protein or its component polypeptides. This technique most commonly used is Edman degradation.

protein synthesis The process by which living cells manufacture proteins from their constituent amino acids, in accordance with the genetic information carried in the DNA of the chromosomes. This information is encoded in messenger RNA, which is transcribed from DNA in the nucleus of the cell. The sequence of amino acids in a particular protein is determined by the sequence of nucleotides in messenger RNA. At the ribosomes the information carried by messenger RNA is translated into the sequence of amino acids of the protein in the process of translation.

proteoglycan (mucoprotein) A type of glycoprotein consisting of long branched heterogeneous chains of glycosaminoglycan molecules linked to a protein core of amino acids.

proteolytic Causing, or capable of causing, the cleavage of proteins into separate subunits.

proteome Entire complement of proteins synthesised by a cell or organism at a given time.

proteomics The study of full set of proteins encoded by a genome.

Proterozoic The eon of geological time extending from the end of the Archaean, about 2500 million years ago, to the start of the present eon about 542 million years ago. Life in the early Proterozoic was dominated by bacteria.

prothallus A small flattened multicellular structure that represents the independent gametophyte generation of clubmosses, horsetails and ferns. In some of these plants a single prothallus bears both male and female sex organs.

p

prothoracic gland *See* ecdysone.

Prothoracicotropic Hormone (PTTH) The insect hormone, released by glands (corpora allata) on either side of the brain, that stimulates release of the moulting hormone, ecdysone, from the prothoracic glands.

prothrombin (Factor II) An inactive precursor of enzyme thrombin. Produced by liver, depend on vitamin-K supply and present in plasma.

Protist (Protoctista) An eukaryote organism that is either unicellular, or multicellular without specialised tissues; sometimes treated as a separate kingdom–Protista. Include simple algae, simple fungi and Protozoa; they display a diversity of forms, ranging from nonmotile or motile independent cells to leaflike or filamentous forms and sheets of cells. They can be autotrophic, saprotrophic, or heterotrophic, or combinations.

protistology The study of protists.

protochordata A chordata without vertebral column.

Protoctista A former kingdom consisting of unicellular or simple multicellular organisms that possess nuclei and cannot be classified as animals, plants, or fungi. The term is no longer considered either descriptively useful or taxonomically valid.

protoderm A plant tissue formed by the apical meristem of shoots and roots that subsequently gives rise to the epidermis.

protogyny

1. Condition in which the carpels of a flower mature before the (stamens) thereby ensuring that self-fertilisation does not occur. e.g. plantain and figwort.
2. Condition in hermaphrodite or colonial invertebrates in which the female gonads or individuals are sexually mature before the male ones.

proton pump A carrier protein or complex of proteins that transport protons (H^+) across biological membranes. Such pumps use energy, e.g. ATP, to establish a high concentration of protons on one side of the membrane compared to the other.

protonema The first stage of the gametophyte of a moss. It develops directly from a spore.

protonephridium Most premitive type of Nephridium. It consists of a system of flame cells. Occurs in Platyhelminthes and rotifers.

protooncogenes Cellular oncogenes (C-onc) present in normal cells, which when activated under certain conditions can cause oncogenic transformation of the cells.

protophloem The primary phloem tissue that arises from the elongating regions of shoot or root apices.

protoplasm The living substance of a cell.

protoplast (energid) The living unit of a cell, consisting of the nucleus and cytoplasm bounded by the plasma membrane.

protostele The simplest type of stele found in lower pteridophytes; consists of a central core of xylem elements surrounded by phloem.

protostome The animal in which the mouth develops from the opening (blastopore) of the embryonic cavity. These constitute one of the two major subkingdoms of animals (Protostomia), the other being the Deuterostomia.

Prototheria A subclass of mammals – the monotremes - that lay large yolky eggs. It contains only the duckbilled platypus and the spiny anteater.

prototroph An organism with the same growth factor requirements as the ancestral strain; said of microbial mutants.

protoxylem The first xylem element which differentiates from pro-cambium. Appear on the inner side, next to pith.

protozoa A group of single-celled heterotrophic often motile eukaryotic organisms, traditionally classified as animals and constituting a phylum or sub-kingdom–Protozoa.

proventriculus

1. Anterior part of the stomach of a bird, where digestive enzymes are secreted. Food passes from the proventriculus to the gizzard.
2. *See* gastric mill.

provirus The intermediate stage in the infection of a host cell by a virus, e.g. a retrovirus, in which the viral genome is integrated into the DNA of the host, where it can undergo successive replications before being transcribed.

proximal Denoting the part of an organ that is nearest to the organ's point of attachment, e.g. the knuckles are at the proximal end of the fingers.

proximal convoluted tubule (first convoluted tubule) The section of a nephron situated between Bowman's capsule and the loop of Henle in the vertebrate kidney. Reabsorption of salt, water and glucose from the glomerular filtrate occurs in this tubule. Also certain substances, including uric acid and drug metabolites, are actively transferred from the blood capillaries into the tubule. Both activities are facilitated by finger-like projections on the inner surface of the tubule, which increase it effective surface area.

psammophyte Plant which occurs only in sandy soil.

pseudoalleles One of two or more closely linked genes on a chromosome. These appear to function as a single allelic pair but occupy distinct, nearly corresponding homologous chromosomes. Such gene pairs produce a mutant effect in the diploid state when located on homologous chromosomes but are being capable of being separated by crossing over during meiosis to produce a wild-type effect when recombined on either of the homologues.

pseudocarp (false fruit) Incorporates, in addition to the ovary wall, other parts of the flower, such as the receptacle, e.g. the fleshy part of the strawberry is formed from the receptacle and the pips on the surface are the true fruits.

pseudocoelomate Animals in which spaces between body wall and internal organs are not lined by the cells of mesodermal origin. Include Rotifera and Nematoda.

pseudogamy Development of ovum into a new individual as a result of stimulation by a male gamete whose nucleus, however, does not fuse with that of the ovum and contributes nothing to hereditary constitution of embryo. Occurs in some nematodes and in some higher plants.

pseudogene It is a sequence of genomic DNA similar to a gene at some other locus, but rendered non-functional by mutations that prevent its normal transcription and translation.

pseudoheart A series of contractile blood vessels in annelid worms that pump blood from the dorsal vessel to the ventral vessle.

pseudomixis The fusion between two vegetative cells or between cells which are not differentiated of as gametes, leads to formation of an embryo.

pseudoparenchyma A tissue that superficially resembles plant parenchyma but is made up of an interwoven mass of hyphae (in fungi) or filaments (in algae), e.g. fruiting bodies (mushrooms, toadstools, etc). of certain fungi and the thalli of certain red and brown algae.

pseudopodium (pl. pseudopodia) A temporary outgrowth of the cell of some protozoans (e.g. *Amoeba*), which serves as a feeding and locomotory organ. It may be blunt or threadlike, form a branching network, or be stiffened with an internal supporting rod. Phagocytic white blood cells also form pseudopodia to engulf invading bacteria.

pseudopregnancy A physiological state resembling pregnancy that occurs in some female mammals such as rabbits, but without the formation of embryos. Caused by the secretion of hormone secreted by corpus luteum.

p-site The ribosomal site most frequently occupied by peptidyl-*t*RNA, i.e. the *t*RNA carrying the growing peptide chain. It is also referred to as the puromycin sensitive site. Puromycin is an antibiotic that shows similarities with a part of aminoacyl–*t*RNA.

psychoactive drugs Any drug that affects the mind or behaviour. There are five main classes of psychoactive drugs.

(a) opiates and opioids (e.g. heroin and methadone)
(b) stimulants (e.g. cocaine, nicotine)
(c) depressants (e.g. tranquilisers, antipsychotics, alcohol)
(d) hallucinogens (e.g. LSD)
(e) marijuana and hashish.

psychology The study of human behaviour.

psychrophilic Microorganisms which thrive at temperatures below 20°C.

pteridophytes Vascular plants with leaves, stems and roots, but lacking both seeds and flowers. Include ferns, horsetails and clubmosses.

Pteropsida In older classifications, it was a subdivision of tracheophytes that contained the ferns and seed plants, or a class of the Pteridophyta containing only the ferns.

Pterosauria The extinct order of flying reptiles – the pterodactyls – that lived in the late Triassic, Jurassic and Cretaceous periods (220-70 million years ago). They had beaked jaws and an elongated fourth finger that supported a membranous wing. They had long jointed tails, no feathers and could probably only fly by soaring.

Pterygota A subclass of insects that typically possess wings, although some are secondarily wing less. Comprises the exopterygotes and endopterygotes.

ptyalin Carbohydrate splitting enzyme of saliva; salivary amylase of man. Responsible for initial starch digestion.

ptyxis The way in which young leaves are rolled in a bud.

puberty *See* adolescence.

pubis One of a pair of bones forming the anterior ventral portion of the tetrapod pelvic girdle. They are sometimes joined to form the pubic symphysis as in most mammals, ostrich and some reptiles.

pudenda External genital organs of a human being, especially of a female.

puff (Balbianiring) Swelling that is seen in certain areas of giant chromosomes.

pulmonary artery A paired artery that carries deoxygenated blood from the right ventricle of the heart to the lungs, in mammals.

pulmonary circulation Part of the circulatory system of birds and mammals that transport deoxygenated blood from the right side of the heart to the lungs and returns oxygenated blood to the left side of the heart.

pulmonary edema An accumulation of fluid in the lung interstitium and air sacs.

pulmonary valve *See* semilunar valve.

pulmonary vein The paired vein that carries oxygenated blood from lung to left atrium of heart.

pulp cavity The central region of a tooth, which is connected by a narrow channel at the tip of the root with the surrounding tissue. It contains pulp connective tissue having blood vessels and nerve fibres.

pulse A rhythmical dilation of the artery caused by the systolic output of the heart.

pulvinus A specialised group of cells with large intercellular spaces that are located at the bases of leaves or leaflets in certain plants. Involved in non-growth nastic movements.

Punnet square Developed by a British geneticist, Reginald C Punnett. A graphical representation to calculate the probability of all possible genotypes of offsprings in a genetic cross.

pupa The third stage, usually dormant, between the larva and the adult of insects, having complete metamorphosis. Pupa of a butterfly is called chrysalis.

puparium The case formed by the hardened last larval skin in which the pupa is formed.

pupil The central aperture of the iris through which light passes to reach retina, in vertebrates and cephalopods.

pupillary reflex The alteration in the size of the pupil in response to a change in the light intensity.

pure line The succession of descendents of homozygous individuals that are identical to each other and continue to breed true.

purine One of the two types of nitrogenous bases occurring in nucleic acids. Purines are composed of fused five and six-membered heterocyclic rings of carbon and nitrogen atoms. Two different purines, adenine and guanine, occur in both DNA and RNA.

Purkinje cells (Purkyne cell) Pear-shaped nerve cells with extensive dendrites in the cerebellar cortex of brain and forming its only axial output.

Purkyne fibres (Purkinje fibres) Modified fibres in the mammalian heart that originate in the bundle of his and spread out in a network over the ventricles. Action potentials generated in the sinoatrial node are conducted extremely rapidly through the ventricles, due to the extensive branching of the Purkyne fibres, causing both ventricles to contract almost simultaneously.

pustule The elevated lesion of skin, containing pus.

putrefaction The condition of decay of organic matter by microbes, usually accompanied by an offensive odour.

pyknotic A nucleus of damaged cell that has decreased in volume and become darker due to some degree of condensation of chromatin.

pyloric valve The sphincter muscle of the pylorus. It is normally closed but opens with building gastric pressure.

pylorus (pyloric sphincter) Exit orifice connecting the stomach with the duodenum.

pyramid of biomass Represents the relationship between producers and consumers in terms of biomass. It can be

- Upright, e.g. in grasslands
- Inverted, e.g. in pond ecosystem

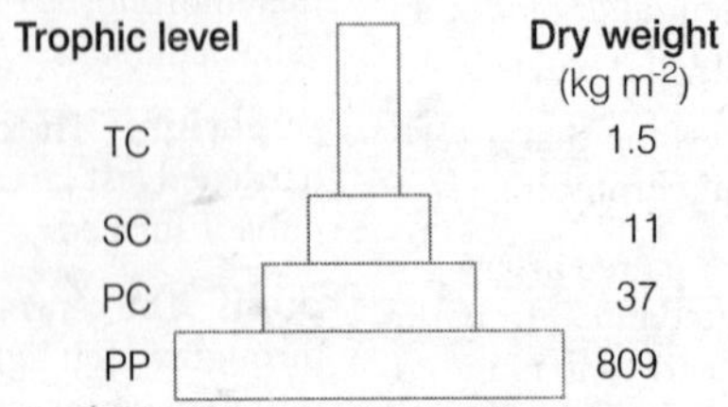

Pyramid of biomass shows a sharp decrease in biomass at higher trophic levels

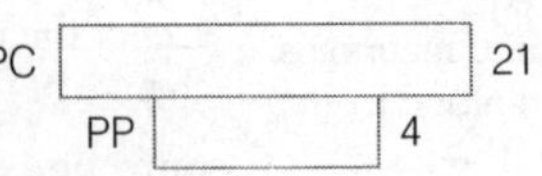

Inverted pyramid of biomass. Small standing crop of phytoplankton supports large standing crop of zooplankton

pyramid of energy The relationship between the producers and consumers in an ecosystem in terms of flow of energy. It is always upright because the energy is lost as heat at each step.

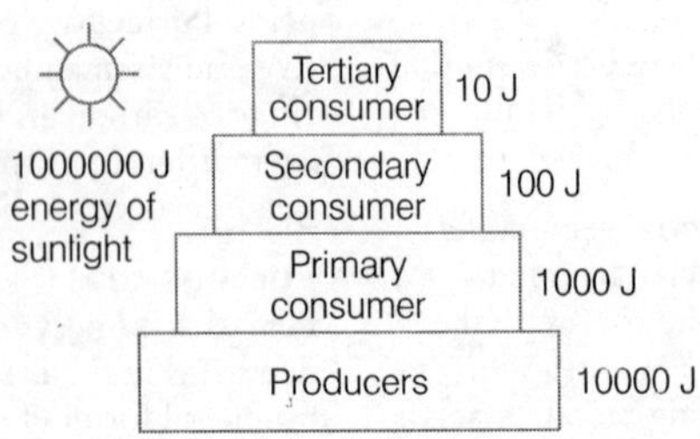

An ideal pyramid of energy. Observe that primary producers convert only 1% of the energy in the sunlight available to them into NPP

pyramid of number Represents the relationship in terms of numerbs, between the producers and consumers in an ecosystem. It can be

- **Upright** Grassland ecosystem
- **Inverted** Tree ecosystem

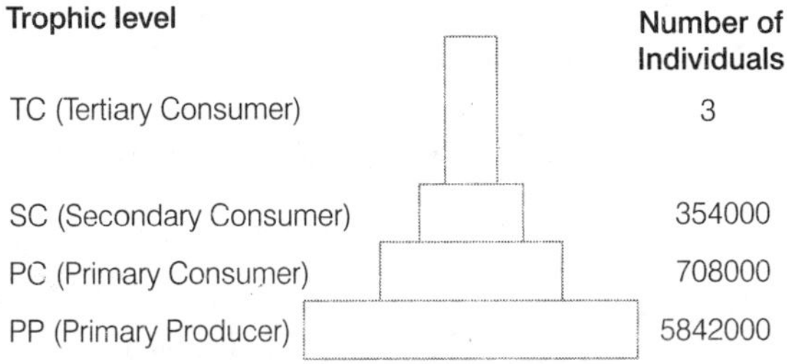

Pyramid of numbers in a grassland ecosystem. Only three top-carnivores are supported in an ecosystem based on production of nearly 6 millions plants

pyramidal cell A type of neuron found in the cerebral cortex of the brain.

pyrenocarp *See* drupe.

pyrenoid Small, spherical, protein bodies found singly or in numbers embedded in chloroplasts of many algae and bryophytes but absent in majority of brown and red algae. In green algae, associated with starch synthesis and surrounded by starch deposits.

pyridoxine (vitamin-B6) The vitamin of B-group, forming a coenzyme (pyridoxal phosphate). Sources include cereal grains, yeast, liver milk etc. Defeciency can cause anaemia, dermatitis, fatigue etc.

pyrimidine One of the two types of nitrogenous bases occurring in nucleic acids. Pyrimidines are composed of a six-membered heterocyclic ring of carbon and nitrogen atoms. The pyrimidines occurring in DNA are cytosine and thymine. In RNA, they are cytosine and uracil.

pyrrole The organic compound having nitrogen that are found in porphyrins.

Pyruvate Kinase (PK) An enzyme catalysing transfer of a phosphate group from PEP to ADP during glycolysis. The result is one molecule of ATP and one molecule of pyruvate.

pyruvic acid (2-oxopropanoic acid) A colourless liquid organic acid, $CH_3COCOOH$. It is an important intermediate compound in metabolism, being produced during glycolysis and converted to acetyl coenzyme A, required for the Krebs cycle. Under anaerobic conditions it is converted to lactate or ethanol.

pyxidium A type of capsule which dehisces by means of a transverse circular split, so that the upper part of the pericarp forms a lid.

Q

QRS complex Component of the electrocar diogram corresponding to the depolarisation of ventricles of human heart.

quadrat The sampling unit chosen at random for the study of composition of vegetation in a selected area, may be a square or a circle.

quadrate Paired bone in upper jaw of bony fishes, amphibia, reptiles and birds which articulates with the lower jawbone. It is absent in mammals (reduced to incus).

quadriceps A group of extensor muscles that cover the anterior surface and sides of the mammalian femur.

quadriplegia Condition of being paralysed in all four limbs (both arms and legs).

quadrupedal Walking on four legs, such as lion.

quadruplet Any of four children produced in a single pregnancy by female.

qualitative variation *See* discontinuous variation.

quantasome Spherical structures present in thylakoid membranes of chloroplasts, about 17.5 nm to 11.0 nm in diameter. These particles are believed to be essential for the light reactions of photosynthesis.

quantitative inheritance (polygenic inheritance; multifactorial inheritance) Pattern of inheritance shown by traits that show continuous variation within a certain range of values (e.g. height in humans). Typically controlled by many genes within a genome.

Quantitative Structure-Activity Relationship (QSAR) A statistical algorithm that quantitatively defines the relationship between the chemical structure of a drug and its effect on an organism. QSAR studies are used to predict the activity or toxicity of new drugs.

Quantitative Structure-Metabolism Relationship (QSMR) A method similar to QSAR, used to predict the metabolism of new drugs.

quantitative trait Any phenotypic trait that shows continuous variation and can be measured quantitatively in terms of length, weight, concentration, test score, etc. Such traits, include height, intelligence and obesity are determined by the cumulative effects of numerous genes at quantitative trait loci.

Quantitative Trait Locus (QTL) Region of a chromosome (genetic locus) containing one or more genes that contribute to determining a quantitative trait. Typically, several or many such genes (polygenes) act together to specify the trait.

quantum hypothesis (in neurophysiology) The hypothesis that nerve impulses are transmitted across, synapses in the nervous system by the release of discrete packets or quanta, of neurotransmitter.

quarantine Process of keeping an animal separate from other animals to prevent the possibility of transmitting disease.

quaternary period Most recent of the three periods of the Cenozoic era, lasting from 2.588 million years ago to the present. It includes two geologic epochs–the Pleistocene and the Holocene. Literally, the fourth age, it is characterised by emergence of humans.

quaternary structure Refers to the ways in which the separate polypeptide chains of a complex protein interact.

queen substance In Hymenoptera, pheromones released by the queen honeybee that attracts and controls the reproductive activities of workers and other queens. Trans-9-keto-2-decanoic acid is the most potent component of the pheromone mixture.

quiescent zone The terminal part in apical meristem of the root apex, where mitotic divisions are absent. They may become meristematic, if other part of meristem is injured.

quills Large feathers on the wings and tail in birds; with a long shaft and broad vane made of barbs and barbules.

quinine An alkaloid obtained from the cinchona tree's bark. Used in the treatment of malaria.

quintipara A woman who has given birth to a viable infant in each of five pregnancies five babies.

quorum sensing A density-dependent mechanism for communication and coordination among populations of bacteria in which the individual bacteria secrete and detect substances called **autoinducers**. These accumulate in the environment depending on the density of the population. Upon reaching a certain threshold concentration, the autoinducer triggers a specific response. Such mechanisms can regulate a wide range of bacterial activities, including bioluminescene, sporulation, virulence and biofilm formation. Some autoinducers are used for intraspecific communication, whereas other affect the behaviour of other species.

quotidian Fever which recurs every day (Quotidian malaria caused by *Plasmodium falciparum*).

R

rabies (common name hydrophobia) An infectious disease primarily affecting mammalian carnivores. Rabies virus (*Lyssavirus*), which is present in the saliva of infected individuals, is communicated to new hosts through bites. The virus travels from the wound to the central nervous system, where it causes paralysis and eventual death, usually by respiratory failure. Until recently, once symptoms set in, this disease has no cure. However, a new treatment, known as the Milwaukee protocol, allows some patients to survive.

race
1. A category that consists of group of individual within a species that are ecologically, geographically, physiologically or cytologically distinct from other members of the species (in classification).
2. A distinct human type with genetically inherited characteristics. Major races are Mongolian, Caucasion and Ethiopian (in anthropology).

raceme Inflorescence with the flowers are borne on short stalks (pedicles) of about equal length arising from an axial rachis.

racemose inflorescence (indefinite inflorescence) Inflorescence in which the growing point of the axis continues to develop and produce lateral branches. Its types include the capitulum, catkin, corymb, raceme, spadix, spike and umbel.

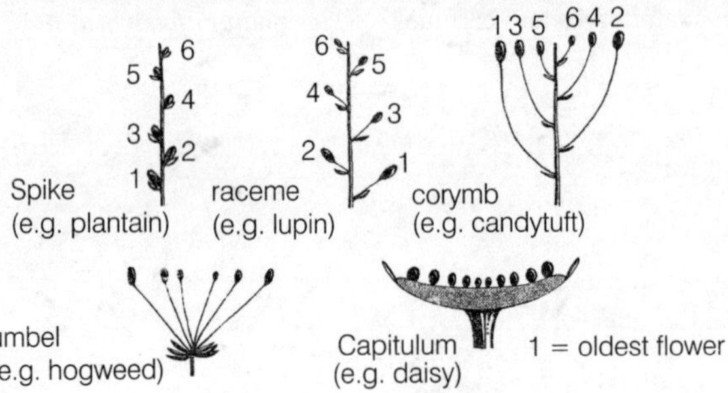

Types of racemose inflorescene

rachis (rhachis)
1. Main axis of a compound leaf or an inflorescence.
2. The shaft of a feather.
3. The backbone.

radial symmetry Similar parts arranged around a common central axis, as in echinoderms and coelenterates. Cutting through the centre of the structure in any direction produces two halves that are mirror images of each other. The stems and roots of plants also show radial symmetry.

radiata A taxonomic rank that includes the coelenterates and ctenophores because of their primary and secondary radial symmetry.

radicle Part of a plant embryo that develops into root system.

radioactive age Age of an archaeological or geological specimen as determined by a process that depends on a radioactive decay.

radioactive waste (nuclear waste) A solid, liquid, or gaseous waste material that contains radioactive atomic nuclei. Such wastes are produced in the mining and processing of radioactive ores, the normal running nuclear power stations and other reactors, the manufacture of nuclear weapons and in hospital and research laboratories.

radiobiology Study of the effects of radiation on living organisms and use of radioactive tracers to study metabolic processes.

radiocarbon dating *See* carbon dating

radiogenic element An element produced by the radioactive decay of an element.

radiograph An image made on photographic film by ionising radiation, e.g. an X-ray image of some portion of a patient's body.

radioimmunoassay (RIA) A very sensitive, extremely specific technique for measuring antigen concentration (e.g. hormone, such as insulin). To carry out the process, an antigen is made radioactive, often with radioactive iodine attached to tyrosine. A measured quantity of this radiolabelled antigen and a known amount of antibody, to that antigen are then mixed and the two bind together. Subsequently, in a separate assay, a serum sample containing an unknown concentration of the same antigen (not radiolabelled) is added. The unlabelled antigen then competes with the radiolabelled antigen for antibody-binding sites. The proportion of antibody bound to unlabelled antigen can then be determined.

radiology The study and use of X-rays, radioactive materials and other ionising radiations for medical purposes, especially for diagnosis (diagnostic radiology) and the treatment of cancer and allied diseases (therapeutic radiology or radiotherapy).

radiopaque (radio-opaque) Medium that is opaque to X-rays and gamma rays, e.g. barium salts, used in diagnostic radiology of the digestive tract.

radius Smaller bone (in lower section of forelimb) that articulates with some carpel bones and the ulna at the wrist and with humerus at elbow, in tetrapod vertebrate.

radula (pl. radulae) A rough tongue-like organ present in many molluscs. Used for rasping food, it has minute chitinous teeth.

rainforests Major terrestrial biome in which trees form the dominant plants and annual rainfall is high (over 200 cm). Restricted to equatorial regions, such as the Amazon basin, Central West Africa and Southeast Asia.

Ramapithecus *See* Sivapithecus

ramet Individual member of a clonal line of individuals. It is capable of separation from other members and independent life.

Ramsar Convention on Wetlands Treaty of various governments that provide the framework for conservation and wise use of wetlands and their resources.

ramus communicans (pl. rami communicantes) A structure containing nerve fibres that runs from a spinal nerve to a ganglion of the sympathetic nervous system.

random genetic drift Changes from one generation to the next, in the population frequency of an allele or trait. This is due to random deviation from the population frequencies expected on the basis of pre-existing frequencies. Such deviations can be quite large in small populations.

random mating (also panmixia and panmixis) Mating in which all potential reproductive pairings between individuals occur equally.

Randomly Amplified Plymorphic DNA (RAPD) DNA primer, made up ot a random sequence of bases (used to amplify segments of an organism's DNA in PCR). The primer effectively 'selects' all regions of DNA that happen to lie within inverted copies of the primer sequence. These regions which will vary in length, are thus amplified by PCR and produce a series of bands when separated by electrophoresis. Because of variation in the sizes of the primer sequences among individuals, the pattern of bands represents a genetic 'fingerprint' unique to each individual. This makes RAPD a useful technique for taxonomic studies, assessing kinship or forensic investigation.

Ranidae It is an anuran family comprised of the true frogs (smooth, moist-skinned frogs with webbed feet and strong hind legs).

rank (category) It is the position of a taxon in classification heirarchy, e.g. class.

raphe

1. A ridge on seed marking the line of fusion between an anatropous (inverted) ovule and the funicle.
2. A groove on the valve of certain diatoms.
3. Any other groove, ridge or suture line marking the line of fusion between two originally separate structures.

raphides Bunches of needle-like crystals of calcium oxalate found in certain plant cell vacuoles.

Raphidophyte A small group of eukaryotic algae that includes both marine and freshwater species. All raphidophytes are unicellular, with large cells but no cell walls.

RAS protein A small G protein that functions as a molecular switch in signal transduction pathways regulating cell growth and other activities. In its active state, it binds the nucleotide guanosine triphosphate (GTP). It its resting (or off state), it is found complexed with GDP. Ras is turned on by removing the GDP and replacing it with a GTP. This nucleotide exchange reaction is controlled and catalysed by proteins called exchange factors.

rate limiting step Slowest step in a metabolic pathway or series of chemical reactions which determine the overall rate of the other reactions of the pathway.

Ratitae (Palaeognathae) A group of the flightless birds, including the ostrich, kiwi and emu. They have long legs, heavy bones, small wings, a flat breastbone and curly feathers.

ray floret (ray flower) One of the marginal florets surrounding tubular florets in the flower heads of certain Compositae plants, e.g. Daisy.

reaction time (latent period) Period of time between the detection of a stimulus at a sensory receptor and the performance of the appropriate response by the effector organ.

reading frame Sequence of bases in messenger RNA that encodes for a polypeptide. Since, each codon of the genetic code consists of three consecutive bases, the reading frame is established according to precisely where translation starts. *See* open reading frame.

readthrough The continuation of transcription of DNA beyond a normal stop signal or terminator sequence, due to failure of RNA polymerase to recognise the signal. It can also occur in translation, when a mutation converts a normal stop codon into one encoding an amino acid.

reannealing The spontaneous realignment of two single DNA strands to reform a DNA double helix.

recapitulation This theory states that some stages of evolution are repeated in the development of an individual organism, i.e. phylogeny is repeated in ontogeny. It is proposed by Ernst Haeckel.

receptacle

1. (Thalamus or torus). Tip of a flower stalk, which bears the floral parts.
2. A swollen part of the thallus of some algae that bears the conceptacles in which the sex organs are situated, e.g. *Fucus*.

receptor

1. A cell or group of cells specialised to detect a particular stimulus and to initiate the transmission of impulses *via* the sensory nerves, e.g. eyes, ears, nose, skin.
2. An area of a plasma membrane consisting of a specially adapted membrane protein, that can bind with a specific hormone, etc., to initiale a change within the cell. Two types with different mechanisms: ionotropic receptors and metabotropic receptors.

recessive allele In a diploid organism, an allele manifested only when two copies of it are present at a locus.

recessive trait A trait which is not expressed in the presence of dominant allele in heterozgous condition. It expresses itself only in homozygous state.

reciprocal cross A cross reversing the roles of males and females to confirm the results obtained from an earlier cross, e.g. if the pollen from tall plants is transferred to the stigmas of dwarf plants in one cross, the reciprocal cross would use the pollen of dwarf plants to pollinate the stigmas of tall plants.

reciprocal translocation A condition in which two non-homologous chromosomes exchange blocks of chromatin (the process produces two new composite chromosomes each of which contain blocks from both of the pre-existing chromosomes).

recombinant DNA The DNA which is experimentally formed by joining portions of two DNA molecules (fragmented by restriction enzymes) by genetic engineering technique.

recombinant DNA technology *See* genetic engineering

recombination The rearrangement of genes during gamete formation. It results form the independent assortment of parental sets of chromosomes and crossing over that occur during meiosis. It results in offspring with combination of characteristics different from that of their parents.

recon The smallest genetic unit that can undergo mutation.

rectal gland Salt secreting gland in the cloaca of cartilaginous fishes.

rectum A short (~12 cm), portion of the large intestine. It connects the sigmoid colon with the anal canal.

red algae Members of class–Rhodophyceae. Characterised by the presence of red pigment (phycocyanin and *r*-phycoerythrin) in their body.

red blood cells *See* erythrocytes.

red tide Sudden toxic proliferation of marine phytoplankton, notably dinomastigotes. It colours the sea red, brown, or yellowish due to the high concentration of the organisms' photosynthetic accessory pigments.

reducing sugar A monosaccharide or disaccharide sugar that possess free aldehyde or ketone group and reduces Cu^{2+} to form a red precipitate with Benedict's solution.

reduction division *See* meiosis

reflex An automatic and innate response to a particular stimulus that is extremely rapid. This is because it is mediated by a simple nervous circuit called a reflex arc.

reflex action An immediate autonomic, e.g. eye blinking in response to dust particle.

reflex arc The nervous pathway taken by nerve impulses in a reflex action.

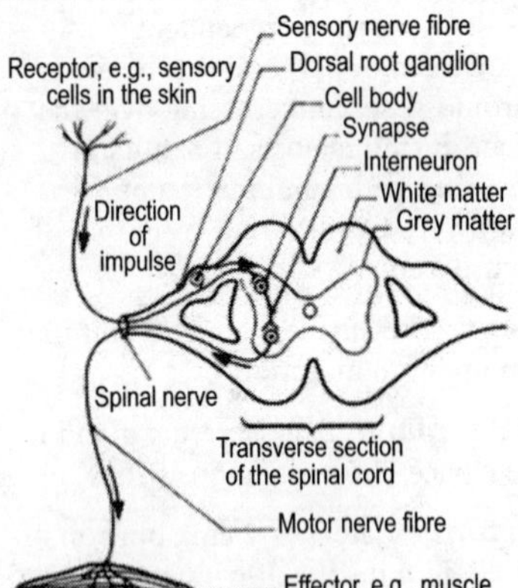

A reflex arc

reforestation Replanting of trees in a deforested area.

refractory period Brief interval following the response of a neuron or muscle fibre during which it is incapable of a second response. It lasts approximately 3 milliseconds and is divided into an absolute refractory period, during which a second impulse may not be generated and a relative refractory period, during which it is possible to generate an impulse only if there is an abnormally strong stimulus.

regeneration Regaining of tissue or organs to replace those lost or damaged by injury. In some Platyhelminthes, a minute fragment can regenerate the whole animal. In mammals, it is limited to wound-healing, regrowth of peripheral nerve fibres, liver and compensatory hypertrophy. Very common in plants, occurring in higher plants by growth of dormant buds, formation of secondary meristems and production of adventitious buds and roots.

regma A dry fruit characteristics of the the Geranium family. It is similar to the carcerulus but breaks up into one-seeded parts, each of which splits open to release a seed.

regulation The processes involved in animal embryonic development that counteract any abnormalities which may arise in the different development stages.

regulator Any organism that can maintain a constant internal environment largely independently of the external environment. This is generally achieved by hemostatic mechanisms. Regulators tends to occupy habitats in which environmental conditions are variable.

regulator gene *See Lac* operon

regulatory enzyme An enzyme that is involved in controlling the different metabolic pathways in the cell by switching them on or off. They include allosteric enzymes and those enzymes controlled by kinases.

regulatory genes Genes that control development by regulating the expression of structural genes responsible for the formation of body components. They encode transcription factors which interact with regulatory sites of other genes causing activation or repression of developmental pathways. e.g. homeotic genes, such as the *Hox* genes of mammals.

regulatory protein (gene-regulatory protein) Any protein that influences the regions of a DNA molecule that are transcribed by RNA polymerase during the process of transcription. These proteins which include transcription factors, help in control the synthesis of proteins in cells.

regurgitation

1. Return of stomach contents into the mouth.
2. Passage of blood back through a heart valve.

reinforcement The action or process of reinforcing or strengthening.

r

Reissner's membrane The membrane in cochlea of inner ear that separates the vestibular canal from cochlear duct.

relative growth rate Defined as the increases in dry mass per unit of plant mass over a specified period of time it is a measurement of the productivity of a plant.

relative refractory period
See refractory period

relaxin A hormone produced by the corpus luteum and placenta. It responsible for the inhibition of uterine contraction. Relaxin stimulates dilation of the cervix and relaxes the pubic symphysis enabling the pelvic girdle to widen.

Release Factor (RF) Any protein that terminates translation of messenger RNA during protein synthesis and releases the completed polypeptide chain from the ribosome.

Release-Inhibiting Hormone (RIH) Any hormone that inhibits the secretion of another hormone. The hypothalamus produces several hormones that inhibit the release of hormones by the anterior lobe of the pituitary (adenohypophysis), e.g. MSH-inhibiting horome prolactin -inhibiting hormone and somatostatin.

releaser *See* sign stimulus

releasing hormone (releasing factor) Any hormone produced by the hypothalamus that stimulates the release of hormone from the anterior pituitary gland. Each hormone has a specific releasing hormone, e.g. a group of organisms that survives as a remnant of a formerly much larger group, in terms of either taxonomic diversity (evolutionary relict) or geographical distribution (geographical relict).

renal Of or pertaining to the kidneys.

renal calculi The stone or insoluble mass of crystallised salts (oxalates) formed within the kidney.

renal capsule *See* Bowman's capsule

renaturation Reconstruction of a protein or nucleic acid that has been denatured such that the molecule resumes its original function. Some proteins can be renatured by reversing the conditions (of temperature, pH, etc.) that caused its denaturation.

renin Hormone produced by the kidney's JGA cells, which causes an increase in blood pressure. It plays a role in the production of angiotensin.

renin-angiotensin mechanism Renin converts angiotensinogen in blood to angiotensin-I and further to angiotensin-II. Angiotensin-II increases glomerular blood pressure by constricting blood vessels and thereby glomerular filtration rate (GFR). It also activates adrenal cortex to release aldosterone. Aldosterone causes reabsorption of Na^+ and water from the distal parts of tubule. This causes an increase in blood pressure and GFR. This is renin-angiotensin mechanism.

rennet The extract of the fourth stomach of the calf, containing rennin.

rennin (chymosin) A milk-clotting enzyme secreted in the stomach of human infants and calves, converting soluble milk protein (caseinogen) into casein which forms insoluble calcium-casein compound.

Renshaw cell An inhibitory interneuron found in feedback circuits in the central nervous system that control the level of excitation of motor neurons. These cells receive input from side branches (collaterals) of motor axons and in turn form inhibitory synapses with the motor-neuron cell bodies.

repetitive DNA The DNA sequence whose base sequence is repeated many times.

replacing bone The bone that is formed by replacing the cartilage of an embryo skeleton.

replication *See* DNA replication

replicon Unit of DNA having replication ability.

repolarisation Restoration of the resting potential in neurons or muscle fibres following the passage of a nerve impulse. It is brought about by diffusion of potassium ions from the neuron and by active elimination of sodium ions.

reporter gene Gene whose phenotypic expression is easy to monitor and is used to study another gene or DNA sequence of interest, e.g. β - galactosidase gene.

repressor Substance produced by regulator gene which binds to the operator gene. It prevents the expression of a gene.

reproduction Process of producing young ones, more or less similar to themselves by living organisms. Asexual and sexual reproduction are two modes.

reproductive cycle (also life cycle) A series of forms that a particular type of organism takes on during the course of reproduction, e.g. the human reproductive cycle involves two forms, a haploid gamete alternating with a diploid organism. Each form produces the other.

reproductive system Organs that are involved in the process of sexual reproduction in an organism. In a flowering plant it is found in the flower and consists of the stamens (male organs) and carpels (female organs). In mammals, the reproductive system consists of the testes, epididymis, spermduct and penis (in the male) and the ovaries, Fallopian tubes and uterus (in the female).

reptile An animal such as a snake, lizard or alligator belonging to the vertebrate class–Reptilia. Reptiles differ from amphibians in having an amnion, a feature held in common with mammals and birds. Due to this trait, oviparous reptiles can lay their eggs on dry land. They are cold-blooded with scaly skin.

repulsion Two linked heterozygous allelic pair in arrangement Ab/aB, i.e. allelic arrangement of two linked heterozygous loci.

reservoir host The host which harbours the parasite and serves as an important source of infection to other hosts.

Residual Volume (RV) The Volume of air remaining in the lungs after maximum expiration. An average human has RV of about 1 L.

resin A naturally occurring acidic polymer secreted by many trees (especially conifers) found either as brittle glassy substances or dissolved in essential oils. Their functions are probably protective.

resistance response A long-term response to any stimulus that potentially threatens the wellbeing of an organism. e.g. release of ACTH (adrenocorticotrophic hormone) in response to stress.

resolving power A measure of the ability of an optical instrument to form separable images of close objects or to separate close wavelengths of radiation. The resolving power of a microscope is usually taken as the minimum distance between two points that can be separated; the smaller the resolving power, the better the resolution.

r

resource In ecology, it is a substance or object in the environment required by an organism for normal growth, maintenance and reproduction.

respiration

1. An act of breathing (external respiration).
2. An oxidation of energy-rich storage molecules, primarily glucose to produce ATP (internal respiration).

respiratory chain *See* electron transport system.

respiratory organ An organ (in animals) across which exchange of carbon dioxide and oxygen takes place. The surface membranes of such organs are always moist, thin and well supplied with blood, e.g. are lungs (air-breathing vertebrates), gills (fish), tracheae (insects) and lung books (arachnids).

respiratory pigment

1. A coloured compound that is capable of reversibly binding with oxygen at high oxygen concentrations and releasing it at low oxygen concentrations, e.g. haemoglobin, myoglobin.
2. Any of the proteins involved in cellular respiration as components of an electron transport chain. They include the cytochromes.

Respiratory Quotient (RQ) The ratio of the volume of CO_2 evolved to the volume of O_2 consumed in respiration.

$$RQ = \frac{\text{Volume of } CO_2 \text{ evolved}}{\text{Volume of } O_2 \text{ consumed}}$$

For carbohydrates, RQ is 1, in case of fats it is less than 1 and in proteins, it is about 0.9.

respiratory substrates Compounds that are oxidised during respiration. Usually, carbohydrates, proteins, fats and even organic acids.

respirometer A device that measures an organism's oxygen uptake. Human oxygen consumption is generally measured by a device known as a spirometer, which can also be used to measure depth and frequency of breathing.

response The physiological or behavioral activity resulting from a change in either the internal or external environment of organisms.

resting potential The difference in electric potential between the outside of a cell in the absence of excitatory or inhibitory potential.

restriction endonucleases (restriction enzymes) Enzymes that make cuts between the ends of a DNA or RNA polymer chain (exonucleases remove nucleotides from the ends of such chains). They cut the internal phosphodiester bonds of the molecule. Each type of restriction enzyme recognises a specific sequence of nucleotides, 4 to 8 base pairs in length (the restriction site) and will cut only where that sequence occurs. Restriction endonucleases are one of the basic tools of recombinant DNA technology.

Restriction Fragment Length Polymorphism (RFLP) The variation between individuals on the basis of difference in the size of the DNA fragments, cut by specific restriction enzymes. Polymorphic nucleotide sequences that result in RFLPs are used as markers on both physical maps and genetic linkage maps. RFLPs usually are caused by point mutation at a cutting site.

restriction mapping A technique for determining the sites at which a length of DNA is cleaved by restriction enzymes.

restriction point A point reached in the G_1 phase of the cell cycle, after which the cell becomes committed to passing through the remainder for the cell cycle, regardless of the external conditions.

restriction site A specific nucleotide sequence where a particular restriction enzyme cuts DNA or RNA. The restriction site of a 'rare cutter enzyme' is itself rare in comparison with other nucleotide sequences of the same length.

rete cord One of the deep, anastamosing strands of cells of medullary cords of vertebrates embryos, which form the rete testis or the rete ovarii.

reticular fibers A type of fiber in connective tissue composed of type III collagen and which forms an intricate interstital network with other similar fibers to serve as a supporting mesh in soft tissues such as liver and bone marrow.

reticulate thickening Secondary wall deposition in the form of an irregular network in tracheids or vessel elements of metaxylem.

reticulocyte An immature red blood cell. It develops in the red marrow in bones from a proerythroblast that gradually accumulates haemoglobin until it is a fully formed reticulocyte.

reticulum The second stomach chamber of ruminants.

retina An interior, light sensitive membrane of the eye. *See* eye.

retinal Derivative of vitamin-A. It forms chromophore component of rhodopsin found in rods of vertebrate retina.

retinoblastoma Cancer of the human retina.

retinol *See* vitamin-A

retrotransposon A type of transposon found in the DNA of various organisms, including yeast, *Drosophila* and mammals that forms copies of itself using a mechanism similar to that of retroviruses.

retrovirus RNA virus that uses reverse transcriptase to create a DNA copy of its RNA. After reverse transcription, the DNA copy is inserted into the host genome and used as a template for the production of additional copies of the virus.

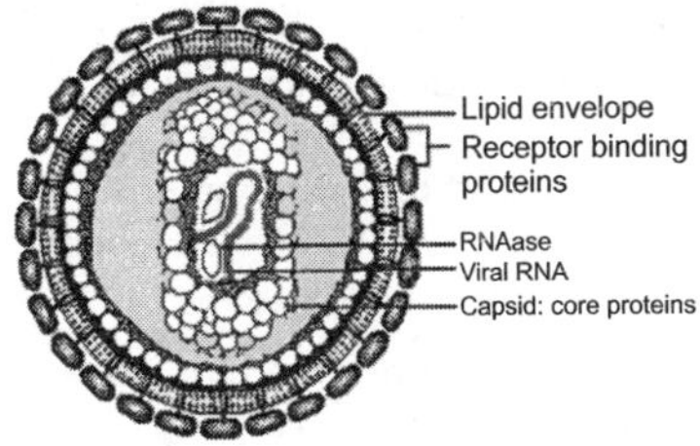

Structure of a retrovirus

Reverse Transcriptase (RT) An enzyme occurring in retroviruses, that create DNA copies of their RNA genomes, an essential step in their cycle of replication. It produces a single strand of DNA using an RNA template. In a natural setting, the template is the genome of a retrovirus.

However, reverse transcriptase is also used in the laboratory to make *c*DNA (complementary DNA sequences) from RNA sequences. The polymerases that normally make DNA and RNA in cells are very accurate and make very few mistakes. Reverse transcriptase on the other hand, makes many.

The RT of the AIDS virus (HIV) makes about one mistake in every 2,000 bases, it copies. This high error rate allows HIV to rapidly mutate and produce drug-resistant strains. Fortunately, the treatments combining several drugs are often effective in combating this problem since the virus cannot mutate to evade all of them at the same time.

reversion The restoration of a mutation to the wild-type genotype or phenotype by a second mutation. The gene or organisms affected in this way is called a revertant.

rf value In chromatography it is the, distance travelled by the solute divided by the distance travelled by the solvent front. Characteristic of a particular molecule.

Rh factor (Rhesus factor) Any of a type of specific antigen whose presence or absence on the surface of red blood cells forms the basis of the rhesus blood group system. Most people possess the Rh factor (Rh+). People who lack the factor are Rh–. If Rh+ blood is given to an Rh– patient, the latter develops anti-Rh antibodies, subsequent transfusion of sRh+ blood results in agglutination with serious consequences. The factor was first recognised in the rhesus monkeys.

rhabdom The transparent crystalline structure found in compound eyes of arthropods.

rhizobium A nitrogen–fixing bacterium that is common in the soil, especially in the root nodules of leguminous plants.

rhizoid One of a group of delicate and often colourless hair-like outgrowths found in certain algae and the gametophyte generation of bryophytes and ferns. They anchor the plant to the substrate and absorb water and mineral salts.

rhizome An underground modified horizontal stem, produced by many plants, e.g. ginger. It enables the plant to survive from one growing season to the next and in some species, it also serves to propagate the plant vegetatively.

rhizomorph A root-like aggregation of hyphae in certain fungi.

Rhizopoda (Sarcodina) A class of protozoans that have pseudopodia for food capture and locomotion, e.g. *Amoeba.*

rhizosphere An area of soil surrounding plant roots. It usually extends a few millimeters from the surface of the root. Rhizosphere may contain 10-15 times more bacteria than the surrounding soil.

Rhodophyta *See* red algae

rhodoplast Plastid giving red algae its characteristic red colour.

rhodopsin (visual purple) Pigment in the rods of retina, sensitive to dim light. Breaks in the light and reforms in the darkness. It consists of opsin (protein component) linked to a retinal (non-protien chromophore).

rhombencephalon *See* hindbrain

rib One of a series of slender curved bones that form a cage to enclose, support and protect the heart and lungs. Ribs occur in pairs, articulating with the thoracic vertebrae of the spinal coloumn at the back and (in reptiles, birds and mammals) with the sternum (breastbone) in front. Movement of the rib cage, controlled by intercostal muscles between the ribs are important in breathing.

riboflavin (Vitamin-B_2) Water-soluble vitamin of the B-complex group. It is constituent of coenzyome FAD and FMN. Deficiency causes inflammation of tongue and lips, mouth sores and conjunctivitis.

ribonuclease (RNAase) An enzyme which catalyses the cleavage of nucleotide in RNA. Each RNAase has specificity for a different cleavage site.

ribonucleic acid (RNA) A long single-stranded polynucleotide containing ribose sugar and the bases adenine, cytosine, guanine and uracil. It is concerned with protein synthesis and is the hereditary material in some viruses. *See* *m*RNA, *r*RNA and *t*RNA cell.

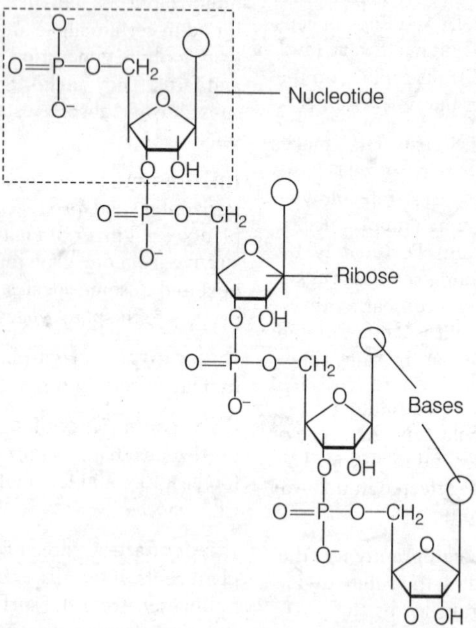

Molecular structure of sugar phosphate backbone of RNA

Ribonucleoprotein (RNP) The complex of protein and RNA that forms during the transcription in eukaryotes. It is involved in packaging and condensation of RNA. Most common is *hn*RNP (heterogenous nuclear RNP).

ribose The five-carbon monosaccharide component of RNA; occurs primarily as D-ribose (in DNA).

ribose-5-phosphate An intermediate of the pentose- phosphate pathway.

ribosomal RNA (rRNA) The most abundant type of RNA. With ribosomal proteins, *r*RNAs forms the structure of ribosomes. It is essential for protein synthesis in all living organisms.

ribosomes Cellular organelles, present in both prokaryotes and eukaryotes. They use *m*RNA templates to synthesise proteins in a process known as translation.

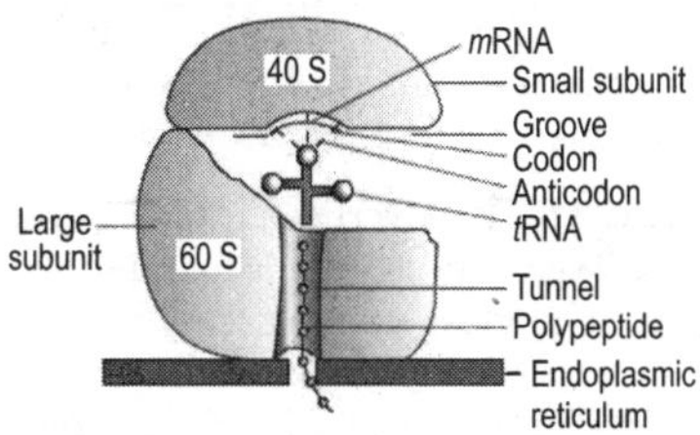

Ultrastructure of ribosome (eukaryote)

It consists of two subunits, one large and one small, each of which comprises a ribosomal RNA and protein. These are described in terms of their sedimentaion coefficients (i.e. rate of sedimentation in an ultracentrifuge), measured in Svedberg unit (S). The prokaryote (70 S) ribosome comprises a 50 S (large) subunit and a 30 S (small) subunit; the eukaryote (80 S) ribosome has large 60 S and small 40 S subunits. Usually, there are many ribosomes in a cell, either attached to the endoplasmic reticulum or free in the cytoplasm.

riboswitch Region of RNA molecule, usually *m*RNA that can bind a specific metabolite and 'switch off' further transcription or translation of the *m*RNA.

ribozyme The RNA molecule that catalyse chemical reactions, e.g. self splicing introns.

ribulose A ketopentose sugar that is involved in carbon dioxide fixation in photsynthesis as a component of ribulose bisphosphate.

rickets Disease in which new bone matrix is not adequately calcified due to the deficiency of vitamin-D or calcium and disorders that cause poor phosphate reabsorption from kidney.

Rickettsia A class of microorganisms that phylogenetically are, described as intermediate between viruses and bacteria. They differ from bacteria in being obligate parasites (cannot grow in the absence of living cells) with one exception.

rigor mortis A state of temporary stiffening of muscles after death. It is due to non-separation of actin-myosin filament caused by non-availability of ATP. Complete after 12 hours, it continues till decomposition starts.

Ringer's solution (also Ringer solution) A solution containing calcium, potassium and sodium chlorides. Used in hospitals to alleviate dehydration and improve circulation and in laboratories to maintain cells and organs alive *in vitro*.

RNA interference (RNAi; post-transcriptional gene silencing) The ability of double-stranded RNA to interfere with, or suppress the expression of a gene with a corresponding base sequence. The phenomenon occurs in many types of organisms, including plants, fungi and animals. There are two pathways, one involves micro RNAs (*mi*RNAs) and other short interfering RNA's (*Si*RANs)

RNA polymerase An enzyme that catalyses the synthesis of RNA polymers. To do this, it reads one strand (the template strand) of a double-stranded DNA molecule. This process is called transcription.

RNA processing (post-transcriptional modification) The modification of newly formed RNA transcripts to produce functional messenger RNA molecules. It occurs in the nucleus of eukaryotic cells and involves removal of the non-coding stretches (introns) from the primary transcript and splicing together of the discontinuous coding sequences (exons, i.e. gene splicing).

RNA splicing A two step reaction in which introns are removed from a primary RNA transcript and exons are joined together to form a mature mRNA.

rod cell One of the visual cells in the retina of vertebrate eye. Contain the pigment rhodopsin and helps in vision in dim light. Absent in the fovea.

rodent Member of the order–Rodentia, e.g. rat and squirrel. Gnawing mammals with a pair of large continually growing chisel-like incisors in upper and lower jaw.

root

1. Structure in vascular plants whose function is anchorage and the uptake of water and mineral salts by osmosis. Roots are partly or completely underground. It may be modified in several ways, e.g. prop root, stilt root, etc (in botany).
2. The portion of a tooth that is not covered with enamel and is embedded in a socket in the jawbone. Incisors, canines and premolars have single roots (in dentistry).
3. The point of origin of a nerve in the central nervous system. There are two roots for every spinal nerve (dorsal root and ventral root) (in anatomy).

root cap (calyptra) Mass of cells covering the root tip and develops as a result of cell division by root meristem at apex.

root hair The unicellular, root epidermal outgrowth; present near root tip; absorbing organs.

Root Mass Ratio (RMR) The dry mass of the roots of a plant divided by the total dry mass of the entire plant. It reflects the proportion of resources that a plant allocates to its root system, which varies with such factors as species, growing conditions and season.

root nodules Small swellings on the roots of leguminous plants. These contain nitrogen-fixing bacteria (e.g. *Rhizobium*). Plants that possess root nodules increase soil fertility by increasing the nitrate content of the soil.

root pressure The upward pressure of water, within vascular plants. It is caused by root cells by pumping water into the plant.

rootstock A vertical, usually short underground stem. It is found in many angiosperms, e.g. rhubarb and strawberry and in certain pteridophytes can refer to a rhizome.

Rotifera (Wheel-animalcules) The phylum of minute Pseudocoelomate animals. Metazoa which swim and feed by ciliated band, the wheel.

Rough Endoplasmic Reticulum (rough ER) A type of ER that has ribosomes attached to its surface. The proteins synthesised on ribosomes are enclosed in vesicles and transported to Golgi from here.

roughage Fibre that forms the bulk in the human diet and essential for the proper working and health of alimentary canal.

round window (fenestra rotunda) A membrane between the middle and inner ear in higher vertebrates. It moves back and forth to compensate for the pressure changes in the perilymph, caused by vibrations of the oval window.

roundworms Belong to class–Nematoda of phylum–Nemathelminthes. Unsegmented, cylindrical and elongated worms without cilia, e.g. hook worm.

royal jelly A highly nutritious food substance secreted from the hypopharygeal and mandibular glands of worker honeybees. It is fed to the newly hatched larvae, for the first three days after hatching. Once they develop into worker bees, their diet changes to pollen and nectar. However, larvae that develop into queen bees continue to be fed exclusively on royal jelly throughout the larval stage.

r-plasmid A plasmid containing gene for resistance to various drugs/antibiotics.

r-selection A type of selection that favours organisms with a high biotic potential (*r* value). Organism that are *r* selected (*r* strategists) are able to colonise a habitat rapidly, utilising the food and other resources before others establish themselves and compete.

rubisco An enzyme that catalyses the crucial step in the Calvin cycle of photosynthesis, i.e. the incorporation of a molecule of carbon dioxide with a molecule of ribulose bisphosphate, forming two molecules of phosphoglycerate. It is the most abundant protein in the plant world.

RuBP (ribulose bisphosphate) A 5-carbon sugar that combines with carbon dioxide at the beginning of the Calvin cycle (dark reaction of photosynthesis).

ruderal A plant species that is characteristic of land with a high level of disturbance (e.g. resulting from environmental disaster or human activity) but rich in resources (water, nutrients, etc). Such species tend to quickly colonise bare patches and reproduce rapidly, but are poor competitors against more robust species.

Ruffini's capsule (Ruffini's corpuscel) A specialised nerve ending that responds to pressure and stretching, providing information about certain types of touch stimuli and movements of joints and limbs. Found in skin and in joint capsules and ligaments, it consists of highly branched unmyelinated nerve ending within a spindle-shaped capsule, which is filled with a mesh of collagen fibres.

rumen The first stomach chamber of a ruminant.

ruminant Member of the suborder–Ruminantia (hooved mammals). Characterised by a four-chambered stomach, it comprises the sheep, cattle, goats, deers and antilopes.

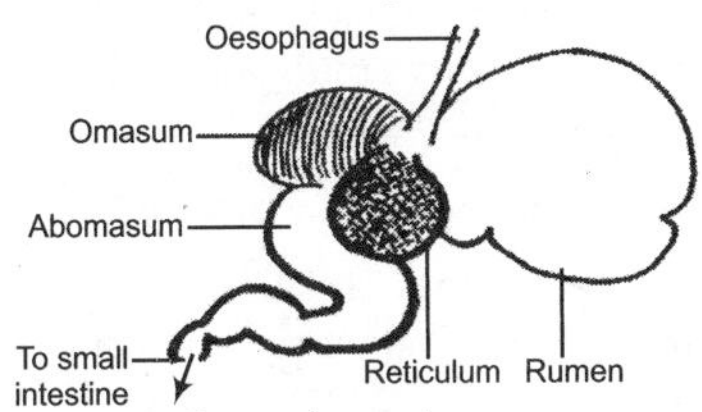

Stomach ruminants

runner Stem that grows horizontally along the soil surface and gives rise to new plants either from axillary or terminal buds, e.g. strawberry.

rusts A group of parasitic fungi of the phylum–Basidiomycota. Many of these species attack the leaves and stems of cereal crops: characteristic rust-coloured streaks of spores appear on infected plants. The life cycles of some rusts may be complex; many form a number of different types of spore and some required two different host plants.

S

saccharide *See* sugar

Saccharomyces A genus of yeasts that is industrially important. S. cerevisiae, of which there are atleast 1000 strains, is used in baking (*see* Baker's yeast), brewing and wine making; it is also used in the production of various proteins and other compounds in biotechnology, including industrial alcohol and for experimental studies in cell biology and genetics. Other yeasts used in the production of beer include S. uvarum carlsbergensis). Reproduction by budding but sexual spores may be formed during adverse conditions.

saccharose *See* sucrose

sacculus (saccule) Chamber of the inner ear from which the cochlea arises in reptiles, birds and mammals. It bears patches of sensory epithelium which is concerned with balance.

sacral Pertaining to the sacrum.

sacral vertebrae The vertebrae that lie between the lumbar and the caudal vertebrae in the vertebral column. They articule with the pelvic girdle and fuse to form a single bone (sacrum).

sacred grooves Tracts of forests set aside, where all the trees and wildlife are given protection. Found in India in Khasi and Jaintia, hills in Meghalaya, etc.

sacroiliac Pertaining to the sacrum and the ilium.

sacrolumbar Pertaining to the sacrum and the loins (or lumbar vertebrae).

sacrum *See* sacral vertebrae

safranin A stain that is used for staining living cells. It is usually used in conjunction with a green or blue counterstain.

sagittal The plane extending longitudinally and dorso-ventrally in the mid-line. Divides bilaterally symmetrical animal into two similar halves.

saline The chemical compound that is a salt or a solution containing a salt.

salinisation The increase in the salt content of the soil (or of a body of freshwater), which can lead to the retardation of plant growth and eventually renders the soil infertile. This problem is particularly acute in hot areas, where water readily evaporates from the soil and it may also occur as a consequence of irrigation.

saliva Watery fluid present in the mouth and secreted by the salivary glands. Saliva contains enzyme ptylalin that begin the digestive process and mucin to lubricate food.

salivary glands Glands that secrete a watery secretion (saliva) in to the buccal cavity. In humans, there are three pairs, i.e. parotid, submandibular and sublingual.

Salmonella Genus of rod-shaped Gram-negative bacteria that inhabit the intestine and cause disease (salmonellosis) in humans and animals. They are aerobic or facultatively anaerobic and most are motile.

Humans may become infected by consuming contaminated water or food, especially animal products, such as eggs, meat and milk or vegetables that have been fertilised with contaminated manure. The bacteria can also be transmitted from human or animal carriers by unhygienic food preparation. Various species of *Salmonella* cause gastroenteritis and septicaemia; typhoid fever and paratyphoid fever.

saltatory conduction The mode of transmission of a nerve impulse along a myelinated nerve fiber, whereby impulse leaps between nodes of Ranvier, considerably speeding up the message.

samara Dry one-seeded indehiscent fruit with the pericarp extended into a membranous wing, i.e. the ash fruit.

sap Sugary fluid found in phloem of plants or the fluid contained in plant vacuoles. It is a solution of organic and inorganic compounds.

saprophyte (saprotroph, saprobe, saprobiont) Plant, fungus or microorganism that feed on the dead or decaying remains of organic matter. They are often called decomposers.

saprozoic The organism that feeds on organic material in solution, rather than on solid organic material.

sapwood (alburnum) The outer region of xylem of older tree trunks. It consists of living xylem cells.

sarcolemma The contractile membrane that surrounds a muscle fibre.

sarcoma Cancer of muscle or connective tissue. Specific types of sarcomas are named for the types of tissue from which they arise.

sarcomere The functional units that make up the myofibrils of striated muscle. Each sarcomere is bounded by two membranes (Z lines), which provide the points of attachment of actin filaments; an other membrane (the **M band** or **line**) is the point of attachment of the myosin filaments. It is divided into various bands reflecting the arrangement of the filaments.

sarcoplasm Cytoplasm of muscle fibres.

sarcoplasmic reticulum A modified form of smooth ER found in striated and cardiac muscles. Release the calcium required for contraction.

satellite DNA Fraction of DNA that forms a separate band from the main body of DNA during gradient centrifugation. It consists of large number of copies of a short nucleotide sequence. It occurs mainly around the centromeres and telomeres of the chromosomes.

savanna Open tropical grassland, usually with widely spaced trees and bushes. Covers much of Africa, South of the Sahara.

scala One of three fluid-filled canals of the cochlea in the inner ear; the **scala media** (cochlear duct), **scala tympani** (tympanic canal) and **scala vestibuli** (vestibular canal).

scalariform thickening A type of secondary wall formation, found in metaxylem tracheids and vessels. Consists of interlaced spiral bonds of thickening.

scales The small bony or horny plates forming the body covering of fish and reptiles. The wings of some insects, notably the Lepidoptera (butterflies and moths), are covered with tiny scales that are modified cuticular hairs. In fishes, placoid, cosmoid and ganoid are the three types of scales.

scapula A large, triangular bone forming part of the shoulder girdle, commonly known as the shoulder blade.

scavenger The organism that feed on dead organic matter e.g. Hyenas

Schiff's reagent A reagent used for testing the presence of aldehydes and ketones; it consists of a solution of fuchsin dye that has been decolourised by sulphur dioxide. Aliphatic aldehydes restore the pink immediately, whereas aromatic ketones have no effect on the reagent. Aromatic aldehydes and aliphatic ketones restore the colour slowly.

schizocarp Dry fruit, formed form two or more carpels that divides at maturity into one seeded achene like segments termed as mericarps, e.g. hollyhock.

schizogeny The separation of plant cells to form intercellular spaces with special function, e.g. resin canals of conifers.

schizogony The multiple asexual fission in Protozoa.

Schwann cell (neurilemma cell) Cell which produce myelin sheath that surrounds a nerve cell. These are in close contact with the axon of the neuron and are separated by gaps called nodes of Ranvier. Named after the German physiologist T Schwann.

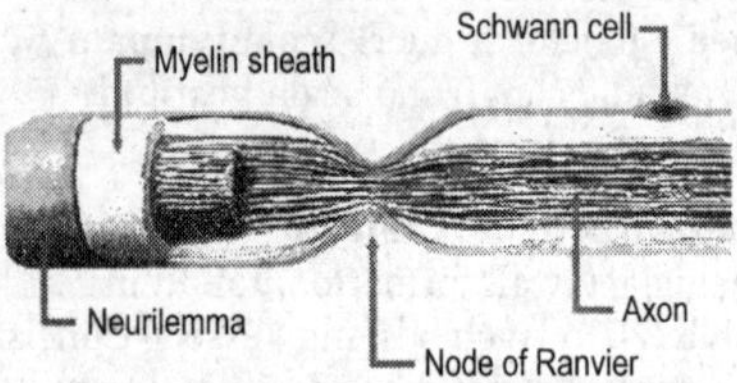

Position of Schwann cell in a neuron

scion In a plant graft, part that is grafted onto stock.

sclera (sclerotic) An outermost coat of the eyeball, extending from the optic nerve to the edges of the cornea in vertebrate eye.

sclereids The spherical, oval or cylindrical, highly thickened dead cells of sclerenchyma with very narrow cavities. These are commonly found in fruit walls of nuts, pulp of fruits like guava, pear, etc.

sclerenchyma main supporting Plant tissue consisting of long, narrow cells with thick and lignified cell walls having a few or numerous pits. They are usually dead and without protoplasts.

scleroderma (systemic sclerosis) A disorder involving a progressive thickening and hardening of the skin. Calcium deposits on the skin are also seen in association with this disease.

sclerophyllous Describing scrub or woodland in which the plants have small leathery evergreen leaves. As an adaptation to conserve water.

scleroprotein The group of proteins found in the exoskeletons of some invertebrates, notably insects, formed by conversion of the relatively soft elastic larval protein by a natural tanning process (sclerotisation). These are secreted and form cross linkages between polypeptides of the proteins, producing a hard rigid covering.

sclerotic The head of an adult tapeworm. A tapeworm attaches itself to the wall of the small intestine with its scolex.

scorpions The type of vision that occurs when the rods in the eye are the principal receptors, i.e. when the level of illumination is low. With scotopic vision colours cannot be iden*tified.*

SCP *See* Single-Cell Proteins.

scrotum (scrotal sac) The external pouch containing the testes, in most mammals. It allows sperm to develop at the optimum temperature, which is slightly lower than body temperature.

scurvy A pathological condition due to the lack of vitamin-C. Symptoms include anaemia, skin discoloration and tooth loss.

scutellum The cotyledon of Gramineae (grasses). Situated towards one side of the embryonal axis.

scyphistoma A polyp stage in the life cycle of scyphozoans which develops from the planula.

scyphozoa A Cnidarian class, in which the medusa is the dominant form, e.g. *Aurelia.*

seaweeds Large multicellular algae living in the sea or in the intertidal zone. They are commonly species of the Chlorophyta. Phaeophyta, and Rhodophyta.

sebaceous gland The epidermal skin gland of mammals, embedded in dermis, always opening into a hair follicle, secreting fatty substance (sebum).

sebum Refers to the substance secreted by sebaceous glands on to the surface of the skin. It is a mild antiseptic material that protects and lubricates the skin and thus helps to prevent desiccation.

Second messenger It is a chemical within a cell that is responsible for initiating the response to a signal from a chemical messenger (such as a hormone, neurotransmitter, or growth factor) that cannot enter the target cell itself, e.g. cyclic AMP.

secondary growth The increase in thickness of plant shoots and roots in dicotyledons and gymnosperms. Tissues involved in secondary growth are two lateral meristems, vascular cambium and cork cambium.

secondary productivity The rate of biomass formation or energy fixation by heterotrophic organisms, such as grazers and decomposers.

secondary sexual characteristics External features of a sexually mature animal that are indirectly significant in reproductive behaviour. The development of such features is controlled by sex hormones (androgens or oestrogens); they may be seasonal (e.g. the antlers of male deer or the body colour of male sticklebacks) or permanent (e.g. breasts in women or facial hair in men). In humans they develop during **adolescence**.

secondary structure (protein) The way in which the chain of aminoacids that make up a polypeptide is coiled or folded. e.g. Alpha helix, β-pleated sheet etc.

secondary thickening *See* secondary growth.

secretin The hormone secreted by the Brunner's glands in the walls of duodenum and jejunum under the stimulus of acidic food coming from stomach. it stimulates alkaline pancreatic secretion and the release of bile.

secretion A process by which substances are produced and discharged from a cell, gland or organ for a particular function in the organism or for excretion.

seed The structure that develops from the ovule after fertilisation. It contains the embryo and nutritive tissue, either as endosperm or food stored in the cotyledons. Angiosperm seeds are contained within a fruit that develops from the ovary wall. Gymnosperm seeds lack an enclosing fruit and are thus termed **naked**.

seed bank A seed bank stores seeds as a source for planting in case seed reserves else where are destroyed.

seed coat *See* testa

seed leaf *See* cotyledon.

seed plant A plant that produces seeds. Most seed plants belong to the phyla. Anthophyta (flowering plants) of Coniferophyta (conifers).

segmentation The animal body plan in which the organism is sub-divided into serially arranged units. see metameric segmentation and cleavage.

segregation The normal biological process, whereby the chromatids of each chromosome pair are separated during meiosis and the two alleles of a gene are randomly distributed to the germ cells.

seismonasty The response to a non-directional shock stimulus in plants.

selection The process by which one or more factors acting on a population produce differential mortality and favour the transmission of specific characteristics to subsequent generations.

selection pressure The extent to which organisms possessing a particular characteristic are either eliminated or favoured by environmental demands. It indicates the degree of intensity of natural selection.

selective breeding The process of breeding to produce offspring that possess the beneficial character of both parents.

selective reabsorption The absorption of some of the components of the glomerular filtrate back into the blood as the filtrate flows through the nephrons of the kidney. Glucose, amino acids and salts can be reabsorbed against a concentration gradient and their transport across the nephron into the capillaries requires energy.

selector gene Any of the zygotic genes expressed subsequently after segmentation genes and who function specifies the fate of the cells in each segment of the fruit fly embryo.

self-fertilisation The union of male and female gametes from the same individual.

self-pollination The transfer of pollen from an anther to a stigma of the same flower.

self sterility The condition found in many hermaphrodites in which male and female gametes of the same individual can not fuse to form a zygote.

Seliwanoff's test A biochemical test to identify the presence of ketonic sugars, such as fructose, in solution. A few drops of the reagent, consisting of resorcinol crystals dissolved in equal amounts of water and hydrochloric acid, are heated with the test solution and the formation of a red precipitate indicates a positive result.

semaphorin One of a class of proteins that act as guidance molecules during the development of nerve cells, immune cells, blood vessels, bone and other tissues.

semelparity The condition of producing offsprings only once during the lifetime of an individual, after which death is inevitable, e.g. Atlantic eel.

semen A thick white fluid containing spermatozoa produced by testes and secretions of prostate seminal vesicles and comper's glands. Semen is produced by male mammals and injected into the female during copulation, a necessary prerequisite of fertilisation. It is slightly alkaline (7.2-7.6) with enzymes that activate sperm after ejaculation.

semicircular canal A part of the ear that is involved in maintaining balance. These are three looped canals that form the part of labyrinth of the vertebrate inner ear. Detect changes in the rate of movement of the head.

semi-conservative replication The general method of DNA replication in which each daughter molecule produced, on replication has one of the original parent strand and one newly constructed strand.

semilunar valve Either of two valves in the heart, found in the pulmonary artery (pulmonary valve) and in the aorta (aortic valve), that prevent the backflow of blood into the right and left ventricles from the pulmonary artery and the aorta, respectively, thus maintaining blood flow in a single direction.

seminal receptacle *See* spermatheca

seminal vesicle

1. One of a pair of small elongated glands in most male mammals that opens into the vas deferens. It secretes a thick alkaline fluid containing substances such as fructose, proteins and various chemicals that contribute to the semen.
2. The organ in lower vertebrates and some inverterbrates used for the storage of sperms.

seminiferous tubule A mass of minute coiled tubules within the vertebrate testis in which spermatozoa are produced. It is lined with sertoli cells.

semiochemical A chemical that affects the behaviour of an organism. e.g. pheromones, which are used for communication between members of the same species and allelochemicals, which act as chemical signals between members of different species.

semipermeable membrane The porous membrane that permits the passage of some substances but not others, e.g. plasma membrane. It permits entry of small molecules such as water but not large molecules, allowing osmosis to occur.

senescence

1. The advanced phase of ageing process, prior to natural death.
2. The deterioration of leaves of a deciduous plant towards the end of growing seasons, resulting in leaf fall.

sense organ A part of the body of an animal that contains receptors that are sensitive to specific stimuli (e.g. sound, light, pressure, heat). Stimulation of these receptors initiates the transmission of nervous impulses to the brain, where sensory information is analysed and interpreted. Example eye, ear, nose and taste bud.

senses Faculties that enable animals to perceive information about their external environment or about the state of their bodies in relation to this environment. Specific receptors are sensitive to pain, temperature, chemicals, etc.

sensillum (pl. sensilla) Various hairlike or peglike sense organs found in insects and other arthropods, comprising a cluster of basal receptor cells whose dendrites extend inside a sheath and hairshaft. They respond to a wide range of chemical and mechanical stimuli, acting as chemoreceptors of both 'smell' and 'taste', touch receptors, mechanoreceptors to detect movement at joints in the skeleton, and auditory organs. They may protrude from the surface, e.g. as setae, or they may be embedded in the cuticle.

sensitisation

1. The alteration of the integrity of a plasma membrane resulting from the reaction of specific antibodies with antigens on the surface of the cell.
2. Initial exposure to a specific antigen such that re-exposure to the same antigen causes a severe immune response in an individual.

S

3. The process whereby an animal gives a stronger than normal response to a stimulus after previous application of a different stimulus, e.g. if the tail of a see slug (*Aplysia*) is given an electric shock, a subsequent tap on the animal's siphon will cause a marked with drawal of the siphon that is greater than that without a preceding tail shock.

sensitivity (irritability) One of the fundamental properties of all organisms, i.e. the capacity to detect, interpret and respond to changes in the environment. Multicellular animals have specialised sense organs and effector organs for this purpose; in unicellular organisms, which lack a nervours system, the reception of and response to a stimulus occur in the same cell.

sensory cell *see* receptor

sensory neuron A neuron sending information from a sensory receptor to the central nervous system and sensory neurons may either link to a receptor or itself be a receptor at one end, as in some pain receptors.

sepal One of the parts of a flower making up the calyx. Considered to be modified leaves with a simpler structure. They are usually green and often hairy.

S

sepsis The presence of pathogenic microorganisms or their toxic products in the bloodstream, leading to infection.

septum A partition separating two chambers, such as that between the nostrils or the chambers of the heart.

sequence analysis The process of characterising sequences of biomolecules, particularly the nucleotides of nucleic acids (DNA or RNA) or the amino acids of proteins.

sequence database A database containing the sequences of biomolecules, which may be the sequences of nucleotides for nucleic acids (DNA or RNA) or of amino acids for proteins. The data also include relevant annotations.

Sequence Tagged Site (STS) A unique short sequence of DNA, typically less than 400 nucleotides long, that can be used as a tag, or marker, in physical mapping of cloned DNA segments.

sere A complete succession of plant communities, which results in the climax community. A sere is composed of a series of different plant communities that change with time. These communities are known as seral stages or seral communities.

serine One of the non-essential amino acids.

seroal stage A stage in ecological succession of communities.

serology The laboratory study of blood serum and its constituents, particularly antibodies and complement, which play a part in the immune response.

serotonin A neurohormone that acts on muscles and nerves and a neurotransmitter found in both the central and peripheral nervous system. It controls dilation and constriction of blood vessels and affects peristalsis and gastrointestinal tract motility.

serous membrane (serosa) The thin, lining membrane that secretes a serous fluid, consists of mesothelium and underlying connective tissue. Specifically, the three membranes lining the pleural, pericardial and peritoneal cavities.

sertoli cells (susten-tacular cells) Cells that line the seminiferous tubules in the testes. These cells provide nourishment to both the developing and mature spermatozoa.

serum *See* blood serum

sesamoid bone A small oval bone that develops as an ossification within a tendon of vertebrate.

sessile

1. Describing animals that live permanently attached to a surface, e.g. see anemones
2. Describing any organ that does not possess a stalk the leaves of the oak (*Quercus robur*) are attached directly to the twigs.

seta (pl. setae)

1. A bristle or hair in many invertebrates, produced by the epidermis and consist either of a hollow projection of cuticle containing all or part of an epidermal cell (as in insects) or are composed of chitin (as in the chaetae of annelid worms).
2. *See* sporogonium

Severe Combined Immuno Deficiency (SCID) A genetic disorder in which the immune system of the patient is impaired as a result of which patient become prone to infectious diseases.

Sewall Wright effect *See* genetic drift

Sex

1. Sexual activity, including specifically sexual intercourse.
2. Either of the two main categories (male and female) into which living organisms are divided on the basis of their reproductive functions.

sex chromosomes Chromosomes that determine the sex of an individual. There are two types in mammals there is a large X chromosome and a much smaller Y chromosome; the female has two X chromosomes making it the homogametic sex, whereas the male has one X chromosome and one Y chromosome, making it the heterogametic sex. In other animal groups, such as butterflies and moths, birds and reptiles, the situation is reversed, males have two matching W chromosomes, while females have one W chromosome and one Z chromosome. Sex chromosomes carry genes governing the development of sex organs and secondary sexual characteristics.

sex determination The method by which the distinction between male and female is established in a species, e.g. in humans, 23 pairs of chromosomes are present, 22 pairs are exactly same in both males and females, these are autosomes. A pair of X-chromosome (XX) is present in females, whereas one X and one Y-chromosome (XY) is present in males.

In males, during spermatogenesis, two types of gametes are produced. 50% of total sperm produced carry X-chromosome and the rest 50% carry Y-chromosomes besides autosomes.

Females produce only one type of ovum with an X-chromosome.

In case, the ovum fertilises with a sperm carrying X-chromosome, the zygote develops into a female (XX) and if ovum is fertilised with Y-chromosome carrying sperm, zygote formed is a male (XY).

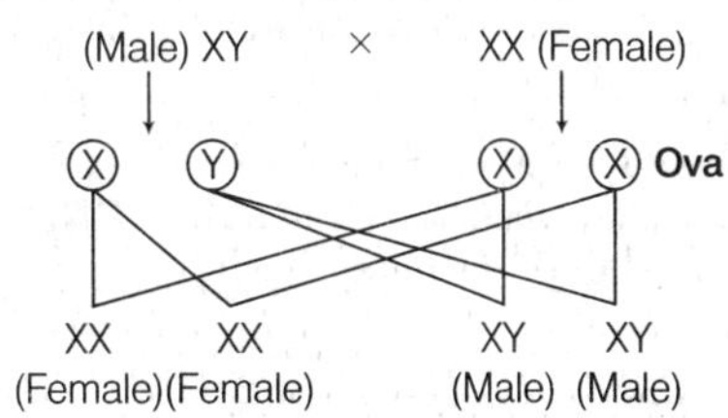

Hence, the genetic make up of sperm, which fertilises an ovum determines the sex of the child. There is 50% chances of having male and 50% chances of having female in each pregnancy. So, woman should not be blamed for the sex of a child.

sex factor A plasmid found in the cytoplasm of certain bacteria that is responsible for initiating conjugation and gene transfer. The sex factor in *Escherichia coli* is known as the F **factor**; it can be transferred from a donor cell (F^+) to a recipient (F^-) by conjugation.

sex hormone Responsible for the development and functioning of reproductive organs. Androgens in male secreted by the testes and estrogen and progesterone in females secreted by the ovary.

sex linkage A linkage involving a locus on the X or Z-chromosome. In certain crosses, sex-linked traits manifest themselves only in the heterogametic sex.

sex-linked traits Traits that are due to alleles found on the X-chromosome. Thus, colour blindness is a recessive sex-linked trait, because, if the male has an allele for colour blindness, it will be expressed in that male, since he does not have another X-chromosome potentially carrying the dominant allele for normal vision. Females can be colourblind, if they are homozygous recessive.

sex ratio The ratio of males to females in a population. In majority of species, this is 1 : 1.

S

sexduction A process whereby a bacterium gains access to foreign DNA brought by F-factor during conjugation.

sexual intercourse (coitus; copulation; mating) The process by which spermatozoa from a male are deposited in the body of a female sexual reproduction.

sexual reproduction A reproduction that involves the fusion of gametes, to form a zygote.

Sexually Transmitted Disease (STD) The disease which is passed from one individual to another during sexual intercourse or other types of sexual activity. Also known as veneral diseases e.g. syphilis.

shikimic acid pathway The key metabolic pathway in the biosynthesis of the aromatic amino acids-tyrosine, phenylalanine, and tryptophan-which occurs in plants, bacteria and fungi, but not in animals.

shine-dalgarno sequence A sequence of five to nine (typically seven) nucleotides preceding the start codon in prokaryotic messenger RNA (*m*RNA) that is recognised by the ribosome as the correct site for binding the *m*RNA molecule prior to the start of translation.

shoot The aerial part of a vascular plant. It develops from the plumule and consists of a stem supporting leaves, buds and flowers.

shoot meristem culture or mericloning Method used in mass production of certain plants like orchids. Apical meristem and one or two leaf primordia are removed under aseptic conditions and placed in the culture medium under light. Adventitious roots and shoots develop and each of them is capable of growing into a new plant.

short-day plant A plant in which flowering can be induced or enhanced by short days, usually of less than 12 hours of daylight, e.g. strawberry and *Chrysanthemum*.

short interfering RNA (siRNA) Small RNA molecules that are produced in cells from double-stranded (ds) precursors by the action of the protein Dicer.

short interspersed element *See* SINE.

Short Tandem Repeats (STRs) Short sequences of nucleotides in DNA, 2-7 bp long, that are repeated in tandem at particular loci throughout the genome. The number of repeats at any one site varies typically between five and thirty. STRs are widely used as the basis for DNA profiling because they are highly variable and their small size makes them readily amenable to detection and amplification by the PCR.

shortsightedness *See* myopia.

shotgun cloning A method for cloning the entire genome of an organism in which the DNA is broken into small fragments at random using restriction enzymes. Each fragment is then incorporated with a cloning vector and introduced into a host organism.

shotgun sequencing A strategy for sequencing large DNA fragments or entire genomes based on sequencing and ordering smaller fragments obtained by shotgun cloning. It relies primarily on computational power.

shoulder girdle *See* pectoral girdle

shunt vessel A blood vessel that links an artery directly to a vein, allowing the blood to bypass the capillaries in certain areas.

shuttle vector Vectors which contain both prokaryotic and eukaryotic replication signals, thus allowing replication of the vector in both kinds of cells.

sialagogue A stimulant of salivation.

sibling species (cryptic species) Two or more groups of organisms that cannot be readily distinguished by their appearance or other traditional taxonomic criteria. However, members cannot interbreed successfully, e.g. many North American species of cricket, occupying the same geographical range, can be distinguished only by their song.

siblings (Sibs) Offsprings from the same cross (both parents in common).

sickle cell anaemia A genetically-controlled defect of haemoglobin synthesis which leads to sickle shape of red blood cells. It is due to a point mutation in the hemoglobin beta gene.

siderocytes Abnormal erythrocytes containing iron granules in the mitochondria.

sieve elements Elongated cells that are placed end to end, make up the sieve tubes within the phloem.

sieve plate An area of relatively large ores present in the common end walls of sieve tube elements.

sieve tube Element of phloem for the transport of food. Consists of thin-walled elongated living cells (sieve elements) having perforations in their cross walls.

sigmoid colon S-shaped portion of the large intestine. It connects the descending colon with the rectum. Also known as the sigmoid flexure of the descending colon.

sigmoid growth curve A population growth curve that is typically s-shaped. It shows an initial phase of exponential growth. The curve levels off when the environment has reached its **carrying capacity**, i.e. when food, space and other conditions can support a given number of individuals without an increase in population numbers.

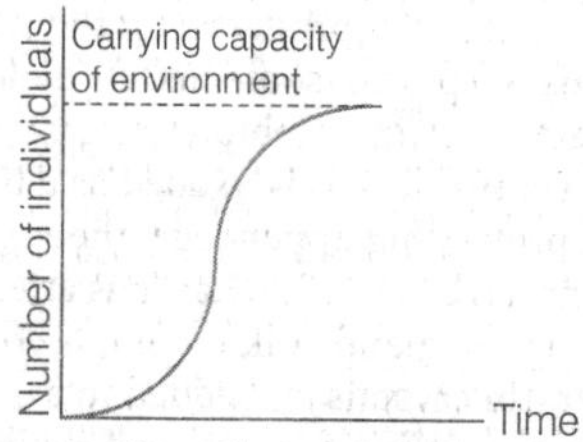

Sigmoid growth curve

S

sign stimulus (releaser) The essential feature of a stimulus, which is necessary to elicit a response, e.g. a red belly (characteristic of courting male sticklebacks) is the sign stimulus necessary to provoke an attack from a rival male.

signal hypothesis A hypothesis to explain how ribosomes become attached to membranes within cells in order to deliver the appropriate proteins to cell organelles, such as mitochondria and chloroplasts, or transport proteins outside the cell membrane.

signal transduction The process in which binding of an extra cellular signal molecule to a cell-surface receptors triggers and response inside the cell. Most signal transduction processes are ordered sequences of intracellular reactions. It is mediated by enzymes and activated by the secondary messengers.

silent mutation The alteration in the genetic code that has no apparent effect on the phenotype of an organism.

silicosis. The lung disorder caused by inhaling gritty dirt.

silicula A capsular fruit typical of the Cruciferae (e.g. candytuft) that is formed from a bicarpellary ovary. It is flattened, short and broad, divided into two locule by a false septum.

S

siliqua (silique) A fruit of some Cruciferae similar of the silicula but longer and thinner, e.g. wall flower.

silk A material produced by the silk glands of spiders, some insects and certain other invertebrates (e.g. centipedes). It is exuded as a liquid by a protruding appendage, the spinneret, but quickly hardens after leaving the gland. Silk is composed of α-keratin crystals embedded in a rubbery matrix of amino acid chains, giving its flexibility and strength.

Silurian period (S) A geologic period of the Palaeozoic era, lasting from 443.7-416 million years ago. During the silurian, the earliest known vascular plants appeared on land. The first coral reefs formed in the oceans and fish with movable jaws made their appearance.

simian Pertaining to monkeys.

SINE (Short Interspersed Element) The class of dispersed moderately repetitive DNA found in eukaryotes, consisting of numerous copies ($>10^5$) of relatively short (<500 bp) sequences scattered throughout the genome. SINEs are not translated into proteins, occur mostly in introns and are nonautonomous retrotransposons, which require a reverse transcriptase from another source in order to replicate via an RNA intermediate, e.g. Alu family.

Single Cell Proteins (SCP) Sources of mixed protein extracted from pure or mixed cultures of algae, yeasts, fungi or bacteria used as a substitute for the protein-rich foods in human and animal feeds.

single circulation The type of circulatory system that occurs in fishes, in which the blood passes only once through the heart in each complete circuit of the body compare double circulation.

single gene disorder A hereditary disorder caused by a mutant allele of a single gene.

Single Nucleotide Polymorphism (SNP) A variation in the base sequence occurring at any given single position in the genome, (e.g. A instead of C). Useful as molecular markers.

sinoatrial (SA) **node** *See* pacemaker

sinus The anatomical cavity, space or channel. Examples are the nasal sinuses in the skull.

sinus venosus A thin-walled first chamber of the heart in fishes, amphibians and reptiles which receives deoxygenated blood from the body. It is absorbed into the right atrium of birds and mammals.

sinusoid A tiny blood vessel or blood filled space in an organ.

siphonaceous (siphonous) Describing algae and other protists in which the thallus is coenocytic and often takes the form of a tubular structure.

Siphonaptera The order of secondarily wingless insects comprising the fleas. The body is laterally compressed and bears numerous backward-directed spines. They live as blood-sucking ectoparasites of mammals and birds, having mouth-parts adapted to piercing their host, injecting saliva to prevent clotting and sucking up the blood, e.g. *Xenopsylla cheopis.*

siphonophora A group of organisms related to jellyfish, coral and sea anemones. While they superficially resemble jellyfish, each apparent organism is really a complex colony of multicellular organisms called polyps. The polyps each have their own specific function in the colony but are genetically identical to each other.

siphonostele (solenstele) A stele in which vascular tissue is in the form of a cylinder surrounding the pith, e.g. fern stem.

siphunculata (anoplura) The order of secondarily wingless insects comprising the sucking lice: blood-sucking ectoparasites of mammals, with piercing and sucking mouth parts forming a snoutlike proboscis.

si RNAs *See* Short Interfering RNA.

sister chromatids Two identical nucleoprotein strands of a single replicated chromosome, which are joined at the centromere.

sister species Either of the two descendant species formed when one species splits during evolution.

site-directed mutagenesis A technique used in genetic engineering for introducing particular changes to the base sequence of a gene at a specific site. It allows precise and specific mutations to be made and is done.

sitology The study of food, nutrition and diet.

sitosterol (also β-sitosterol) Plant sterol with a molecular structure similar to that of cholesterol.

Sivapithecus (Ramapithecus) A genus of extinct primates that lived about 2-8 million years ago. Fossil remains have been found in India and Pakistan, the Near East, and East Africa. They chewed from side to side and had fairly short muzzles, i.e. humanoid features. How they are now regarded as ancestral to the Asian great apes (e.g. orang-utans), not the hominids.

skeletal muscles *See* voluntary muscle

skeleton The hardened framework of an animal body serving for support and protecting soft parts; it may be external (exoskeleton) or internal (endoskeleton).

skin The outer layer of the body of a vertebrate composed of two layers, the epidermis and dermis, with a complex nervous and blood supply. It may bear a variety of specialised structures, including hair, scales and feathers. It plays an important role in protecting the body from mechanical injury, water loss and the entry of harmful agents. It is also a sense organ, containing receptors sensitive to pain, temperature and pressure.

In warm-blooded animals it helps regulate body temperature by means of hair, fur, or feathers and sweat glands.

skull The bony framework of the head. It is composed of two set of bones (cranial and facial) that forms total of 22 bones [bones of middle ear (6 bones) and hyoid bone (1) are also included in bones of skull.

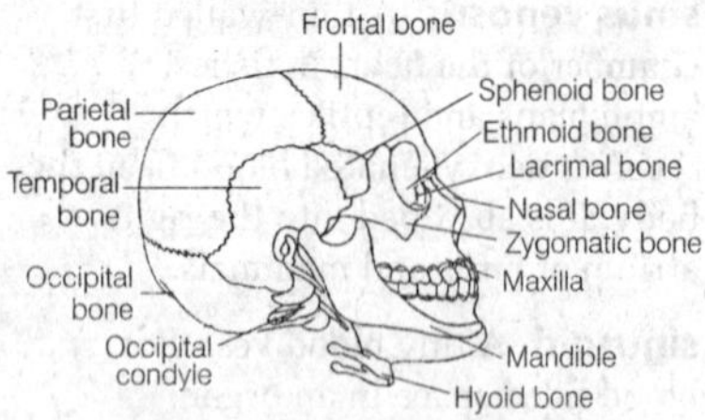

Diagrammatic view of human skull

sliding filament theory It states that contraction of muscles takes place by the sliding of thin and thick filaments that past over each other with the help of cross-bridge to reduce the length of the sarcomere.

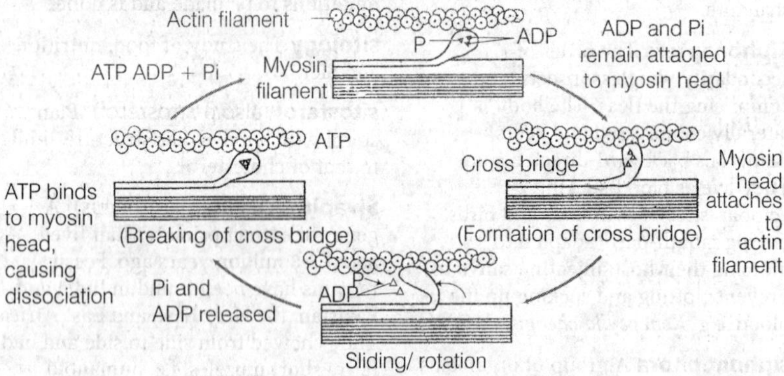

Stages in cross-bridge formation, rotation of heat and breaking of cross-bridge

slime moulds Saprophytic protists. They are extremely resistant and survive for many years even under adverse conditions.

small intestine The portion of the alimentary canal between the stomach and the large intestine. It is subdivided into the duodenum, jejunum and ileum. It plays an essential role in the final digestion and absorption of food.

smooth endoplasmic reticulum (smooth ER) A series of membranous networks involved in the final modification of proteins and the synthesis of lipids ER that lacks ribosomes. Contrast rough ER.

smooth muscles *See* involuntary muscle.

smuts A group of parasitic fungi of the phylum Basidiomycota. Many of these species attack the ears of cereal crops, replacing the grain by a mass of dark spores.

snakes *See* squamata

sociobiology The branch of biology that studies the evolutionary basis of social behaviour.

sodium chloride (common salt) A colourless crystalline solid, NaCl, that is soluble in water. It has a key role in biological systems in maintaining electrolyte balances and used as a food preservative.

sodium pump A mechanism by which sodium ions are transported out of a eukaryotic cell across the plasma membrane. The process requires energy in the form of ATP, being a form of active transport.

softwood A wood of gymnosperms, e.g. pine. Only this wood functions in water conduction.

soil The layer of unconsolidated particles derived from weathered rock, humus, water, and air that forms the upper surface over much of the earth and supports plant growth.

soil erosion The removal and thinning of the soil layer due to climatic and physical processes, such as high rainfall, which is greatly accelerated by certain human activities, such as deforestation.

soil horizon Distinct horizontal layers that can distinguished in a vertical section (profile) of soil. Four basic horizons are common to most soils; an uppermost A horizon (or **topsoil**) containing the organic matter; an underlying B horizon (or **subsoil**), which contains little organic material and is strongly leached; a C horizon consisting of weathered rock; and a D horizon comprising the bedrock.

solenocytes *See* flame cells.

solenoid structure Refers to the supercoiled arrangement of DNA in the eukaryotic nuclear chromosomes.

soma *See* cell body

somaclones Plants produced by tissue culture techinque of somatic cells.

somatic Relating to the all body cells, other than reproductive cells.

somatic cell The cell in the body except gametes and their precursors.

somatic embryogenesis A process where a plant or embryo is derived from a single somatic cell or group of somatic cells.

somatic hybridisation The hybridisation by the fusion of protoplasts of somatic cells (not of sex cells).

somatic hybrid A hybrid plant produced by the fusion of two distinct species of plants. The process in called somatic fusion or protoplast fusion.

somatic mutation (also somatic cell mutation) The mutation in a non-reproductive cell.

somatic neural system Type of peripheral nervous system. Relays impulses from CNS to skeletal muscles.

somatostatin (Growth Hormone Inhibiting Hormone; GHIH) A hormone, secreted by the hypothalamus, that inhibits the release of growth hormone from the anterior pituitary gland.

somatotrophic *See* Growth hormone.

somites Blocks of mesoderm along side the neural tube of a chordate embryo.

sorosis A type of composite fruit incorporating a spike as seen in the pine apple.

Southern blotting (also Southern blot) A method developed by E M Southern for transferring DNA fragments, separated in electrophoretic gels, onto membrane filters. Used to detect specific fragments by complementary radioactive probes.

spadix A type of inflorescence found in the family. Araceae, e.g. cuckoopint. It is a modified spike with a large fleshy axis on which are borne small hermaphrodite or more usually, unisexual flowers.

spathe A large bract that encloses the spadix.

spawn

1. To lay eggs in water (said of an aquatic animal).
2. Eggs laid in water by an aquatic animal.

speciation Evolutionary processes by which new kinds of organisms evolve from preexisting kinds or evolve from currently existing kinds.

species A group of individual organisms with fundamental similarities.

species diversity The diversity at the species level, *e.g.,* the Western Ghats have a greater amphibian species diversity than the Eastern Ghats.

spectroscopy The study of methods of producing and analysing spectra using spectroscopes and other instruments. It is employed in a wide range of biological research areas, such as biochemistry and toxicology, for the identification of metabolites and other compounds of biological significance.

Spemann's organiser A cluster of cells in the dorsal mesoderm that determines the development of the spinal cord and brain in the early amphibian embryo.

sperm Small, motile, usually flagellated male sex-cell (non-flagellated in *Ascaris*).

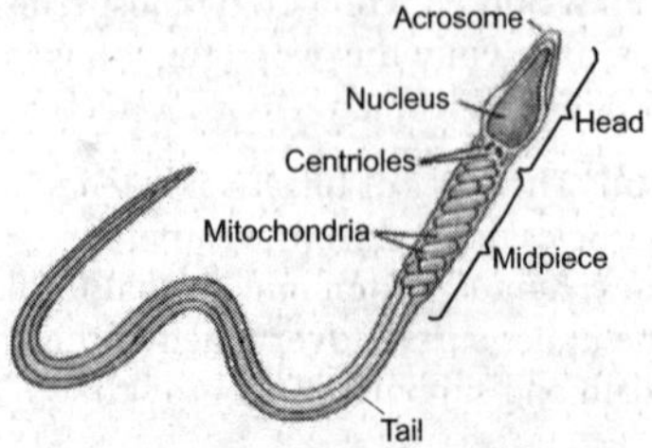

Structure of human sperm

sperm competition Competition between sperm from different males to reach and fertilise the egg cell of a single female. e.g. in rodents.

spermatheca (seminal receptacle) A sac or receptacle in some female or hermaphrodite animals (e.g. earthworms) in which sperm from the mate is stored until the eggs are ready to be fertilised.

spermatid One of the four haploid cells produced by meiosis that subsequently develop into spermatozoa without further division.

spermatium (pl. spermatia) A type of nonmotile male sex cell that is produced by the red algae (Rhodophyta) and by certain fungi (e.g. the rusts).

spermatocyte A diploid cell in the testis that divides by meiosis to give rise to four spermatids (see spermatogenesis).

spermatogenesis The production of sperms in males.

(i) In testis, the immature male germ cells (spermatogonia) produce sperms by **spermatogenesis** that begins at puberty.
(ii) **Spermatogonia** (sing, spermatogonium) present on the inside wall of seminiferous tubules multiply by mitotic division and increase in numbers.
(iii) Each spermatogonium is diploid and contains 46 chromosomes. Some of the spermatogonia called **primary spermatocytes** periodically undergo meiosis.
(iv) A primary spermatocyte undergoes first meiotic division leading to two equal, haploid cells called **secondary spermatocytes**, which contains only 23 chromosomes each.
(v) The secondary spermatocytes undergo the second meiotic division to produce four equal, haploid **spermatids**.
(vi) The spermatids are transformed into spermatozoa (sperms) by the process called **spermiogenesis**.

(vii) Sperm heads become embedded in the Sertoli cells and are finally released from the seminiferous tubules by the process called **spermiation**.

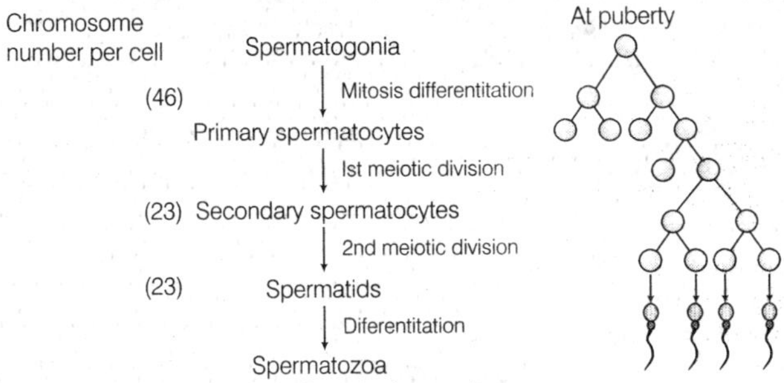

Spermatiogenesis

spermatogonium (Pl. spermatogonia) Diploid cells in the walls of the seminiferous tubules of the testis that give rise to the primary spermatocytes.

Spermatophyta In traditional classifications, a division of the plant kingdom containing plants that reproduce by means of seeds. In modern systems seed plants are grouped into separate phyla, the most important of which are the Anthophyta and Coniferophyta.

spermatozoan (pl. spermatozoa) A male haploid gamete capable of active movement by means of a undulipodium, which is produced by the testes. It consists of a head section containing a haploid nucleus and an acrosome, which allows the sperm to penetrate the egg at fertilisation; a middle section containing mitochondria to provide the energy for movement; and a tail section, which contains a single undulipodium.

spermatozoid *See* antherozoid

S-phase The portion of interphase during which DNA content of the cell doubles and chromosomes are replicated.

Sphenophyta (Arthrophyta) A phylum of Tracheophyte plants, the only living members of which are the horsetails (*Equisetum*). Horsetails have a perennial creeping rhizome supporting erect jointed stems bearing whorls of thin leaves.

spherosome (oleosome) A small spherical organelle of plant cells, about 0.8-1.0 μm in diameter, bounded by a single membrane and capable of storing and synthesising lipid.

sphincter A ring of smooth muscle that can close off a tubular structure such as an arteriole or part of the colon.

sphygmic Pertaining to the pulse.

sphygmomanometer A instrument comprising an inflatable cuff and a pressure gauge for determining the blood pressure.

spike A type of racemose inflorescence in which stalkless flowers arise from an undivided floral axis, as in plantain and *Orchis*. In the family Gramineae (Poaceae; sedges and grasses) the flowers are grouped in clusters called **spikelets** which may be arranged to form a compound spike (as in wheat).

spinal column *See* vertebral column.

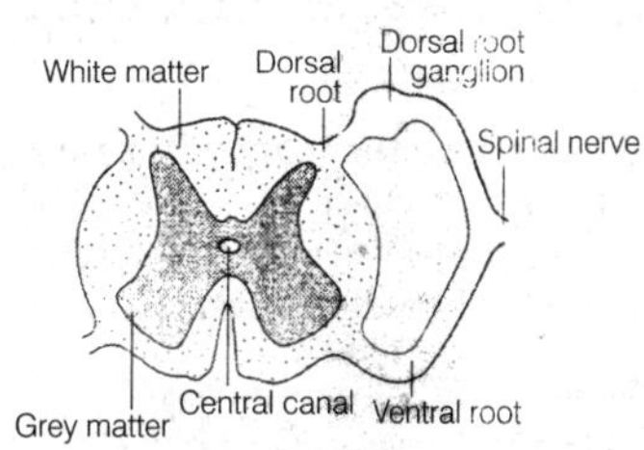

spinal cord A column of nervous tissue that in vertebrates runs along the back and that in bony animals is enclosed within the vertebral column. In humans, it gives rise to all the nerves of the trunk and limbs. It consists sists of a hollow core of grey matter (H-shaped in cross section) surrounded by an outer layer of white matter; the central cavity contains cerebrospinal fluid. The white mater contains numerous longitudinal nerve fibres organised into distinct tracts: **ascending tracts** and descending tracts.

spinal nerves Pair of nerves that arise from the spinal cord. Each nerve arises from a dorsal root and a ventral root and contains both sensory and motor nerve fibres. In humans, there are 31 pairs.

spindle A structure formed from microtubules in the cytoplasm during cell division that moves chromatids or chromosomes diametrically apart and gathers them in two clusters at opposite ends (poles) of the cell. Broad in the middle and narrowing to a point at either pole, its construction is directed by a microtubule-organising centre, the centrosome.

spine

1. *See* vertebral column.
2. A hard pointed protective structure on a plant that is formed through modification of a leaf, part of a leaf, or a stipule.

spinneret A small tubular appendage from which silk is produced in spiders and some insects.

spinocerebellar tracts Pathways of neurons that transfer sensory information from the spinal cord to the cerebellum of the brain.

spinous process The prominence at the posterior extremity of a vertebra.

spiracle

1. External openings of insect tracheae for breathing.
2. Dorsally situated, reduced evolutionary remnant, first gill slit in many fishes.

spiral valve A spirally arranged fold of epithelium in the intestine of all fishes except teleosts. It delays the passage of food and provides an increased surface area for the secretion and absorption.

spirillum The rigid spiral-shaped bacterium. Generally, are Gram-negative aerobic and highly motile, bearing flagella either in tufts or singly. They occur in soil and water, feeding on organic matter.

spirochaete The nonrigid corkscrew-shaped bacterium that moves by means of flexions of the cell, produced by a series of rotatory axial fibrils underlying the cell's outer sheath. e.g. *Treponema.*

spirometer The instrument used for determining the volume of air inspired.

spiroscope A device for measuring respiratory capacity.

splanchnology Study of the internal organs.

spleen A vertebrate organ, lying behind the stomach, that is basically a collection of lymphoid tissue. Its functions include producing lymphocytes and destroying foreign particles. It acts as a reservoir for erythrocytes and can regulate the number in circulation. It is also the site for the breakdown of worn-out erythrocytes and it stores the iron they contain.

splenic flexure The angular bend in the large intestine between the transverse and descending colons.

spondylitis The inflammation of and resulting damage to the vertebrae.

sponge *See* Porifera

spongy bone The inner layer of bone, found at the ends of long bones and is less dense than compact bones.

spongy mesophyll Parenchyma cells found in plant leaves that are irregularly shaped and have large intercellular spaces.

spontaneous generation A supposed process that has never actually been observed in which living things arise from non-living matter.

sporangiophore The simple or branched stalk that bears one or more sporangia.

sporangium (pl. sporangia) The plant reproductive body in which asexual spores are formed.

spore

1. An asexual reproductive cell in a plant or fungus that does not participate in fertilisation.
2. A dormant, relatively impervious cell in prokaryotes that is resistant to destruction by heating.

spore mother cell (sporocyte) A diploid cell that gives rise to four haploid spores by meiosis. See also megaspore mother cell; microspore mother cell.

sporic meiosis In sporic meiosis, there are both haploid individuals and diaploid individuals within the life cycle. Meiosis produces haploid spores, which then divide to produce an individual that is made entirely of haploid cells.

sporocyst

1. Larval form in many trematode flatworms that develops in the body of its molluscan intermediate host.
2. A secondary cyst that develops and produce one or several sporozoites.

sporocyte *See* spore mother cell.

sporogenous tissue A group of compactly arranged homogenous cells in young anthers which occupy centre of each microsporangium.

sporogonium The sporophyte generation in mosses and liverworts. It is made up of an absorptive **foot**, a stalk (**seta**), and a spore-producing **capsule**. It may be completely or partially dependent on the gametophyte.

sporophore (fructification) The aerial spore-producing part of certain fungi, e.g. the stalk and cap of a mushroom.

sporophylls Leaf-like appendages bearing sporangia in ferns. It may form distinct compact structures called strobilli or cones in mosses, horsetails etc.

sporophyte In plants, a multicellular stage that produces haploid spores by meiosis.

sporulation (also sporogenesis) The production of spores.

sputum The ejected matter mixed with mucous and sometimes pus.

squamata The most diverse order of Class-Reptilia. It includes the lizards and snakes. Most lizards have four legs and a long tail, eardrums, and movable eyelids. Snakes are limbless reptiles that lack eardrums; the eyes are covered by transparent immovable eyelids and the articulation of the jaws is very loose, enabling a wide gape to facilitate swallowing prey whole.

squamous epithelium A type of epithelium that is made up of flattened cells and therefore presents a small distance over which substances have to pass, e.g. lining of alveoli in the lungs and Bowman's capsule in the kidney.

stabilising selection (normalising selection) A natural selection that acts to maintain the constancy of a species over successive generations. It involves selection against the extremes of the range of phenotypes for a particular characteristic.

staining A technique in which cells or thin sections of biological tissue that are normally transparent are immersed in one of more coloured dyes (**stains**) to make them more clearly visible through a microscope. If the colour comes from the negative ion (organic anion), the stain is described as **acidic**, e.g. eosin. If the colour comes from the positive ion (organic cation), the stain is described as **basic**, e.g. haematoxylin. **Neutral stains** have a coloured cation and a coloured anion; e.g. Leishman's stain.

stamen The part of the flower that produces pollen. Basically, a micro (or small) sporophyll. It is the stamen part where the pollen grains develop in the anther. The narrow stalk supporting the anther is the filament.

staminode A sterile stamen.

standing crop Total mass of living material at a particular time in each trophic level. Measured as the mass of living organisms (biomass) or the number in a unit area.

S

stapes (stirrup) The third of the three middle-ear bones in mammals.

Staphylococcus A genus of Gram positive cocci usually present on human skin and mucous membranes. Under the microscope, it can be seen as rounded structure and form clusters like bunches of grapes. Either by penetration or toxin production, bacteria in this genus cause many different diseases, both in human beings or in other animals (the toxins are a frequent source of food poisoning).

staple (food) A staple food, sometimes simply referred to as a staple. It is a food that is eaten routinely and in such quantities that it constitutes a dominant portion of a standard diet in a given population.

starch The storage polysaccharide (consist of D-glucose polymers, amylose and amylopectin) of most plants. Starch polymers are cleaved by the enzyme amylase.

start codon (initiation codon) The triplet of nucleotides on a messenger RNA molecule at which the process of translation is initiated. In eukaryotes the start codon is AUG, which codes for the amino acid methionine; in bacteria the start codon can be either AUG, coding for N-formly methionine, or GUG, coding for valine.

statoblast A chitinous bud that is produced asexually by members of the Bryozoa. It stores food and is able to remain dormant for long periods, withstanding drought and extreme temperatures.

statocyst (otocyst) A balancing organ found in many invertebrates. It consists of a fluid-filled sac lined with sensory hairs and contains granules of calcium carbonate, sand, etc. (statoliths). As the animal moves the statoliths stimulate different hairs, giving a sense of the position of the body or part of it. e.g. The semicircular canals in the ears of vertebrates.

statocyte A plant cell containing statoliths and thought to be involved in the perception of gravity.

statolith

1. *See* statocyst.
2. A membrane-bound group of starch grains in plant cells that is believed to act as a sensor to gravity.

stele The vascular tissue (xylem and phloem) of tracheophyte plants, together with the endodermis and pericycle (when present). The arrangement of stelar tissues is very variable. In monocotyledons the vascular bundles are scattered throughout the stem whereas in dicotyledons (and gymnosperms) they are arranged in a circle around the pith.

stem A longitudinal axis to plants upon which are borne the leaves, buds and reproductive organs of the plant.

stem cells Undifferentiated, primitive cells with the ability, both to multiply and to differentiate into various types of cells.

stem tuber A swollen structure produced by stolons and runners which remains dormant during adverse conditions and later grows into a new plant when the conditions become favourable for growth.

stenohaline Organisms that are unable to tolerate small variations of salt concentrations in environment.

stenosis The abnormally narrow opening.

stenostomia The narrowing of the mouth.

stenothermal Having the capacity to tolerate only a narrow range of temperatures.

stereocilium (pl. stereocilia) A relatively short, nonmotile cilium, some 20 to 300 of which form an array on the apical surface of a sensory hair cell. Their shaft contains numerous fine longitudinal actin filaments.

sterigma (pl. sterigmata) A small spore-bearing stalk found in certain fungi. In the Basidiomycota, sterigmata develop as outgrowths of a basidium, each bearing a basidiospore at its tip.

sterile

1. (of organisms) Unable to produce offspring.
2. (of objects, food, etc.) Free from contaminating microorganisms.

sterilisation

1. The process of destroying microorganisms that contaminate food, wounds, surgical instruments, etc. Common methods of sterilisation include heat treatment and the use of disinfectants and antiseptics.
2. The operation of making an animal or human incapable of producing offspring. Men are usually sterilised by va-sectomy and women by tubectomy.

sternebra (pl. sternabrae) The individual segment of the sternum, in early life, that later fuse to form body of sternum.

sternum (breast bone)

1. A flat bone at the front of the chest that articulates with clavicle of the pectoral girdle and with most of the ribs.
2. The ventral portion of each segment of the exoskeleton of arthropods.

steroid A class of lipids characterised by a four carbon ring structure. e.g. Male hormones (androgens) and female hormones (oestrogens) are steroids.

sterols A class of complex cyclic alcohols with a tetracyclic structure. Cholesterol is the most widely known sterol.

sticky end (cohesive end) A single unpaired strand of nucleotides protruding from the end of a double stranded DNA molecule. It is able to join with a complementary single strand, e.g. the sticky end of another DNA molecule, thus forming a single large double-stranded molecule. Sticky ends provide a means of annealing segments of DNA in genetic engineering, e.g. in the packaging of vectors.

stigma

1. The sticky tip of a carpel; the stigma is the part of a flower that receives pollen grains.
2. *See* eye spot.

stilt root An enlarged form of prop root, seen in some mangrooves that helps to support plants in unstable soil.

stimulus A change in the external or internal environment of an organism that elicits a response in the organism. The stimulus does not provide the energy for the response.

stipe

1. The stalk that forms the lower portion of the fruiting body of certain fungi, such as mushrooms, and supports the umbrella-shaped cap.
2. The stalk between the holdfast and blade (lamina) of certain brown algae, notably kelps.

stipule The outgrowth from the petiole or leaf base of certain plants. Those of the garden pea are like leaves and photosynthetic organs. Those of the lime tree are scalelike and protect the winter buds, whereas those of the false acacia (*Robinia*) are modified as spines.

stolon The sub-aerial modification of stem, grows horizontally, roots at nodes.

S

stoma (pl. stomata) Microscopic pores in the epidermis of plants; stomata allow gas exchange with the atmosphere. Each stoma is bordered by two semicircular guard cells, whose movements (due to changes in water content) control the size of the aperture.

stomach The part of the alimentary canal in vertebrates that lies between the oesophagus and the duodenum. It is a large sac with thick muscular walls and is closed at each end by a ring of muscle.

stomium A region of thin-walled cells in certain spore-producing structures that ruptures to release the spores, e.g. in the sporangium of the fern *Dryopteris* the stomium ruptures when the annulus dries out.

stomodaeum The invagination of ectoderm of the animal embryo that forms the anterior region of the alimentary canal—the mouth, pharnyx and oesophagus in mammals.

stool The excrement discharged from the bowls.

stop codon (termination codon) The triplet of nucleotides on a messenger RNA molecule at which the process of translation ends. It is recognised by proteins called release factors, which bind to the A site of the ribosome. This effectively stops the formation of a polypeptide chain at that point. The three stop codons are UGA, UAA, and UAG.

stratification

1. The arrangement of the components of an entity in layers. It is a feature of sedimentary rocks and soils. It is also seen in stratified epithelium and thermal stratification can occur in some lakes.
2. The practice of placing certain seeds between layers of peat or sand and then exposing them to low temperatures for a period, which is required before they will germinate.

stratified epithelium The epithelium that is made up of a number of layers of cells and is present in areas of the body that are subject to wear and tear. Stratified epithelium containing keratin forms the outer layer of the skin; stratified epithelial cells that do not contain keratin are found in moist areas, such as the mouth and vagina.

stratosphere The layer of the Earth's atmosphere that lies above the troposphere and extends to about 50 km above the Earth's surface.

stratum corneum The outermost, horny layer of the mammalian epidermis. Consist of several layers of flat, keratinised non-nucleated dead cells.

stratum germinativum The epidermis's innermost layer, which is composed of columnar epithelial cells that divide to replace the outer layers as they wear away.

stratum granulosum The thin layer of the epidermis lying between the stratum spinosum and stratum lucidum. Keratinocyte become granular in this region.

stratum lucidum A transparent layer of the skin. It lies between the stratum granulosum and stratum corneum.

stratum spinosum The layer of epidermis between stratum granulosum and stratum basale.

strepticemia The presence of streptococci in the blood.

Streptococcus *(pl. streptococci)* A genus of Gram positive cocci, of which most strains are harmless. However, some are both common and among the most dangerous of human pathogens. e.g. speumoniae.

stretch receptor A specialised cell or group of cells found in muscles or tendons that is sensitive to mechanical stress.

stretch reflex (myotatic reflex) The reflex initiated when a muscle is stretched; e.g. kneejerk reflex.

striated muscle *See* voluntary muscle.

stridulation The production of sounds by insects rubbing one part of the body against another.
As seen in Orthoptera (grasshoppers, crickets, cicadas), in which the purpose of the sounds is usually to bring the sexes together mainly but also used in territorial behaviour, warning, etc.

strobilus

1. A type or composite fruit that is formed from a complete inflorescence. It produces achenes enclosed in bracts and when mature becomes cone-shaped. e.g. hop fruit.
2. *See* cone.

stroma The tissue that forms the framework of an organ; e.g. the tissue of the ovary that surrounds the reproductive cells, or the gel-like matrix of chloroplasts that surrounds the grana.

structural gene A gene that encodes a product (*e.g.* a polypeptide, *t*RNA or *r*RNA). *See* operon.

structural protein A protein that contributes to cell or tissue structure.

style The stalk of a carpel, between the stigma and the ovary. In many plants it is elongated to aid pollination.

styloid process A slender, pointed process on a bone at the attachment point of certain muscles (in particular, the styloid process of the temporal bone).

subarachnoid space The space between the middle arachnoid membrane and the inner pia mater, which surrounds and protects the brain and spinal cord in vertebrates. It is crossed by delicate tissue strands and filled with cerebrospinal fluid, which cushions the central nervous system against external shocks.

subclavian artery In mammals, a large paired artery that carries oxygenated blood to the forelimbs. It arises on the left side from the aorta and on the right side from the innominate artery.

subcutaneous tissue The tissue that underlies the dermis. It is made up of loose fibrous connective tissue, muscle and fat, which in some animals (e.g. whales and hibernating mammals) forms an insulating layer or an important food store.

suberin A waxy, waterproof, substance found in the layers of dead cork cells that sheath woody stems and mature roots.

suberisation Infiltration of plant cells with suberin, as in formation of cork tissue.

sublittoral

1. Designating or occurring in the shallow-water zone of a sea, over the continental shelf and below the low tide mark.
2. Designating or occurring in the zone of a lake below the littoral zone, to a depth of 6-10 metres.

submucosa A layer of areolar connective tissue that lies beneath the mucosa, e.g. in the stomach and intestine.

subsidiary cells Epidermal cells, present in the vicinity of the guard cells, specialised in their shape and size.

S

subspecies A group of individuals within a species that breed more freely among themselves than with other members of the species and resemble each other in more characteristics.

substrate

1. The substance upon which an enzyme acts in biochemical reactions.
2. The material on which a sedentary organism (such as a barnacle or a plant) lives or grows. The substrate may provide nutrients for the organism or it may simply act as a support.

succession The sequence of communities that develops in an area from base situation to stable mature climax community. It has a number of intermediate stages called sere.

succulent The xerophyte plant that store water within its tissues. They have fleshy leaves or stems.

succus entericus The digestive juice secreted by glands in the duodenum of vertebrates. Contains numerous enzymes, e.g. erepsin, sucrose, lactose and enterokinase.

sucker (turion) A lateral branch arising from the underground root or stem at its node. It grows at the expense of the parent plant.

sucrose The enzyme that hydrolyses sucrose into fructose and glucose, produced in small intestine.

sucrose (cane sugar; beet sugar; saccharose) A sugar comprising one molecule of glucose linked to a fructose molecule. It occurs widely in plants and is particularly abundant in sugar cane and sugar beet (15-20%), from which it is extracted and refined for table sugar.

sugar (saccharide) The group of water-soluble carbohydrates of relatively low molecular weight and typically having a sweet taste. The simple sugars are called monosaccharides. More complex sugars comprise between two and ten monosaccharides linked together: disaccharides contain two trisaccharides three, and so on.

sulpha drugs *See* sulphonamides.

sulphonamides Amides of sulphonic acids. Many have antibacterial action and are also known as **sulpha drugs**, including sulphadiazine, sulphathiazole, and several others.

They act by preventing bacteria from reproducing and are used to treat a variety of bacterial infections, especially of the gut and urinary system.

sulphur bacteria The group of unrelated bacteria that utilise sulphur, sulphide, or sulphate in their metabolism. e.g. photoautotrophic green sulphur bacteria and purple sulphur bacteria.

sulphur cycle The cycling of sulphur between the biotic (living) and abiotic (nonliving) components of the environment.

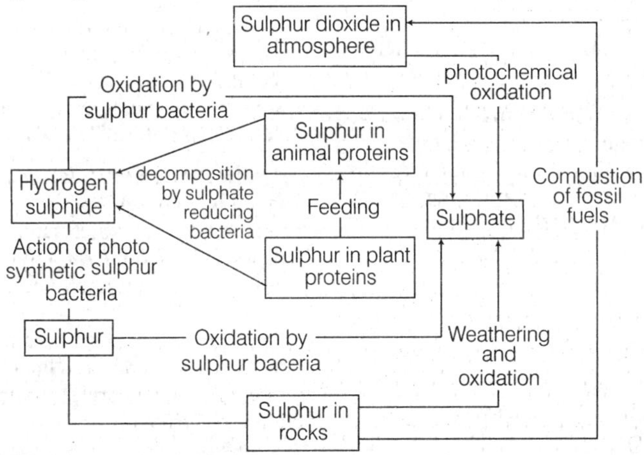

Oxidation by sulphur bacteria

supercoiling A form of DNA in which the double helix is further twisted about itself, forming a tightly coiled structure. This is the form generally adopted by DNA in nature and enables it to condense sufficiently to be packaged into living cells.

supergene A collection of closely linked genes that tend to behave as a single unit because crossing over between them is very rare.

superior Describing a structure that is positioned above or higher than another structure in the body, e.g. in flowering plants the ovary is described as superior when located above the other organs of the flower.

supernormal stimulus

1. A stimulus that produces a more vigorous response that the normal stimulus eliciting that particular response in animal behaviour.
2. (in neurophysiology) It is a stimulus that is more intense than a normal stimulus and is capable of inducing a response in a nerve fibre during the relative refractory riod.

supernumerary chromosome *See* accessory chromosome.

supination The rotation of the lower forearm so that the hand faces forwards or upwards with the radius and ulna parallel.

surgery

1. A manual procedure, usually involving incision or excision and the creation of a suture, to find out whether disease is present or to remove or repair a part of the body in order to relieve pain, correct a defect, cure disease, or prolong life.
2. The branch of medicine dealing with such procedures.

suspension culture A type of culture in which intact cells are maintained in suspension in the culture medium so that they are distributed evenly within it.

suspensor The chain of cells that anchors a plant embryo in the surrounding gametophyte tissue. In flowering plants the suspensor attaches the embryo to the embryo sac and extends to push the embryo into the endosperm.

suspensory ligaments Ligament that supports a body part, especially an organ, e.g. suspensory ligament of lens (Zinn's membrane) connecting the ciliary body with the crystalline lens of the eye.

suture The line marking the junction of two body structures. Examples are the immovable joints.

sweat A watery fluid secreted by the sweat glands. Contains small amount of sodium chloride, potassium ions and urea.

sweat glands Epidermal glands embedded deep in the dermis of the skin of mammals secreting a hypotonic solution of the salts and other small molecules present in blood. Cools the body by evaporation.

swim bladder (air bladder gas bladder) The air-filled sac lying above the alimentary canal in bony fish that regulates the buoyancy of the animal.

syconus A type of composite fruit formed from a hollow fleshy inflorescence stalk inside which tiny flowers develop. Small drupes, the 'pips', are produced by the female flowers.

symbiont The organism that is a partner in a symbiotic relationship.

symbiosis A relationship between two distinct types of organisms, in which they live together. Symbiosis is of various types as parasitism, mutualism and commensalism etc.

symmetry The regularity in the arrangement of the parts of an organism. *See* bilateral symmetry; radial symmetry.

symmorphosis A hypothesis, pro-postulating that biological systems adhere to an 'economy of design' giving a close match between their various structural and functional parameters. Hence, no single parameter in the system has unnecessary excess capacity beyond the requirements of the system.

sympathetic nervous system One of the two divisions of the autonomic nervous system, which supplies motor nerves to the smooth muscles of internal organs and to heart muscle. Their endings release mainly nor-epinephrine hormones which increases heart rate and breathing rate, raises blood pressure and slows digestive processes.

sympathetic tone The condition of a muscle when the tone is maintained predominantly by impulses from the sympathetic nervous system.

sympatric speciation A process by which new species evolve within the range of its parent species without any sort of geographical isolation. This type of speciation is well documented in plants where it typically involves polyploidy.

symphysis

1. A line of fusion between two bones that were once separate.
2. A joint, lined by cartilage, between two bones (as between two vertebrate).

symplastic system The system of interconnected protoplasts in plants.

symplesiomorphy The ancestral trait that is shared by two or more modern groups. e.g. simple leaves of some modern flowering plants are probably inherited from simple-leaved ancestors. But this does not mean that all simple-leaved modern plants are closely related.

sympodium The composite primary axis of growth in such plants as lime and horse chestnut. After each season's growth the shoot tip of the main stem stops growing growth is continued by the tip of one or more of the lateral buds.

symporter A transport protein that couples the movement of a substance across a cell membrane with the simultaneous movement of ions in the same direction.

synapse The physiological junction between two neurons or between a neuron and a muscle cell across which nerve impulses can be transmitted. Synapses occur between the knob like axon endings of one neuron and the dendrites or cell body of another.

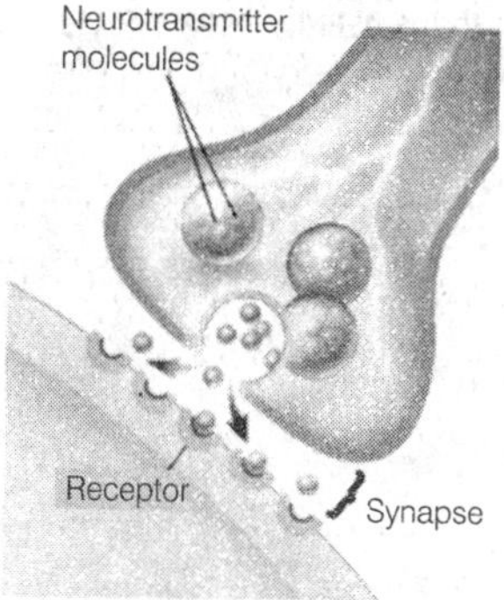

A nerve cell showing where neurotransmitters are exhanged

synapsis *See* pairing

synaptic knobs These are bulbous structures at the end of an axon, each of which contains vesicles of neurotransmitter substance.

synaptic plasticity The change in efficacy or connections of the junctions (synapses) between neurons in the nervous system.

synaptic vesicle It is one of the many minute vesicles in a synaptic knob where neurotransmitters are stored.

synaptonemal complex It is a complex of proteins that binds together the members of each homologous pair of chromosomes during pairing in prophase of the first division of meiosis. Formation of the complex begins at the telomeres, which at this stage are attached to the nuclear envelope and proceeds along the length of the chromosomes, 'zipping' them together so that matching regions are aligned.

syncarpous The condition when carpels are fused in flower, e.g. in mustard and tomato.

syncytium (pl. syncytia)
A multinucleated mass of cytoplasm, a multinucleated cell, e.g. cells of striated muscle.

syndrome The group or recognisable pattern of symptoms or abnormalities that indicate a particular trait or disease. e.g. Down's syndrome.

synecology The study of all the living and relevant non-living components of a natural community and their relationships with each other.

synergid cells Two haploid cells located near the egg cell at the micropylar end of the embryo sac, in flowering plants. They do not participate in the fertilisation process and abort soon afterwards.

synergism (summation)

1. The phenomenon in which the combined action of two substances (e.g. drugs of hormones) produces a greater effect than would be expected from adding the individual effects of each substance.
2. The combined action of one muscle (the synergist) with another (the agonist) in producing movement.

syngamy *See* fertilisation.

syngeneic Genetically identical.

synovial fluid A form of lymph which bathes the synovial membrane covering the ends of a bone in a movable joint. It is lubricating in nature.

synovial joint The fluid-filled synovial cavity present between the articulating surfaces of the two bones. Helps in locomotion. Examples are ball and socket joint, hinge joint, pivot joint, gliding joint and saddle joint.

synovial membrane The membrane that forms the capsule surrounding a joint. It consists of tough connective tissue with a high proportion of white collagen fibers and it secretes the viscous lubricating synovial fluid.

syntenic group A block of genes occurring in the same order in two different types of organisms.

syrinx The sound-producing organ in birds, present at the bifurcation of trachea into bronchi (i.e at base of trachea).

systematics Study of the diversity of living organisms, their relationship to one another and their classification.

systemic arch Blood vessel found in adult tetrapods that carries usually oxygenated blood from the heart to the dorsal aorta. It is derived from the fourth aortic arch. Amphibians and reptiles retain both arches, birds only the right arch and mammals only the left arch.

systemic circulation Part of the circulatory system of birds and mammals that transports oxygenated blood from the left ventricle of the heart to the tissues in the body and returns deoxygenated blood from the tissues to the right atrium of the heart.

systole Portion of the cardiac cycle during which the cardiae muscles contract.

syzygy Fusion of two or more organs or portions of the body in certain protozoans usually before sexual reproduction.

T

tachycardia An increase in heart rate. The heart rate is normally approximately 70 beats per minute. During exercise, there is an increase in sympathetic activity and tachycardia occurs and the heart rate increases to up to 110 beats per minute. Immediately after exercise, at rest, the heart rate decreases to around 60 beats per minute (bradycardia), due to an increase in parasympathetic activity.

tachymetabolism A relatively high metabolic rate. It is characteristic of warm blooded animals.

tactic movement *See* taxis.

tactile Of, or pertaining to the sense of touch.

tadpole The young aquatic larva of any amphibian.

tagma (pl. tagmata) Section of the body of an arthropod with distinct function and structure, formed by the fusion of mesodermal sometime. The basic tagmata are the head, thorax and abdomen, but the form of the tagmata (known as tagmosis) varies between arthropod groups, each group having its characteristic tagmosis, e.g. many crustaceans have a cephalothorax and abdomen, while arachnids have a prosoma and opisthosoma.

taiga Coniferous forest with thin, acidic soil. A biome characteristic of the northern regions of Eurasia and North America, where summers are short and winters are long and cold.

tallow Rendered fat of animals, particularly cattle and sheep.

talus (ankle bone) Articulates distally with the calcaneus and navicular bone of the tarsus and proximally with the fibula and tibia.

tandem repeats (also tandem duplications) Multiple copies of the same nucleotide base sequencing lying in series; often used as markers in physical mapping.

tannin Heterogenous group of phenol derivatives. Strongly astringent substances dissolved in cell sap; common in bark, unripe fruits, leaves, galls. Used in tanning of leather and ink-making.

tap root The root system having a prominent main root, that grows vertically downwards bearing smaller lateral roots. Develop from radicle of the embryo.

tapetum

1. In anther, a layer of cells lining the locule and absorbed when pollen grains mature.
2. A reflecting layer in choroid of eye of vertebrate nocturnals contain crystals of guanine.

tapeworms *See* Cestoda.

taphonomy The study of biological, chemical and physical processes which change organisms after death leading to their preservaton as fossils. Iniatially, it leads to decay or removal of soft parts by microbes.

tarsal (tarsal bone) One of the bones that form the ankle in terrestrial vertebrates.

tarsus An ankle or corresponding part of the hindlimb in terrestrial vertebrates. It consists of a number of small bones (tarsals). The number of tarsal bones varies with the species: e.g. humans.

taste

1. Sense that enables the difference in flavour of different substances.
2. Flavour of a substances.

taste bud Small sense organ in most vertebrates, specialised for the detection of taste. The taste bud containing gustatory receptors have distinct distribution patterns on the upper surface of the human tongue, sensitive to four types of taste, i.e. sweet, salt, bitter or sour.

TATA box (Hogness box) Analogous to Pribnow box in prokaryotic promotors. It is a sequence of nucleotides which serve as site for attachment of RNA polymerase in promotor region of eukaryotic gene.

taxis (taxis response; tactic movement) Behaviour consisting of a directed movement of the cell or a microorganism towards or away from a stimulus.

taxon (pl. taxa) A taxonomic group of any rank in the hierarchical classification of organisms.

taxonomy In biology, the science dealing with the classification and identification of organisms.

TCA cycle *See* Krebs' cycle

T-cell (T-lymphocyte) Lymphocyte involved primarily in cell-mediated immunity. Like B-cells, T-cells originate in the bone marrow, but during embryological development, they migrate to the thymus where they mature and differentiate. Sub-types of T-cell play different roles in immune response and are characterised by their surface antigens.

T-cell receptor A protein molecule found attached to the plasma membrane of a T-cell that recognises foreign antigen displayed on the surface of host cells and activates an immune response by the T-cell. A single T-cell typically has about 30000 such receptors.

tectorial membrane Thin elastic membrane above the rows of the hair cells in internal ear.

tectum A region of the brain specifically the dorsal part of the mesencephalon (midbrain).

tegmen

1. Thickened and leathery forewing in the Orthoptera (grasshopper) and Dictyoptera (cockroach).
2. Inner layer of seed coat.

teleostei The major superorder of the bony fishes, that have been dominant, since the Cretaceous period (about 70 million years ago). Teleosts have colonised an extensive variety of habitats and show great diversity of form. The group includes the eel, seahorse, plaice and salmon.

telocentric chromosome Chromosome having its centromere at a terminal position.

telodendron Fine terminal branches of an axon. They have knobbed ends called end bulbs or buttons, that releases neurotransmitters.

telolecithal An ovum with yolk concentrated at one pole, e.g. eggs of reptiles, birds, fish and monotremes.

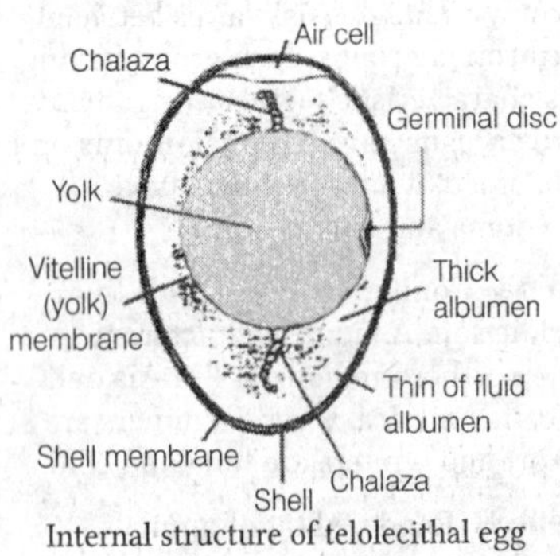

Internal structure of telolecithal egg

telome theory The theory which states that leaves of ferns and seed plants evolved by modification of terminal branches of stem (telomes).

telomerase A ribonucleoprotein enzyme that adds individual bases to produce repeating units of one DNA strand at the telomere.

telomere Region at the tip of a chromosome that consists of repeating sequences of nucleotide bases. Functions ensuring that each cycle of DNA replication have been completed. The telomere is restored to its original length by an enzyme called telomerase.

If a particular cell does not have telomerase, the cell can only undergo a set number of cell divisions until dying out. Scientists believe this process is related to the ageing process of multicellular organisms including humans.

telophase One of the stage in cell division. In mitosis, the separated chromatids collect at the poles of spindle, nuclear membrane forms around each, forming two daughter nuclei. In meiosis-I complete chromosomes from the pairs that separated at first anaphase form the daughter nuclei. The number of chromosomes in these nuclei is therefore, half the number in the original one. In the second telophase, daughter nuclei are formed from chromatids (as in mitosis).

template Molecule used as a pattern for the synthesis of another new molecule.

temporal bones A symmetrical pair of bones which lie on either side base of the skull.

tendon A thick strand or sheet connects a muscle to bone or cartilage. It consists of a mass of parallel, non-elastic, white collagen fibres continuous with those of the muscle sheath and the periosteum of the bone. It pulls the bone, causing movement at joint.

tendril Modification of stem leaf, leaflet or petiole into a slender, thread-like structure. Used for support in climbing.

ten per cent law States that only 10% of the energy entering a particular trophic level in a food chain is available for transfer to the next higher trophic level. It was introduced by Raymond Lindeman (1942).

tentacle A soft flexible appendages in aquatic invertebrate animals that are used principally for feeding. Tentacles are possessed by many cnidarians, some echinoderms and by cephalopod molluscs, in which the bear rows of suckers.

tepal The part of perianth not differentiated into calyx and corolla.

teratogen An environmental factor that acts on a fetus to cause congenital abnormality, e.g. X-rays nutritional deficiencies, drugs (e.g. thalidomide), toxic chemicals and virus infections (e.g. rubella).

teratology Study of plants and animal physiological abnormalities.

tergum A plate, stiffened by chitin, that forms the protective dorsal covering to each segment of the thorax and abdomen of an insect.

terpenes Group of unsaturated hydrocarbons present in plants, consist of isoprene units. Its derivatives (terpenoids) include ABA and gibberellin (plant hormones) and pigments (chlorophyll and carotenoid) used in photosynthesis.

territory An area containing a resource that is actively defended by an animal, from other animals (of the same species) or in some cases of different species. Note that a territory is not the same thing as a home range, in that, a home range is where the animal may tend to roam.

tertiary consumer A carnivore that feeds on other carnivores, below it in a food chain.

tertiary period Formerly, it was older geological period of the Cenozoic era. It began about 65 million years ago and extended to the about 2 million years ago. It is characterised by the rise of the modern mammals and the development of shrubs, grasses and other flowering plants. It has been replaced by the Palaeogene and Neogene periods.

tertiary structure of protein The three dimensional shape of the coiled or pleated polypeptide.

test cross A mating designed to test, if an organism with the dominant phenotype is homozygous or heterozygous (carrying a recessive allele). This is done by mating the individual with a homozygous recessive individual. Today, many genetic systems biochemical tests which eliminate the need for test crosses.

1. For example, F_1 hybrid (Tt) heterozygous of a pure tall plant (TT) and a pure dwarf plant (tt) is crossed with a pure dwarf plant.

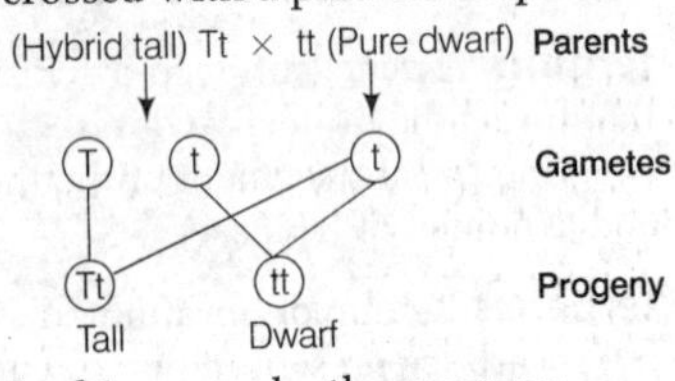

In this example, the progeny consists of tall and dwarf plants in the ratio of 1 : 1. Thus, monohybrid test cross ratio is 1:1.

2. In case of both homozygous parents, i.e. TT, the progeny obtained will have all tall plants.

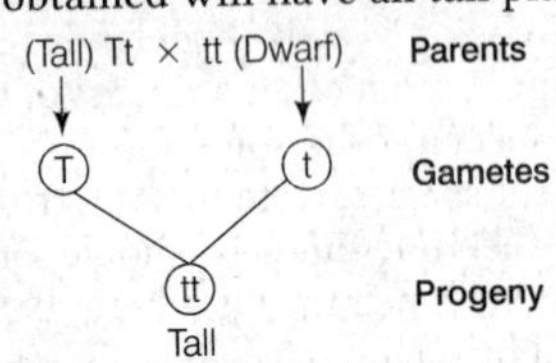

testa (seed coat) The outer hard, dry layer of seed coat. It may be lignified or fibrous.

testacea

1. Rhizopoda having a shell.
2. Formerly, a name for shellfish of any kind.

testis (testicle) A male reproductive organ in animals. In vertebrates, two are normally present, that produces spermatozoa and androgens. In mammals, including humans, the testes are present in the scrotum.

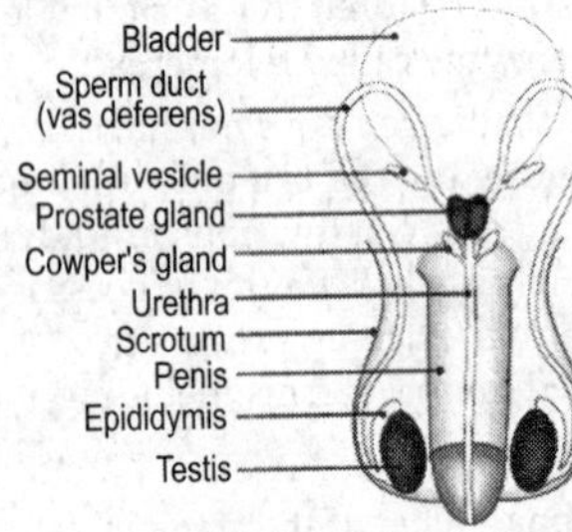

Position of testes in human

testosterone In mammals, the primary androgen, an anabolic steroid; secreted primarily in the testes of males (the adrenals also secrete small amounts).

tetanus

1. A sustained maximal contraction of a voluntary muscle. It results from a rapid succession of nerve impulse to the muscle.
2. A disease caused by the bacterium *Clostridium tetani*, which generally enters the body *via* an open wound. A toxin produced by the bacterium irritates the nerves, which induce muscle spasms that begin in the jaw muscles (giving the disease the informal name of lockjaw).

tetany Rapid spasms in muscles due to the low Ca^{2+} in body fluid.

tetrarch Primary xylem of root having four protoxylem strands.

tetrad Group of four haploid cells or nuclei at the end of second meiotic division.

tetraploidy A condition in which a cell has or an organism has, four complete sets of chromosomes (4*n*).

tetrapod The amphibian, reptile, bird or mammal having four limbs or being descended from such an animal.

tetraspore In life cycle of red algae, any one of the four spores produced by meiosis in an independent diploid sporophyte. These are produced within tetrasporangium and germinate to produce sexual haploid generation.

thalamus

1. A part of the forebrain that relays sensory and motor impulses to and from the appropriate centers in the cerebral hemispheres (in zoology).
2. A receptacle of flower (in botany).

thalassaemia A blood disease caused by a substitution mutation in the DNA sequence that controls the synthesis of the α or β-chains of haemoglobin. It results in anaemia and growth retardation.

Thallophyta Most primitive division of plant kingdom which includes algae, fungi and lichens. The body is simple and not differentiated into root, stem and leaves.

thallus A simple vegetative plant-body showing no differentiation into root, leaf and stem and lacking a true vascular system.

theca (pl. thecae) A thick outer covering present in some unicellular protists.

thecodont Teeth embedded in socket of jaw bone as in mammals.

thelytoky The phenomenon of reproduction in certain animals in which fertilised eggs give rise to males and unfertilissed eggs to females. Found among aphids and certain other insects and some mites. The males are diploid, whereas females are haploid and transmit only the maternal genome their production represents thelytokous parthenogenesis. *Compare* arrhenotoky.

therapeutic drugs Relating to therapeutics, the branch of medicine that is concerned specifically with the treatment of disease. The therapeutic dose of a drug is the amount needed to treat a disease.

thermocline Steep temperature gradient that exists in the middle zone (the metalimnion) of a lake and gives rise to thermally induced vertical stratification of the water.

thermogenesis The process of generation of heat by the body due to increased metabolic rate, e.g. shivering.

thermonasty A nastic movement in response to change in temperature.

thermoneutral zone The range of environmental temperatures over which the heat produced by a 'warm-blooded' animal remains fairly constant neither generate extra heat to keep warm nor expend metabolic energy on cooling mechanisms. Animals adapted to cold environments tend to have broader thermoneutral zones than ones living in hot environments.

thermoperiodism A phenomenon shown by some plants in which there is a response to alternating periods of low and high temperatures.
e.g. *Chrysanthemum* and tomato.

thermophile An organism that thrives at temperatures, above which life normally can not be found (usually 50 degrees celsius or above). Some thermophiles in hot springs and in deep sea thermal vents live at near the boiling point of water.

thermoreceptor A group of specialised cells in the skin, sensitive to temperature.

thermoregulation The regulation of body temperature by any means (physiological or behavioural). Some animals, particularly mammals and birds, can maintain a fairly constant internal body temperature.

therophyte Annual plant that complete their life cycle in a short period when conditions are favourable and survive harsh condition as seed. It is a plant life form in Raunkier's system of classification.

thigmotropism An orientation toward or away from a touch stimulus. Found in the tendrils of many vines.

thoracic cavity The space within the thorax, which contains the heart, lungs, and rib cage, in vertebrates.

thoracic duct The main collecting vessel of the lymphatic system. It runs longitudinal in front of the backbone and drains its lymph into the superior vena cava.

thoracic vertebrae The vertebrae of the upper back which articulate with the ribs. They lie between the cervical vertebrae and the lumbar vertebrae and are distinguished by a number of articulating facets for attachment of the ribs. In humans there are 12 thoracic vertebrae.

thorax

1. Portion of the body between the neck and diaphragm.
2. In insects, the second of the three body segments, between the head and abdomen.

thorn A hard side stem with a sharp point at the tip, replacing the growing point. In some plants, the its development and subsequent suppression of the growing points may be a response to dry conditions, e.g. thorns of gorse and hawthorn.

thread cell (nematoblast; cnidoblast) A specialised cell found only in the ectoderm of the Cnidaria. It contains a nematocyst (a fluid-filled sac with a long hollow coiled thread). When a small sensory projection (cnidocil) on the surface of the thread cell is touched (e.g. by prey) the thread is shot out and adheres to the prey, coils round it, or injects poison into it. Numerous thread cells on the tentacles of jellyfish produce their sting.

threonine An essential amino acid.

threshold An electrical potential across a neuron cell membrane that is just sufficient to trigger an action potential.

thrombin An enzyme that catalyses the conversion of fibrinogen to fibrin. See blood clotting.

thrombosis Formation of thrombus or clot in a vessel, causing its obstruction.

thylakoid A flattened membranous sac that comprises part of the granum in plant chloroplast. The pigment system and electron transport system for the light dependent reactions of photosynthesis are part of the thylakoid membrane.

thymidine A nucleoside consisting of one thymine molecule linked to a D-deoxyribose sugar molecule.

thymine (T) One of the four nucleotide bases, a pyrimidine found in DNA, but not RNA; pairs with adenine.

thymocyte An immature lymphocyte in the thymus that develops to T-cell. Progenitor cells enter the thymus from bone marrow to become thymocytes which proliferate and undergo distinct developmental stages marked by changes in their CD surface proteins.

thymosin A peptide hormone secreted by thymus gland. Plays major role in the differentiation of T-lymphocytes which provide cell-mediated immunity.

thymus gland A body organ located in the chest behind the sternum only in vertebrates. Formerly, considered functionless, but now known to play an important role in the immune system. In the thymus, lymphocyte precursors become thymocytes and then mature into T-cells.

thyrocalcitonin *See* calcitonin

Thyroglobulin (TGB) Glycoprotein made in the thyroid gland. TGB is the precursor of the thyroid hormones, thyroxine and triiodothyronine. Iodine binds to the tyrosine residues in thyroglobulin, which is then hydrolysed into **iodotyrosines** that combine to form triiodothyronine (T_3) or thyroxine (tetraiodothyronine or T_4).

thyroid

1. Secretory gland at the front of the neck in vertebrates. It surrounds the front of the upper trachea. It is made up of two lobes connected by an isthmus and is grossly enlarged in goitre. It secretes two iodine-containing thyroid hormones, thyroxine (T_4; see formula) and tri-iodotyronine (T_3), which are formed in the gland from thyroglobulin;
2. Thyroid extract (derived from the thyroid glands of cattle), used as a therapeutic drug.

thyroid hormone Hormones produced by the thyroid gland which increase cell metabolism. Most important is thyroxine.

thyrotropin (Thyroid Stimulating Hormone TSH) A hormone produced by the anterior pituitary gland, that stimulates the thyroid gland to release thyroxine. Level of thyroxine controls thyrotropin release by a negative feedback mechanism.

Thyrotropin-Releasing Hormone (TRH) The secretion of thyroid-stimulating hormone is controlled by Thyrotrophin-Releasing Hormone (TRH) from the hypothalamus. The release of TRH depends on many factors, including the levels of TSH, glucose and thyroxine in the blood and the rate of metabolism in the body.

thyroxine (T_4) Principal hormone of the thyroid gland. Together with T_3, control the rate at which body uses energy.

tibia Larger of the two bones in the lower leg. It lies at the inner side of the leg while the smaller fibula lies toward the outside.

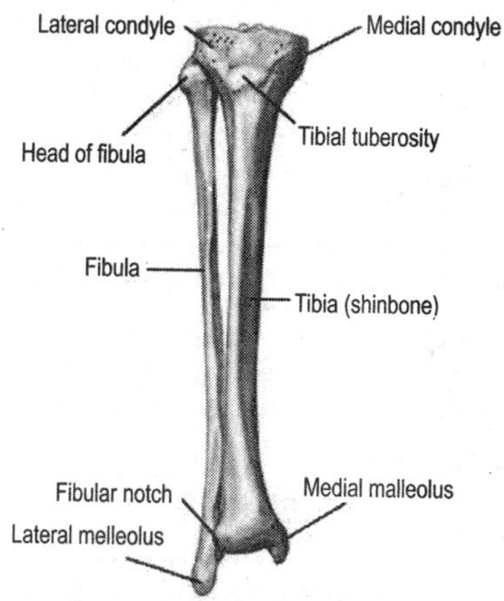

Position of tibia and fibula bone in human leg

tidal volume The volume of air inspired or expired during normal respiration. In an average human, approx 500 cm^3.

tight junction (zonula occludens) Structure that encircles many animal epidermal cells holding adjacent cells tightly together with no intercellular space. Tight junctions prevent material at the epidermal surface from penetrating between the cells, particularly at sites of absorption, e.g. gut and kidney.

tiller Shoot that develops at the base of a plant stem from an axillary bud, are often produced in response to injury of the main stem, as occurs when a tree is lopped.

tinsel A type of eukaryotic flagellum containing numerous mastigonemes (hair-like projections) along the shaft.

Ti-plasmid (Tumour inducing plasmid) A plasmid of *Agrobacterium tumefaciens.* It is a segment of it is transferred to cells of the plant host. This T-DNA (transfer DNA) segment which comprises the genes responsible for the tumorous growth, becomes integrated into the genome of infected plant cells. Possession of the Ti plasmid has made *A. tumefaciens* an important tool in genetic engineering for the introduction of foreign genes into plant tissue. The turnour inducing genes are usually replaced with the gene of interest and a marker gene (e.g. the antibiotic resistance gene) is added to enable selection of transformed cells.

tissue A group of similar cells bound together by intercellular material that carries out a common set of functions.

tissue culture Part of living body taken out and grown in a test tube under sterile conditions in special nutrient media.

tissue engineering The production of synthetic or semisynthetic tissue that can be used in place of human tissue in surgery. Different kinds of tissue have been developed or are currently being researched, including skin, bone, cartilage, cornea and spinal tissue, using various combinations of biopolymers, cultured cells and growth factors.

tissue fluid The fluid, consisting of water, ions and dissolved gases and food substances, that is formed when blood is ultrafiltered from the capillaries into the intercellular spaces. It surrounds the body cells, facilitating the exchange of nutrients and waste materials.

tissue typing The process of identifying the Major Histocompatibility (MHC) antigens in humans encoded by the HLA system in both the recipient and the potential donor before organ transplantation. If the donor and recipient tissue types do not match exactly, the transplanted organ will be rejected.

titre

1. Number of infectious virus particles present in a suspension.
2. Measure of the amount of antibody present in a sample of serum, given by the highest dilution of the sample that results in the formation of visible clumps with the appropriate antigen.

TLC Thin Layer Chromatography.

tobacco Obtained from the leaves of plant *Nicotiana tabacum* contains an alkaloid nicotine–a poisonous substance which stimulates the central nervous system of the user.

Tobacco Mosaic Virus (TMV) A rigid rod-shaped RNA-containing virus. It causes distortion and blistering of leaves in a wide range of plants, especially the tobacco plant and is transmitted by insects when they feed on plant tissue. It was the first virus to be discovered.

tocopherol *See* vitamin-E

tolerance

1. The ability of an organism to withstand extreme variations in environmental conditions.
2. Build-up of resistance to drugs or other chemicals (such as pesticides), which occurs after prolonged use or application. Increasingly large doses of the chemical are required to produce the desired effect in the organism.
3. *See* immunological tolerance.

tone (tonus) State of sustained tension in muscles that is necessary for the maintenance of posture.

tongue A muscular organ situated in the buccal cavity of vertebrates. In mammals, it consists of smooth and striped muscle covered with stratified epithelium, beneath which is loose connective tissue. It plays an important role in manipulating food during chewing and swallowing and in terrestrial species it bears numerous taste buds on its upper surface.

tonoplast (vacuole membrane) Outer membrane of a vacuole in plant cell.

tonotaxis Movement along a gradient of osmotic pressure.

tonsil The mass of lymphoid tissue in mouth or pharynx of tetrapods. In humans there are the palatine tonsils at the back of the mouth, lingual tonsils below the tongue and pharyngeal tonsils (adenoids) in the pharynx. They are concerned with the production of lymphocytes and form part of the gut-associated lymphoid tissue protecting the gut against infection.

topoisomerase An enzyme which can remove supercoiling and tension in duplex DNA by creating transition breaks in one or both strands of the sugar phosphate back bone.

topological isomers (topoisomer) An isomer of, a large complex molecule that is distinct from other similar isomers by its topology, such as degree of twisting or interlocking of ring structures. For example, topoisomers of DNA may be distinguished according to the degree of supercoiling of the double helix.

tornaria A ciliated larva produced by some acorn worms *see* Hemichordata.

totipotency

1. Ability of individual plant cells to form new plant when isolated, provided it is given suitable nutrients, hormones, etc.
2. Describing embryonic cells at a stage before their fate is irreversibly determined, when they have the ability to develop into any differentiated cell given the appropriate stimulation.

toxicology Study of toxins.

toxin Poison, produced by an animal or plant, that elicits the production of an antibody (antitoxin) when introduced into body tissue, typically by injection.

trachea (windpipe) In air breathing vertebrates, it is the portion of single respiratory tube connecting the bronchial tubes with the larynx.

tracheal system (trachea) A system of air tubes and their branches which help in respiration in insects and some other arthropods.

tracheids Specialised dead cells in xylem tissue with thick and lignified walls which are normally elongated and pointed at both ends. Found in all vascular plants.

tracheophyta Vascular plants, including pteridophyte and spermatophyte.

trait A distinguishing or marked feature; a peculiarity.

transamination Transfer of amino group from one amino acid to the keto group of a keto acid to form a new amino acid and keto group. Enzyme transaminase catalyses this reaction.

transcribed spacer Portion of a primary *r*RNA transcript discarded during the production of functional *r*RNAs.

transcript RNA sequence complementary to and produced from a DNA template.

transcriptase An enzyme which brings about transcription.

transcription The process in living cells in which the genetic information of DNA is transferred to a molecule of messenger RNA (*m*RNA) as the first step in protein synthesis. It takes place in the cell nucleus or nuclear region and is regulated by transcription factor by the action of RNA polymerase enzymes. It form a complementary strand of *m*RNA from the DNA template (in eukaryote). The subsequent removal of the noncoding sequences from this primary transcript (see RNA processing) to form a functional *m*RNA molecule.

transcription factor Protein that binds to RNA regulatory regions and helps control gene expression.

transcriptome Full complement of RNA transcripts present in a given tissue at a particular time.

transducin G protein that acts as a second messenger in the response to light by vertebrate photoreceptors. When a photon of light.

transduction

1. Transfer of part of the DNA from a bacterium to another by a bacteriophage. Process does occur naturally mainly known as a technique in recombinant DNA technology.
2. Conversion of stimuli detected in the receptor cells into electric impulses, which are transported by the nervous system.

transect Straight line across an expanse of ground along which ecological measurements are taken, continuously or at regular intervals.

transfection Introduction of foreign DNA into a eukaryotic cell.

transfer cell A type of plant cell specialised for the short-distance high volume transport of materials. These cells have numerous knobs and ridges on the inner surface of primary cell wall. These cells are found mainly in salt secreting glands and in region where sugar is loaded or unloaded into sieve element of phloem.

transfer RNA (tRNA) The RNA molecule that adds specific amino acids during translation. Transfer RNAs also serve as primers for reverse transcription.

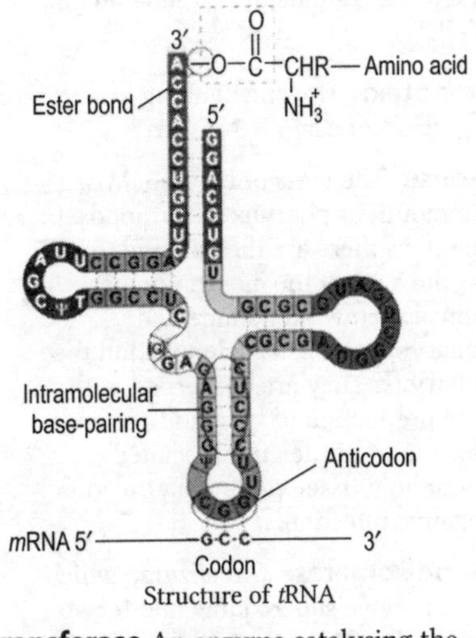

Structure of *t*RNA

transferase An enzyme catalysing the transfer of a functional group (e.g. a methyl or phosphate group) from a donor molecule to an acceptor molecule.

transferrin A plasma protein (a β-globulin) that can bind to iron and transport it to the liver for storage or to the cells of bone marrow, where it is used for the formation of haemoglobin. One transferrin molecule can bind to two iron ions.

transformation

1. An incorporation of naked DNA into the cells of bacteria causing an alteration in the bacteria cell's phenotype.
2. Conversion of a normal cell into a malignant cell which can be brought about by mutation of genes, sometimes through the action of carcinogens or oncogenic viruses.

transgene A gene that is taken from one organism and inserted into the germ line of another organism so that it is replicated as part of the genome and present in all the recipient's cells. The resulting organism is described as transgenic.

transgenic animal An experimentally produced animal created by the introduction of new DNA into the organism's germ line.

transient polymorphism Polymorphism in which a particular form is in the process of spreading through a population causing relative proportion of phenotypes to alter.

transition Substitution mutation in which a pyrimidine is replaced by another pyrimidine or a purine base is replaced by another purine base.

translation Process in living cells in which the genetic information encoded in *m*RNA in the form of nucleotide triplets is translated into a sequence of amino acids in a polypeptide chain during protien synthesis.

translocation

1. Process of transportation of sucrose or food through the phloem (in botany).
2. Chromosomal mutation in which a section of one chromosome is broken and become attached to another chromosome resulting in loss of genetic information from the first one (in genetics).

transpiration Loss of water from the leaves of a plant; creates an osmotic gradient that draws nutrient-laden water up from the roots.

transplantation Transfer of a tissue or organ from one part of the animal to another part or from one individual to another.

transport proteins Proteins found in the membrane of cells and organelles that are responsible for the movement of many substances into or out of the cell. Sometimes, these proteins are passive gateways for substances to diffuse into or out of the cell. Other transport proteins use energy to actively pump molecules into the cell or to the outside.

transposase Enzyme capable of catalysing the insertion of transposons.

transposon Segment of DNA capable of shifting from one position in the eukaryote genome to another. Often transposons code for enzymes that facilitate the transposon's movement. Sometimes called jumping genes.

transverse tubules (T tubules) A series of infoldings of the plasma membrane of muscle fibres that envelope each myofibril at the Z-line. The T-tubules transfer action potentials from the sarcolemma to the sarcoplasmic reticulum, which releases calcium ions into the cytosol causing contraction of the muscle fibres.

transversion A substitution mutation in which a pyrimidine base is replaced by a purine base or *vice-versa.*

trematoda A class of parasitic flatworms, commonly known as flukes. They have suckers and hooks to anchor themselves to the host and their body surface is covered by a protective cuticle. The whole life cycle may either occur within one host or required one or more intermediate hosts to transmit the infective eggs or larvae.

Treponema A genus of spirochetes, among which are the causative agents of syphilis (*Treponema pallidum*) and yaws (*Treponema pertenue*).

triacylglycerols (also triglycerides) Molecules that are the typical form in which fatty acids are stored in cells. A triacylglycerol is formed by esterification of one fatty acid to each of the three hydroxyl groups of a glycerol molecule. Triacylglycerols are insoluble in water and in salt solutions, so they form droplets in cells. For body tissues, these droplets are a source of energy. They are stored in adipose cells.

Triassic period The first period of the Mesozoic era. Pangaea began to breakup during this time. During the period marine animals diversified: molluscs were the dominant invertebrates - ammonites were abundant and bivalves replaced the declining brachiopods. Reptiles were the dominant vertebrates and included turtles, phytosaurs, dinosaurs and the marine ichthyosaurs.

tribe A number of species or genera having certain structural characteristics in common.

triceps Muscle that causes extension of the arm. Runs parallel to humerus and works antagonistically with biceps.

trichinosis A disease resulting from ingestion of the nematode *Trichina spiralis.* The usual source of infection is undercooked pork.

trichloroacetic acid An acid commonly used to precipitate proteins in the course of biochemical extractions.

trichocysts A thread-like barbs of ciliated protozoans that can be fired for defence or to capture prey.

trichome (hair) Uni or multicellular outgrowth arising solely from the epidemis. Root and leaf hairs and glandular hairs are type of trichome.

tricuspid valve Right atrioventricular valve, which has three cusps.

trilobites A group of marine arthropods abundant in the Palaeozoic; similar in appearance to the extant horseshoe crab (although received wisdom holds that they became entirely extinct at the end of the Palaeozoic era).

trioecious Existing in three distinct sexual types, i.e. males, females, and hermaphrodites.

triple fusion In double fertilisation, one male gamete fuses with the egg under the process of syngamy and forms zygote while, the other male gamete fuses with two polar nuclei. Three nuclei are involved in this fusion process.

triplet code *See* codon; genetic code

triploblastic Describing an animal having a body composed of three embryonic cell layers; the ectoderm, mesoderm and endoderm. Most multicellular animals are triploblastic; the coelenterates which are diploblastic are an exception.

triploidy Condition in which a cell or an organism three complete sets of chromosomes ($3n$).

trisomy Condition of having three copies of a particular chromosome instead of the normal two ($2n + 1$).

Tritanopia A rare form of colour blindness in which person cannot distinguish between blue and green due to lack of cones sensitive to blue light.

trochanter

1. Several bony knob on the femur of vertebrates to which muscles are attached.
2. Second segment of an insect's leg, between the coxa and the femur.

trochlear nerve (cranial nerve IV) A pair of nerves that emerges from the dorsal midbrain in vertebrates to supply the superior oblique muscle of each esye ball. It contains chiefly motor nerve fibres.

trochophore The small, translucent, free-swimming larva characteristic of marine annelids and most group of molluscs. Trochophores are spherical or pear-shaped, girdled by a ring of cilia.

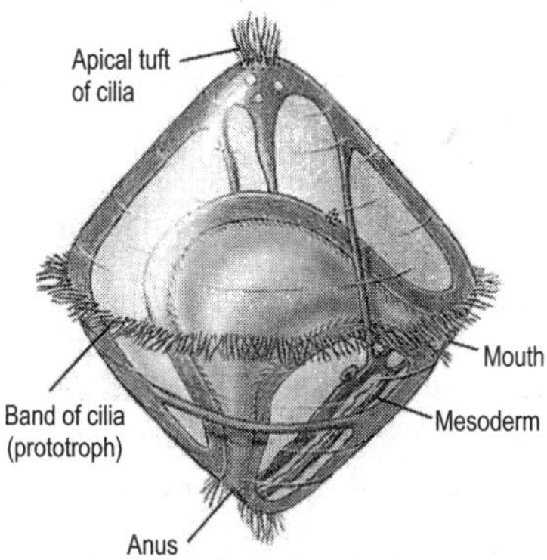

Structure of a trocophore larva

trophic level Position that an organism occupies in a food chain. For example, green plants are the primary producers; herbivores are primary consumers (and secondary producers). A carnivore that eats only herbivores is a secondary consumer and a tertiary producer. Many animals feed at several different trophic levels.

trophoblast Outer layer of cells of a blastocyst.

tropism The directional growth of a plant organ, toward or away from a stimulus. In contrast to a taxis, tropism do not involve movement of the whole organism relative to the stimulus.

tropomyosin Protein found in actin filament in skeletal muscle that regulates contraction of the muscle filaments.

troponin A structural protein of muscle, involved in regulation of contraction. Site of calcium binding.

troposphere First zone of the atmosphere that is about 10-12 km from the surface of the earth.

true fruits Fruits which develop from the ovary, e.g. sweet pea.

trumpet cells Elongated cells which are joined end to end to transport photosynthate through the body of brown algae. The end walls are perforated by pores. Trumpet cells and sieve elements are example of convergent evolution.

trypanosomiasis Disease of human and animals caused by the various protozoans of the genus–*Trypanosoma.*

trypsin An enzyme, splitting proteins and peptides in alkaline solution. Secreted by vertebrate pancreas. Trypsinogen is activated by enterokinase enzyme of succus entericus to trypsin.

tryptophan An essential amino acid.

tube feet A sac-like protrusion of body wall in Echinoderms used for feeding, locomotion and respiration.

tuber A fleshy, swollen underground stem that serves to store nutrients and as a means of vegetative reproduction. Examples are potatoes, yams and Jeruselem artichokes.

tuberculosis (TB) A common, often fatal disease caused in human beings primarily by the *Mycobacterium tuberculosis.* Most often, TB attacks the lungs, but it can affect almost any part of the body. It is extremely contagious. The infection spreads through the air, and from surfaces contaminated with infected mucous and saliva. Frequently Person's infected cattle and milk spread the disease to human beings. The usual symptoms of pulmonary tuberculosis are chronic cough with bloody sputum, night sweats, fever and loss of weight. A wide range of other symptoms can occur depending on the region of the body affected. The BCG vaccine decreases the risk of getting infection.

tubulin Protein that forms the primary component of microtubules. Each unit of tubulin is a dimer with an alpha and beta subunit.

tumour The new, autonomous growth forming an abnormal mass that performs no physiological function.

Tumour Necrosis Factor (TNF) Either of two proteins that act as cytokines, eliciting numerous responses in virtually all types of cells by activating cellular signalling pathways and promoting the expression of genes. They are involved in many aspects of the body's immune defences.

tumour-suppressor gene (anti-oncogene) A class of genes that suppress the development of cancers in living cells. The products of tumour-suppressor genes are typically involved in monitoring replication of DNA and progress of the cell cycle and in promoting repair of damaged DNA.

tundra An arctic or alpine vegetation biome beyond timberline.

tunicates *See* Urochordata

Turbellaria A class of free-living Platyhelminthes comprising the planarians, which occur in wet soils, freshwater and marine environments. Their under surface is covered with cilia, used for gliding over stones and weeds. Planarians can also swim by means of undulations of the body.

turgid Swollen and firm from fluids. Used in reference to plants kept in a hypotonic solution allowing water to enter and cause the plant cells to swell.

turgor State of cell in which the cell wall is rigid and stretched by increase in volume of vacuole and protoplasm, due to the absorption of water.

turion Winter bud, covered with scale leaves and mucilage, that is produced by certain aquatic plants, such as frogbit. Turions become detached and remain dormant on the lake or pond bottom during winters before developing into new plants.

turner syndrome A condition in women which only a single X-chromosome (X-monosomy) is present. They are anatomically female but sterile. The ovaries are usually absent or small.

twitch A single contraction of a muscle.

tylose Ingrowth from a parenchyma cell into an adjacent tracheid or vessel through a paired pit. Tyloses are often found in injured tissue, older wood and below an abscission layer and can completely block the conducting vessel.

tympanic membrane (Eardrum, tympanum) The membrane that separates the outer ear from the middle ear. It vibrates in response to sound waves and transmits these vibrations *via* the ear ossicles of the middle ear to cochlea of the inner ear absent in amphibians and some reptiles.

tympanites Abdominal distention resulting from accumulation of gas in the intestine or peritoneal cavity.

tympanum A membrane that separates outer ear from the middle ear.

tympanum cavity Cavity of the middle ear.

type Material used to define a species. It is usually a dried specimen stored in a herbarium but may also be a drawing. The term 'type' is also used to describe the representative genus of a family, etc., e.g. the genus *Solanum* is the type genus of the family –Solanaceae.

tyrosine One of the non-essential amino acid.

tyrosine kinase A kind of protein kinase, phosphorylating tyrosine residues in its protein substrate. They commonly form intracellular components of cell-surface receptors to growth factors and homones, being activated when ligand binds the extracellular domain. Some form parts of signalling pathways involved in cell fate determination in both invertebrate and vertebrate, development.

U

ubiquinone (coenzyme Q) Any of a group of quinone derived compounds. It is a coenzyme that is an essential component of the electron transport chain, serving as electron carriers.

ubiquitin A small protein present in an eukaryotes. Binding to a protein may initiate the protein's degradation.

ulcer The open sore on the surface of a body organ.

ulna The larger of the two bones in the forearm of vertebrates, it lies on the side opposite the thumb.

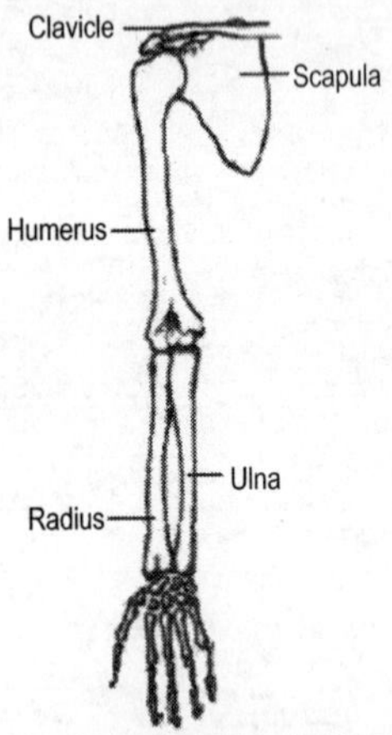

Position of ulna and other bones of upper arm of human

It articulates with the outer carpals at the wrist and with the humerus at the elbow.

ulnocarpal Relating to the ulna and carpus.

ulocace An ulcerating inflammation of the gums.

ulorrhea Bleeding from the gums.

ultracentrifuge A high-speed centrifuge, operating at upto million revolution per second. It is used to measure the rate of sedimentation of colloidal particles or to separate protein or nucleic acid molecules. The rate of sedimentation depends on the molecular weight of the molecule.

ultradian rhythm A biological rhythm whose periodicity is greater than a day. For example, the body temperature of hibernating animals often shows periodic transient increases corresponding to brief periods of arousal.

ultrafiltration Filtration of urine under high pressure in kidney.

umbel An inflorescence in which the flower stalks (pedicels) radiate from a common point. It is a type of racemose inflorescence.

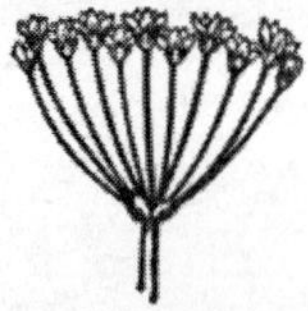

Umbel inflorescence

umbilical cord The cord connecting the placenta to the embryo or foetus in mammals. It contains the umbilical artery and vein. It is severed at birth to free the newly born. The navel (or umbilicus) marks the point of its former attachment.

umbilicus

1. A shell in mollusc, the axially aligned hollow cone-shaped space within whorls of a coiled mollusc shell.
2. A genus of over 90 species of flowering plant in family Crassulaceae, e.g. Venus's navel.
3. The navel in mammals.

Uncoupling Protein (UCP) Any of a family of proteins that uncouple electron transport from oxidative phosphorylation in mitochondria. This leads to release of energy derived from aerobic tissue respiration chiefly as heat instead of being used to make ATP.

undernutrition A condition in which insufficient amount of food is taken for an extended period of time.

undulipodium (pl. undulipodia) Movable, whip-like extension of the plasma membrane of certain eukaryotic cells, such as spermatozoa. An undulipodium contains cytoplasm and a shaft composed of microtubules, known as axoneme where the undulipodium attaches to the cell, there is a basal body called the kinetosome which is connected to the axoneme with motor proteins. The flagella of prokaryotes are similar in appearance and function to the undulipodia of eukaryotes, but differ from them with respect to structure.

ungulate A hoofed mammal, such as cow. Grouped into Artiodactyla and Perissodactyla.

unguligrade Animals which walk and run on tips of digits covered with modified nail called hoof, e.g. pigs, horses and rhinoceros.

uniaxial Having or relating to a single axis.

unicellular A single-celled organism, tissue or organs such as bacteria.

uniformitarianism An important assumption in science that states that the natural laws and processes that operate today have operated in the same manner in earlier times, back to close to the beginning of the universe. It does not mean that conditions have been unchanging in past times.

unilocular Having, consisting of, or characterised by only one loculus or cavity; single chambered

uninucleate Having a single nucleus.

uniparental inheritance A non-mendelial form of inheritance that includes transmission of genotypes from one parent to all progeny. All the genes in offsprings will originate from only mother or only the father. It is observed in eukaryotic organelles as mitochondria and chloroplasts.

uniparous Producing only one offspring at birth.

unipolar neuron The neuron which has one main process (the axon) extending from its cell body. These include many sensory and many vertebrate motor neurons and interneurons.

uniporter A transport protein that moves molecules across a membrane only down their concentration gradient (in one direction only).

Uniramia (Mandibulate) A phylum of arthropods that contains the centipedes millipedes and centipede like animals and insects, as described in some classification systems. Members are characterised by uniramous appendages.

uniramous appendage A type of appendage that is a characteristic of insects and other members of the uniramia. It consists of an unbranched series of segments.

unisexual An individual organism producing either male or female gamete.

unit membrane The limiting membrane of cells and organelles like that of mitochondria, endoplasmic reticulum, nucleus, etc., as seen under electron microscope as a three layered structure. It consists of a layer of lipids sandwiched between two protein layers.

uracil One of the four nucleotide bases, a pyrimidine, found in RNA (another pyrimidine, thymine, replaces uracil in DNA).

urea (carbamide) The diamide of carbonic acid. A soluble nitrogenous waste; found in blood and lymph. It is the primary nitrogenous component of urine in ureotelic animals, e.g. mammals and amphibians.

urea cycle (ornithine cycle) The series of biochemical reactions. Which converts highly toxic ammonia and carbon dioxide to much less toxic urea in liver of mammals. The urea is ultimately excreted in solution as urine.

urease An enzyme that splits urea into ammonium carbonate.

u

uremia The term used for urea in the blood. These are the symptoms of profound kidney malfunction.

ureotelic The organisms excreting urea as main breakdown product of amino acids. These include cartilaginous fish, some bony fish, adult amphibians and mammals.

ureter Either that carries urine the tube leading from the kidney to the bladder in vertebrates. The internal surface (lumen) of the human ureter is star-shaped.

urethra A tube leading urine from the bladder to the exterior of the body in mammals. In males it also serves as a channel for sperm.

uric acid The ultimate product of purine catabolism; main nitrogenous component of urine in birds and reptiles.

uricase An enzyme that catalyzes oxidation of uric acid. Usually occurs in liver and kidney.

uricotelic Animals which excrete nitrogenous waste as uric acid in the form of pellet or paste with a minimum loss of water, e.g. reptiles, birds, land snails and insects.

uridine A nucleoside consisting of uracil, attached to a D-ribose ring.

urinary system The collection of organs and tissues that perform osmoregulation and excretion. In mammal it consists of two kidneys each linked to the bladder by a ureter.

urine The waste fluid produced by the kidneys and stored in the urinary bladder prior to excretion in animals. In addition to water, the major constituents of urine are one or more of the end products of nitrogen metabolism-ammonia, urea, uric acid and creatinine. It may also contain various inorganic ions, the pigments urochrome and urobilin, amino acids and purines. The habitat of a particular species mainly decide the composition aquatic animals produce copious volumes; terrestrial animals need to conserve water and produce much less (about 1.0-1.5 L/Day in humans).

uriniferous tubule A convoluted tube forming bulk of each nephron of vertebrate kidney.

Urochordata (Tunicata) A group of marine chordates. These are sedentary with gill slits reduced nervous system. Notochord absent. Body enclosed in cellulosic tunic that protects it.

urochrome The pigment which gives a characteristic yellow colour to the urine.

urogenital Of or pertaining to the genital and urinary organs.

urology The study and treatment of disorders of the urogenital organs.

urostyle A long unsegmented bone that represents a number of fused vertebrae in frogs and toads. It forms the posterior part of body.

uterus (womb) An organ of gestation, present only in female mammals. It holds and nourishes the developing embryo.

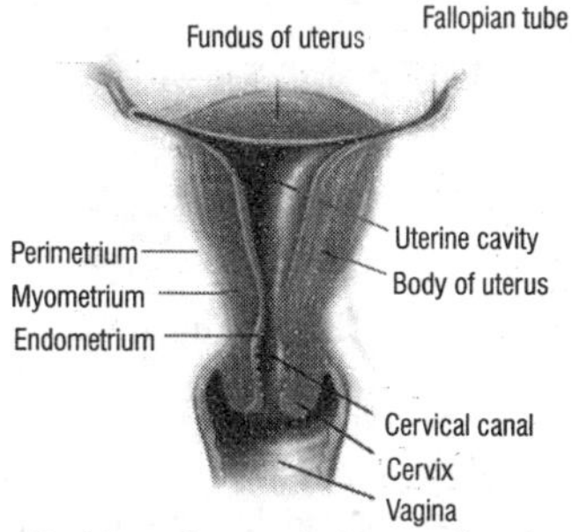

Position of uterus in human female

utriculus (utricle) The upper large chamber of the vertebrate inner ear bearing semicircular canals. Responsible for the detection of changes in the direction and speed of movement.

uvea The choroid, ciliary body and iris taken together. It is the pigmented layer of the eye lying beneath the sclera and cornea.

V

vaccination The introduction of an inactive or weakened pathogen into the body in order to stimulate an immune reaction that will allow the body to respond quickly to the pathogen, in case of an actual infection.

vaccine A preparation eliciting an immune response when injected into the body. Vaccines contain dead or weakened pathogens or their products.

vacuole A space within the cytoplasm of a living cell filled with air, water or liquid cell sap or food particles. In plant cells there is usually one large vacuole bounded by a single-layered tonoplast animal cells usually have serveral small vacuoles.

vagal tone The effect on the heart when only the parasympathetic never fibres are controlling the heart rate. It slows the heart rate from approximately 70-60 beats per minute.

vagina

1. It is mucomembranous, thin-walled passage forming the birth canal; the passageway leading out from the uterus to the exterior of the body.
2. Any sheath-like part.

vagus nerve Tenth cranial nerve that supplies branches to many major internal organs. It carries motor nerve fibres to the heart, lungs and viscera and sensory fibres from the viscera.

valine An essential amino acid.

Variable Number of Tandem Repeats (VNTRs) **locus**

1. The locus which is hypervariable due to tandemely repeated DNA sequences. The variability is perhaps by unequal crossing over or slippage during replication.
2. In humans, these are 1-5 kb long. These have proved valuable in DNA profiling as they contribute to repetitive DNA.

variation Phenotypic and/or genotypic differences between individuals of a population. They may be the result of environmental conditions, e.g. water supply and light intensity affect the height and leaf size of a plant. Differences of this kind acquired during the lifetime of an individual, are not transmitted to succeeding generations since the genes are not affected. Genetic variations are inherited. The most important sources of these are mutation and recombination.

variegation The irregular variation in colour of plant organs(e.g. leaves, flowers) through suppression of normal pigment development in certain areas.

vas deferens One of a pair of sperm-bearing tube leading from the epididymis to the urethra (in mammals) or to out side.

vas efferens In reptiles, birds and mammals, one of a number of small ducts that convey spermatozoa from the seminiferous tubules of the testis to the epididymis. In invertebrates, vasa efferentia conduct spermatozoa directly from the testis to the vas deferens.

vasa recta Thin-walled blood vessels that branch from the efferent arterioles leaving each glomerulus in the vertebrate kidney U-shaped loops adjacent to the loop of Henle that eventually drain into the renal vein are formed by vasa recta.

vascular Containing or composed of conductive tubes.

vascular bundle (fascicle) A fluid-conducting plant tissue that consists of cells which comprise the xylem and phloem. Separate vascular bundles are linked by interfascicular cambium.

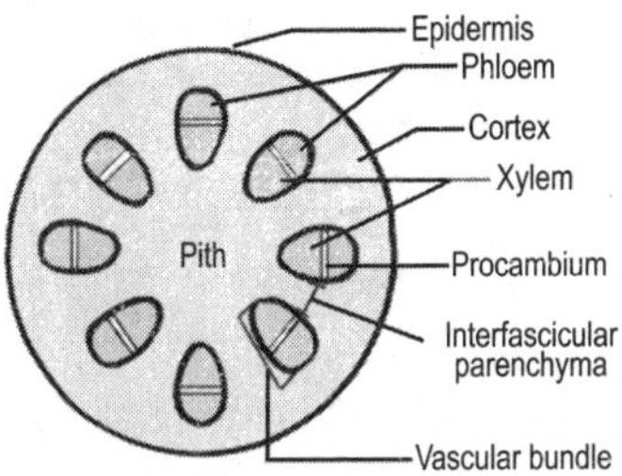

Vascular bundle and other structures in a section of plant stem

vascular cambium Occurs in the stem and root. It divides to produce secondary xylem and secondary phloem. In mature stems, it is extended laterally to form a complete ring.

vascular cylinder The central core of the vascular tissue in a plant root. It is surrounded by parenchymal ground tissue.

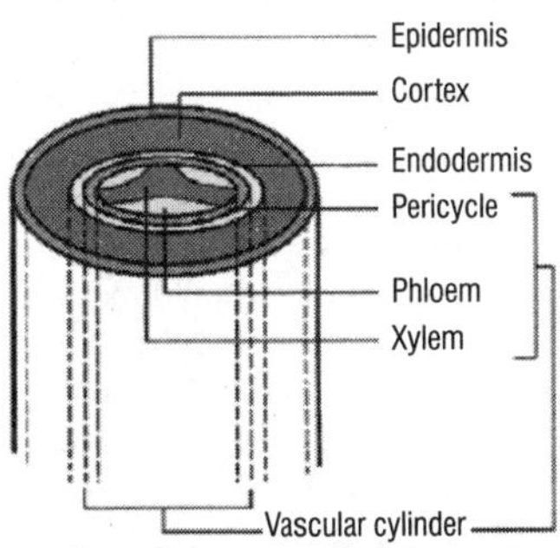

A section of plant stem showing position of vascular cylinder

vascular plants Plants with vascular tissue (i.e. all plants other than mosses and their relative species).

vascular system

1. In animals, a specialised network of vessels for the circulation of fluids throughout the body. All animals, except simple invertebrate groups possess a **blood vascular system,** which enables the passage of respiratory gases, nutrients, excretory products and other metabolites into and out of the cells.
2. In plants, the system of vascular tissue.

vascular tissue In plants, the tissue that conducts water and nutrients through the plant body in tracheophytes consists of xylem and phloem.

vasectomy An operation severing the vas deferens for the purpose of sterilisation or permanent contraception in males.

vasoconstriction The reduction in internal diameter of small blood vessels due to the contraction of smooth muscles in their walls. It results from stimulation by vasoconstrictor nerve fibres or from the secretion of epinephrine in response to decreased blood pressure, low external temperature, pain, etc.

vasodilation The increase in the internal diameter of blood vessels, mainly arterioles or capillaries. It is mediated by the action of nerves on the smooth muscle fibres of the arteriole walls and results in a decrease in blood pressure.

vasomotor centre The region of the medulla oblongata of the brain that mediates the diameter of the blood vessels (*via* the action of vasomotor nerves) and hence controls blood pressure.

vasomotor nerves Nerve fibres of the autonomic nervous system that control the diameter of blood vessels. They transmit impulses from the vasomotor center in the medulla oblongata of the brain to the smooth muscle in the vessel wall, causing them to constrict or dilate.

vasopressin *See* antidiuretic hormone

vector

1. An animal (mostly insect) that transmits disease from an infected individual to another uninfected individual.
2. Cloning vector.

vegetal pole (vegetative pole) The end of an egg, opposite the animal pole, containing most of the yolk. Farthest from egg nucleus.

vegetative cell The cell of pollen grains which is bigger, has abundant food reserves and a large irregularly shaped nucleus.

vegetative reproduction (vegetative propagation) A type of asexual reproduction in which a complete new plant is produced from any vegetative part (e.g. gemmae, rhizome) of the plant body.

vein

1. A thin-walled vessel carrying blood towards the heart. Most veins carry deoxygenated blood (pulmonary vein is an exception).
2. Vascular bundle in a leaf.
3. Any of the chitinuous tube of an insect's wing that strengthen it.

velamen A spongy tissue meant for absorbing moisture surrounding the aerial roots of epiphytic plants. It is comprised of dead empty cells.

velum A torn membrane which encircles the stipe, close to the attachment of pileus, in mushroom or toadstool.

velvet worms *See* Onychophora

vena cava Either of two main veins that convey deoxygenated blood from the body to the right atrium of the heart in tetrapods. The anterior (precaval superior) vena cava is a paired vein serving the head and forelimbs. The posterior (postcaval or inferior) vena cava is a single vein serving most of the body and the hindlimbs.

venation

1. An arrangement of veins in a leaf. Dicotyledons have reticulate type while leaves of monocotyledons have parallel venation.
2. The arrangement of the veins in an insect's wing often important in classification.

venom A toxin produced by an animal which is usually injected by a bite or sting into victim, for the purpose of causing harm to it.

venter The swollen base of an archegonium that contains the egg cell (oosphere) for development.

ventilation The process that maintains continuous exchange of gases across respiratory surfaces. Also called external respiration, this is achieved by respiratory movements; in air-breathing vertebrates, it is movement of air into and out of the lungs.

ventilation centre The group of neurons in the medulla oblongata of the brain that controls the process of ventilation. It responds to an increase in the amount of carbon dioxide in the blood by increasing the rate of breathing. Within the ventilation centre are subcentres that control inspiration (inspiratory centre) and expiration (expiratory centre).

ventral

1. The surface resting on, or facing substratum (in)
2. In humans, toward the front (anterior surface).
3. In animals, toward the underside or belly.

ventral aorta The blood vessel in fish and embryo of tetrapods that carries deoxygenated blood from the ventricles of the heart to aortic arches. In adult tetrapods, it is equivalent to the ascending limb of the aorta. In adult fish, it branches into afferent branchial arteries supplying gills.

ventral cavity The major body cavity in vertebrates.

ventral root The part of a spinal nerve that leaves the spinal cord on the ventral side and contains motor fibres.

ventricle

1. A chamber of the heart that receives blood from an atrium and pumps it into the arterial system. Amphibians and fish have a single ventricle, but mammals, birds and reptiles have two, pumping deoxygenated blood to the lungs and oxygenated blood to the rest of the body respectively.
2. Any of the four linked fluid-filled cavities in the brain of vertebrates. One in the medulla oblongata, two in the cerebral hemispheres and one is in the posterior part of the forebrain. They contain cerebrospinal fluid filtered from the blood by the choroid plexus.

ventrolateral

1. In humans, toward, on or in the anterior and sides.
2. In animals, toward, on or in the belly and sides.

venules Small blood vessels that connect veins with capillaries.

vermiculture Refers to composting by the activity of earthworms. It includes cultivating worms in order to breakdown organic wastes.

vermiform appendix *See* appendix.

vernalisation The promotion of flowering by exposure of a plant to low temperatures.

vertebrae (also vertebras, sing. vertebra) Any of the bones that make up the vertebral column. In mammals, each consists of a main body, (centrum) from which arises a **neural arch** through which the spinal cord passes and **transverse processes** projecting from the side. There are five groups of vertebrae, specialised for various functions and varying in number with the species. In humans, there are 7 cervical vertebrae, 12 thoracic vertebrae, 5 lumbar vertebrae, 5 fused sacral vertebrae, and 5 fused caudal vertebrae (forming the coccyx).

vertebral column (backbone; spinal column; spine) A flexible bony column in vertebrates that extends down the long axis of the body. It provides the main skeletal support and also encloses and protects the spinal cord, it also provides attachment for the muscles of the back. The vertebral column consists of a series of bones separated by cartilaginous intervertebral discs.

vertebrate A chordate animal with a backbone; a member of the subphylum–Vertebrata (includes mammals, birds, reptiles, amphibians and fishes).

verticillaster (false whorl) A type of inflorescence in which two oppositely placed cymes are inserted on either side of a stem giving the impression of a whorl. Typical of the deadnettle family.

vesicle A membrane-bound fluid filled sac that transports macromolecules into and out of a eukaryotic cell and between the organelles within the cell.

vessel

1. A substance conducting tube or duct within a living organism, especially blood or lymphatic vessel (in zoology).
2. A tube within the xylem composed of joined vessel elements. Vessels facilitate the efficient movement of water from the roots to the shoots and leaves of a plant (in botany).

vessel elements Wide cells that join end-to-end to form water-carrying tubes within xylem in angiosperms.

vestibular apparatus A complex system in inner ear, located above the cochlea. Composed of three semi-circular canals and otolith organ consisting of the saccule and utricle. Responsible for the maintenance of balance of the body and posture.

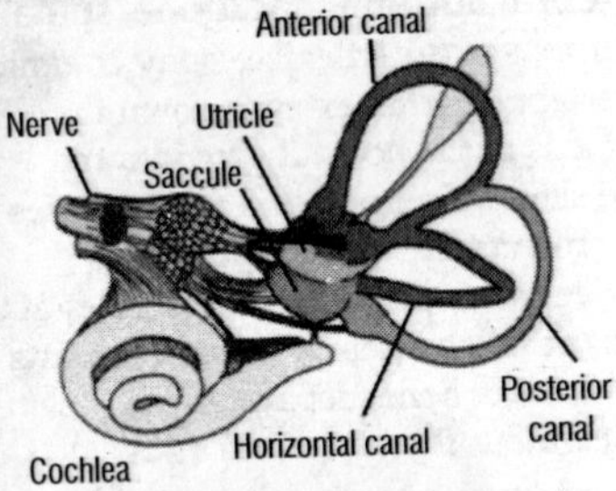

Structure of vestibular apparatus in human ear

vestibular canal A canal in the cochlea of the inner ear that connects with the oval window. It contains perilymph, through which pressure waves are transmitted from the oval window *via* Reissner's membrane to the endoylmph within the cochlear duct.

vestibule A chamber that leads to a body cavity or that links one cavity to another. For example, a vestibule leads from the vulva into the vagina int the mammalian reproductive system.

vestigial organ Any part of an organism that has diminished in size during its evolution because the function it served decreased in importance or became totally unnecessary, e.g. the human appendix and the wings of the ostrich.

viable count A measure of the number of living cells within a culture.

Vibrio A genus of comma-shaped bacteria including the causative agents of cholera (*Vibrio cholerae*) and brucellosis (*Vibrio fetus*).

villi (sing. villus) Short, filamentous processes that occur on certain membranous surfaces of some tissues and organs. Such as, the interior of the intestine or the surface of the chorion. Serve to increase the surface area of the organ.

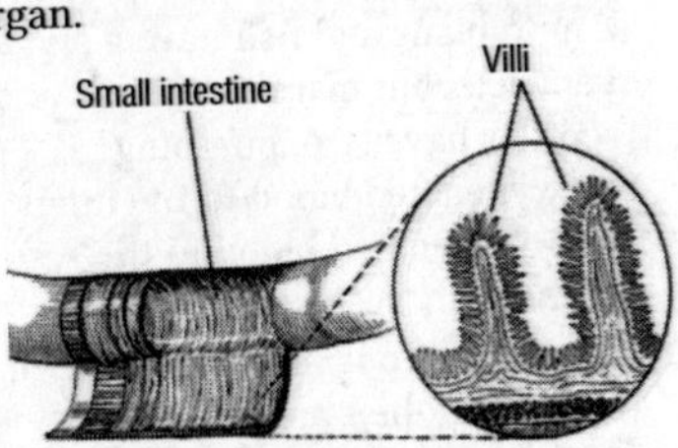

Position of villin in small intestine

vinegar A liquid containing 3-6% acetic acid obtained by the oxidation of ethanol by the action of bacteria on wine, beer, etc.

VIP (Vasoactive Intestinal Peptide) A neuropeptide that acts as a neurotransmitter or regulator of a cell function in many organs and tissues, including the central and peripheral nervous systems, heart, lungs, kidneys, and immune system.

It also acts as a peptide hormone, being secreted by endocrine cells of the upper part of the small intestine in response to the entry of partially digested food from the stomach. VIP along with secretin stimulates the pancreas to produce a thin watery secretion containing bicarbonate.

viral Of or pertaining to viruses.

virion An extracellular inert phase of a virus. Virion consists of a protein coat surrounding one or more strands of DNA or RNA. Virion may be polyhedral or helical and vary greatly in size, from 20-600 nm in diameter.

viroids Tiny naked single-stranded RNA plant pathogens, usually about 300 nucleotides long; thousands of times smaller than the smallest virus.

virology The study of viruses.

virulence The relative pathogenicity of an organism. It depends on the invasiveness and toxicity of the pathogen. It may vary between strains of the same organism.

virulent phage A bacteriophage that infects a bacterial cell and undergo immediate lysis of the host cell.

virus The surface area of an organ. It is the ultramicroscopic, non-cellular parasite that can reproduce only within a host cell. Viruses consist of nucleic acid covered by protein. Some animal viruses are also surrounded by membrane (enveloped viruses). Inside the infected cell, the virus uses the translational system of the host to produce progeny virus. They carry enough genetic material to encode a few proteins. Many viruses such poliovirus and rhinovirus, carry the bare minimum—just enough to specify their own structure and get synthesis started once they get inside the cells.

virusoid The subviral particle that infects plants. It consists of a small naked single-stranded circular RNA molecule.

viscera Internal organs, especially those present within the thorax or abdomen.

visceral arch One of a series of bony or cartilaginous skeletal arches in fish and other vertebrate embryos that occur in the lateral wall of the pharynx and support the tissue behind the mouth and between the gill slits. Often modified for different functions.

visceral muscles Located in the inner wall of hollow visceral organs of the body. For example, alimentary canal, reproductive tract and others.

visceral sensory neuron A sensory neuron located in sympathetic and parasympathetic nerves in the heart, lungs and other organs.

visceroptosis The downward displacement of any or all body organs of the abdominal viscera.

visual cortex (striate cortex) An area of the cerebral cortex in brain, where the neural impulses are analysed and the image formed on the retina is recognised based on earlier memory and experience.

visual purple *See* rhodopsin

vital capacity The amount of air that can be expired by the most forcible expiration, after the deepest inspiration. The average vital capacity of human is about 4.5L.

vital stains Non-toxic colouring materials that can be used in dilute concentrations to stain living material without damaging it, e.g. Janus green, tryptan blue.

vitamin An organic compound that cannot be produced by the body and is needed in trace amounts to maintain normals health. There are some 14 generally recognised major vitamins: the water-soluble vitamin-B complex (containing 9) and vitamin-C and the fat-soluble vitamins-A, D, E and K. Most B vitamins and vitamins-C occurs in plants, animals and microorganisms; they function typically as coenzymes. Vitamins-A, D, E and K occur only in animals, especially vertebrates and perform a variety of metabolic roles.

vitamin-A (retinol) A fat-soluble vitamin (a derivative of the yellow pigment, carotene) occurring chiefly in yellow and green vegetables. Deficiency causes night-blindness.

vitamin-B-complex A group of water-soluble vitamins which tend to occur together. Obtained from whole grain of cereals and from meat and liver. These include thiamine (B_1), riboflavin (B_2), nicotinic acid (niacin), pantothenic acid (B_5), pyridoxine (B_6), cyanocobalamin (B_{12}), biotin, lipoic acid and folic acid (see intrinsic factor).

vitamin-C (ascorbic acid) A water-soluble vitamin; widely required in metabolism. Chiefly available in citrus fruits and green vegetables deficiency results in scurvy.

vitamin-D A fat-soluble vitamin found in fish liver oil, butter, milk, cheese, egg yolk and liver. It increases absorption of calcium and phosphorus from the intestine and calcification process in bone. Deficiency causes rickets in children and osteomalacia in adults. See calciferol and cholecalciferol.

vitamin-E (tocoferol) A fat-soluble vitamin that is the main antioxidant of cell membranes and other lipid-rich tissue components. Deficiency leads to muscular dystrophy, liver damage and infertility. Good sources are cereal grains and green vegetables. It prevents the oxidation of unsaturated fatty acids in cell membranes by removing lipid peroxyl rad icals; so maintaining their structure.

vitamin-K Any of several related fat-soluble compounds, including vitamins-K_1 (phylloquinone) and K_2 (menaquinone), that are essential for the function of various proteins involved in blood coagulation (e.g. prothrombin) and bone metabolism. Deficiency leads to extensive bleeding. A form of it is manufactured by intestinal bacteria. Green vegetables and egg yolk are good sources.

vitelline membrane Membrane that surrounds the ovum of animals. Membrane around yolk of hen's egg.

vitreous chamber The space between the lens and the retina is filled with vitreous humour.

vitreous humour A jelly-like material which fills the cavity of vertebrate eye behind the lens.

viviparity A form of reproduction in animals in which the developing embryo obtains its nourishment directly form the mother *via* a placenta or by other means. Occurs in some insects and other arthropods, in certain fishes, amphibians and reptiles and in the majority of mammals (in zoology).

(i) A form of asexual reproduction in certain plants, such as the onion, in which the flower develops into a bud-like structure that forms a new plant, when detached from the parent.

(ii) The development of young plants on the inflorescence of the parent plant, as seen in certain grasses and the spider plant (in botany).

vocal cords Two strong bands of elastic tissue that are stretched across the opening of larynx. These vibrate by the movement of air past them and provide tones of speech.

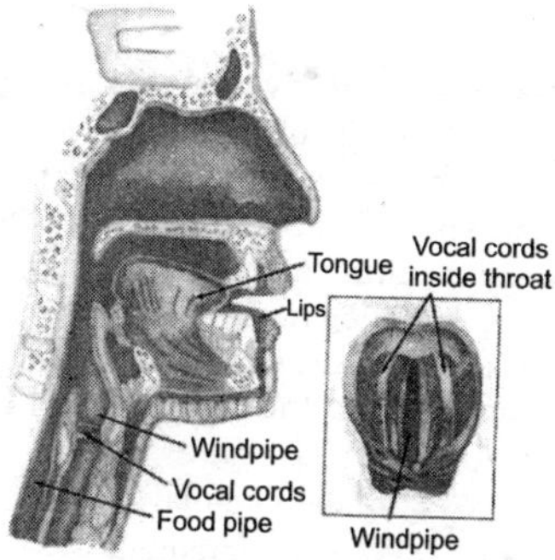

Position of vocal cords in human

Volkmann's contracture It occurs when there is lack of blood flow in the forearm. Due to injury to the muscles of the forearm leading to deformity of the hand, fingers and wrists. Also known as Volkmann's ischaemic contracture.

voluntary muscle (skeletal striped or striated muscle) The muscle that is under our will and is generally attached to the skeleton. An individual muscle consists of bundles of long muscle fibres, each bounded by a sarcolemma and containing sarcoplasm, sarcoplasmic reticulum, and many nuclei. The myofibrils runs through it (see myofibril).

volution Polyphosphate granules that are used as food reserves by certain bacteria, including the cyanobacteria.

volva A thin covering of hyphae over the entire young fruiting body of mushroom.

vomeronasal organ (Jacobson's organ) An organ of smell found in many terrestrial vertebrates. These are paired blind-ending sacs opening into the oral or nasal cavities, lined with olfactory epithelium, whose cells express receptor proteins for airborne chemical.

vomit Ejection of stomach contents through the mouth and nose.

vomitus The material ejected from the stomach through the mouth and nose.

vulva The external genital opening of the female mammal.

V

W

waggle dance A type of communication in honeybees, characterised by tail-wagging movements. It indicates the direction of food source at a distance of more than 100 M.

wall pressure The pressure exerted by the cell wall on the cell content. It is equal in force and opposite in direction to turgor pressure.

Wallerian degeneration The characteristic sequence of changes that occurs in a vertebrate sensory or motor peripheral neuron after its axon has been severed.

wandering cells Cells of connective tissue and blood which actively migrate *in vivo*.
e.g., macrophages, amoebocytes.

Warburg effect The inhibition of CO_2 fixation in photosynthesis by high concentration of oxygen.

warm-blooded *See* endotherm

warning colouration (aposematic colouration) The conspicuous colouring or markings by which a noxious or dangerous animals can be recognised by potential attackers. It is common in insects, e.g. the bright-black and yellow stripes of abdomen in many wasps, warns of it's sting.

water cycle *See* hydrological cycle

water holding capacity The amount of water a soil can hold against the gravitational pull.

water logging An excessive irrigation or more amount of water in the fields.

water potential Symbol-Ψ. It is the chemical potential of water in a biological system compared to the chemical potential of pure water at the same temperature and pressure. It is measured in kilopascals kPa = 0. Pure water has the value kPa = 0. For a plant cell, the water potential of the cell Ψ cell is given by $\Psi_{cell} = \Psi_s + \Psi_p$.
where ψ_s = Solute potential
ψ_p = Pressure potential.
If cell's water potential is greater than its surroundings, the cell will lose water to its surroundings.

water vascular system In echinoderms, a system of water filled canals which supplies hydraulic power for the tube feet, primarily for locomotion.

Watson-Crick model The double-stranded twisted ladder-like molecular structure of DNA as determined by James Watson and Francis Crick at Cambridge, England, in 1953. It is commonly known as the double helix.

wax A lipid consisting of long chain fatty acids. Waxes are used by organisms to form water proof layers or structures, e.g. beeswax.

weberian ossicles A paired chain of three or four small bones in certain fish (e.g. carps and cat fishes) that connect the air bladder with the auditory capsule. They conduct pressure changes from the air bladder to the inner ear.

weed Unwanted plants in human controlled setting, especially fields and gardens. Compete with the main plant and affects its growth.

Weismannism The theory of continuity of the germ plasm that proposes the contents of the reproductive cells (sperms and ova) are passed on unchanged from one generation to the next, unaffected by any changes undergone by the rest of the body. It rules out any possibility of the inheritance of acquired characteristics and has become fundamental to Neo-Darwinian theory.

Western blotting (protein blotting) A technique analogous to Southern blotting used to separate and identify proteins instead of nucleic acids. Protein mixture is separated by electrophoresis and blotted onto a nitrocellulose filter. Antibodies specific to the proteins of interest are applied and bind to their target proteins. A second antibody, specific to the first antibody, is then applied. This carries a radioactive label, so enabling it to be located by autoradiography, wherever it binds to the antibody-protein complex. See Southern blotting.

wet zone Regions receiving extremely heavy rainfall. Average rainfall in this zone is well over 200 cm.

whalebone *See* Baleen.

whales *See* Cetacea

White Blood Cells (WBCs) *See* Leucocytes

white matter Tissue of central nervous system (brain and spinal cord) of vertebrates mainly formed of nerve fibres whose myelin gives it glistening white appearance. Glia and blood vessels also present.

Widal test A test to confirm typhoid fever.

wild type The form of an allele possessed by most members of a population in their natural environment. Wild-type alleles are usually dominant.

wildlife sanctuary An area where capturing or killing of any species of wild animals is prohibited.

wilting The limpness of plant tissue that occurs when there is insufficient cell sap to keep the cells rigid and which may be caused by the lack of water or disease.

wind pollination *See* anemophily.

windpipe *See* trachea

wings Forelimbs modified for flying, e.g. in birds.

wisdom tooth Last molar in each side of each jaw. Generally appears at the age of 17-25 years.

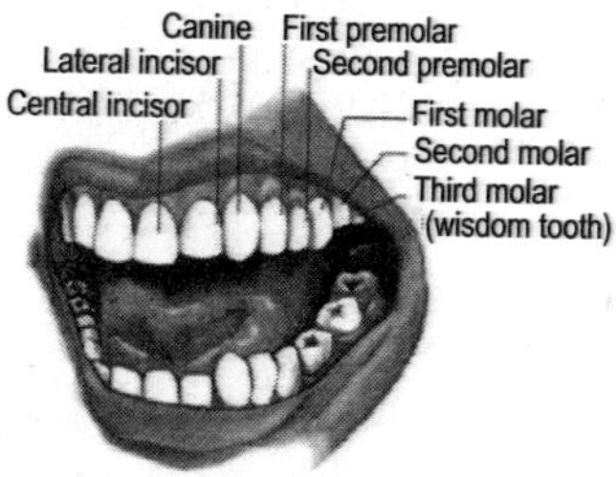

Position of wisdom tooth and other teeth in mouth

WNT protein A family of proteins that are involved in signal transduction pathways crucial in regulating cell proliferation, differentiation, and survival, both during embryological development and in later life.

wobble hypothesis The hypothesis, which states that the nucleotide at the 5′-end of a *t*RNA anticodon plays a less important role than the other two in determining which *t*RNA molecule will align by base-pairing with a codon triplet in *m*RNA. Helps to explain pattern of degeneracy observed in the genetic code which is restricted to the 3′-end of codons.

W

Wolffian duct One of a pair of ducts in fish and amphibians that transports urine from the kidney to the cloaca. In reptiles, birds and mammals, it is functionally replaced by the ureter and persists only in the male, forming the epididymis and vas deferens.

womb *See* uterus

wood A hard, fibrous tissue found in many plants. It is composed of cellulose fibres (which resist tension) and lignin (which withstands compression).

Woronin body A rounded granular body, double membranous found in the hyphae of filamentous ascomycote fungi.

One or more Woronin bodies are closely associated with the pore in each cross-wall (septum) of the hyphae.

If the hypha is injured, the Woronin body is swept by the flow of cytoplasm to plug the pore and minimise further damage.

X

xanthochromia The yellowing of the skin.

xanthophyll The yellow carotenoid pigment present in chloroplasts and chromoplasts.

xanthophyll cycle A cyclic series of interconversions involving certain xanthophylls, which plays a vital role in the photoprotection of plant chloroplasts. Under intense illumination, violaxanthin is converted, *via.* an intermediate, to zeaxanthin.

Xanthophyta A phylum of yellow-green algae. Mostly unicellular, non-motile, others are colonial, filamentous or tubular, coenocytic. Cell wall materials mainly cellulose and pectic substances; in some species, impregnated with silica.

xanthoproteic test A standard test for proteins. Concentrated nitric acid is added to the test solution, a yellow precipitate produced either immediately or on gentle heating indicates a positive result.

xanthous Yellow.

X-chromosome Paired sex determining chromosome in many animal species. In humans it is two in females, whereas males have one X-chromosome.

xenia Changes due to the effect of foreign pollen in visible characters of the endosperm.

xenobiotic Man-made compounds which are not found in natural condition. They include drugs, pesticides and carcinogens. Their detoxification mainly occur in liver.

xenogamy The transfer of pollen grains form anther to the stigma of a different plant.

xenograft (xenotransplant) Tissue or organs from an individual of one species transplanted into or grafted onto an individual of other distinct species, e.g. pig heart valves are often transplanted into humans.

xerarch succession Occurs in dry areas and the series progress from xeric to mesic conditions.

xeric The condition characterised by an inadequate supply of water. Such condition exist in arid habitats, extremely cold habitats and salt marshes. The halophyte or xerophyte are adapted to live in these conditions.

xeromorphic Structurally adaptations in certain plants to withstand dry conditions.

xerophthalmia A pathological condition of the conjunctiva of the eye, due to lack of vitamin-A. The eye fails to produce tears, thus leading to corneal ulceration and finally blindness.

xerophyte Plant capable of living under xeric conditions. Succulents, such as cacti and agaves, have thick fleshy stems or leaves to store water.

xerosere The characteristic sequence of communities reflecting the developmental stages of a plant succession that begins in an arid environment.

X inactivation The inactivation of all but one of the X-chromosomes in cells of female mammals to prevent an abnormal does of X-linked genes compared with cells of male individuals, which carry a single X-chromosome.

xiphocostal Concerning the ribs and xiphoid process.

xiphoid process The smallest and most inferior portion of the sternum protruding caudally between the two halves of the rib cage. It is usually ossified in adult human.

X-ray A technique used to study the molecular structure of substances. In this method, X-rays are passed through the substance under analysis. On the basis of the resulting patterns of diffraction, inferences are made.
X-rays are electromagnetic radiation of shorter wavelength than ultraviolet radiation and longer wavelength than gamma radiation. The range of wavelengths is 10^{-11}-10^{-9} m.

X-ray crystallography The technique for analysing the structure of molecules such as DNA or crystals. It is based on the measurements of diffraction pattern of X-rays as they pass through a crystal of the substance.

xylem The vascular tissue in plants that carries water and minerals upward from the roots. It is composed of vessels, tracheids and xylem parenchyma.
Vessel elements are joined end to end and the end walls are perforated to allow passage of water. In dicot plants, the protoxylem is replaced by secondary xylem formed by vascular cambium. The walls of xylem cells are thickened with lignin. Wood is mostly made up of secondary xylem.

xylophagous Wood-eating.

Y

Y-body Y-chromosome has a brightly fluorescent band on its long arm. It appears as a fluorescent spot under ultraviolet light in the interphase nuclei that are stained with quinacrine dyes. Helps in determining the number of Y-chromosomes in a cell.

Y-chromosome An unpaired sex determining chromosome. It is present only in males in humans.

yeast A group of unicellular fungi in the phylum–Ascomycota. They occur as single cells or as groups or chains of cells. They reproduce asexually by budding and sexually by producing ascospores.
Yeasts of the genus–*Saccharomyces* ferment sugars and are used in the baking and brewing industries.

Yeast Artificial Chromosome (YAC) The linear vector derived from a circular plasmid found in *Saccharomyces cerevisiae* and capable of accomodating DNA inserts of upto 1000 kb. These have a centromere, enabling them to attach to the mitotic spindle of their yeast host and undergo normal segregation during cell division. These behave like minichromosomes.

yeast two-hybrid screen A technique for detecting an interaction between two proteins by means of expression of a reporter gene in yeast cells. It can be used in high-throughput screening of libraries of genes or potential genes from any species (e.g. mouse or human) to identify pairs of genes that encode interacting proteins.

yellow body *See* corpus luteum.

yellow sensitive cones The cones containing photopigment most sensitive to the light of 570 nm wavelengths.

yolk Highly nutritive food found in all eggs except alecithal eggs. It consists mainly protein or phospholipids and fats. Eggs of oviparous animals contain relatively large yolk.

yolk plug A plug formed by large yolk cells, found protruding slightly from the blastopore in the gastrula of amphibians.

yolk sac One of the protective membranes surrounding the embryos of birds, reptiles and mammals. The embryo derives nourishment from the yolk sac *via* a system of blood vessels. In birds and reptiles, the yolk sac encloses the yolk; in most mammals, a fluid replaces the yolk.

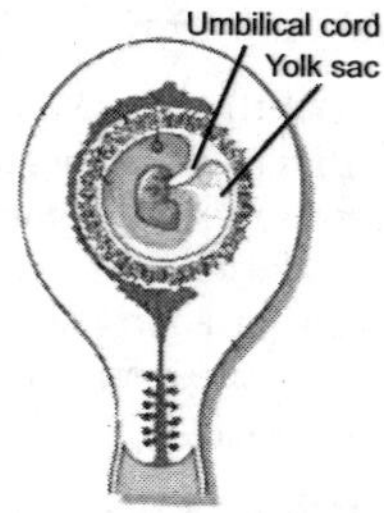

Position of yolk sac in an embryo

yolk stalk A narrow duct-like part which connects the mid-gut and the yolk sac of the developing embryo. Also called omphalomesenteric duct.

Y-organ Either of a pair of glands in the head of certain crustaceans that secrete the hormone ecdysone and play a role in the moulting process.

Z

zaid crops A crop grown between Rabi and Kharif crop season, during the months of February-March to May-June, e.g. black gram, green gram, etc. Grown mainly in the summer season, require warm dry weather and longer day length for flowering.

Z-DNA The left-handed form of DNA having 12 bp per turn. It is found under physiological conditions in short GC segments that are methylated.

zeatin A naturally occurring cytokinin identified in 1963 in grains of maize.

zinc finger Structural motif characteristic of certain proteins that bind to DNA (notably transcriptioin factor). It consists of finger-like folds of amino acids at the base of which two cysteine and two histidine residues lies. These four residues bind a single zinc ion in tetrahedral array.

Z-lines (Z-band) Dense areas in myofibrils that mark the beginning of the sarcomeres in striated muscle. The centre point of each thin (actin) filament is anchored to the Z-line and exerts a pull on the membrane when the fibres contracts.

zona pellucida A non-cellular, transparent glycoprotein layer immediately surrounding the mammalian ovum. Disappears before implantation in the uterus. It develops as a jelly coat around the primary oocyte and is surrounded by the granulosa cells.

zona radiata Zona pellucida which exhibits radial striations. It is caused by radiating canals within the membrane.

zonation The distribution of the different species of community into separate zones which are created by variations in the environment. Competition between species may also contribute to zonation.

zone of differentiation In plant roots, the region where new cells develop into distinct cell types.

zone of elongation In plant roots, the region where new cells grow and lengthen before differentiating.

zone fossil *See* index fossil

zoochore A plant whose seed, spores or fruits are dispersed by animals.

zoogeography A branch of biogeography concerned with the geographic distribution of animals.

zooid An individual member of a colony of invertebrate animals, especially an individual of the phylum–Bryozoa.

zoology Scientific study of animals.

Zoomastigota (Zoomastigina) A phylum of parasitic or free-living heterotrophic Protozoa that possess one or more undulipodia (flagella) for locomotion in older classification.

zoonosis (pl. zoonoses) An infectious disease of non-human vertebrates that can be transmitted to humans, e.g. rabies and anthrax.

zoophagous Eating animals as a diet (carnivorous).

zoophyte A plant-like animal such as sea anemone.

zoosporangium A sporangium that produces zoospores as in some Phycomycetes, green and brown algae.

zoospore (swarm spore) An asexual motile spore produced by a zoosporangium. It has one or more flagella. Produced by many algae and certain protists.

zygodactyly A condition of having two toes facing forward and two backwards, as in some birds.

zygoma Refers to zygomatic bone, a bone of human skull commonly referred as cheekbone or malar bone.

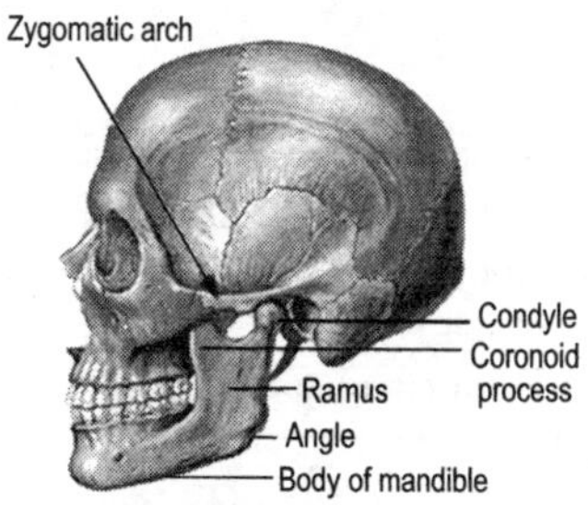

Position of zygomatic arch in human skull

zygomatic arch A structure in the human skull formed primarily by the parts of zygomatic bone and temporal bone.

zygomatic process A bony protrusion of human skull, mostly composed of zygomatic bone but also contributed by the frontal bone, temporal bone and maxilla.

zygomorphic Bilateral symmetry in flower, e.g. *Antirrhinum.*

Zygomycetes Any member of phylum of saprophytic or parasitic fungi including the bread moulds. Their hyphae lack cross walls and they can reproduce asexually by sporangiospores or sexually by means of zygosphores.

zygospore A diploid zygote that results from the fusion of two hyphae of different mating types in some algae and fungi (Zygomycetes). The zygospore forms sporangia, in which meiosis takes place and spores form.

zygote A fertilised egg produced by the union nucleus of the ovum or ovule with nucleus of sperm or pollen grain. A zygote goes on to develop into a multicellular organism by repeated cell divisions (mitosis).

Zygote Intra Fallopian Tube (ZIFT) In *vitro* fertilisation method in which zygote or early embryo (up to 8 blastomeres) is transferred into the Fallopian tube of woman.

zygotene The second stage in prophase-I of meiosis. It is characterised by the pairing (synapsis) of homologous chromosomes. Bivalents are formed.

zygotic meiosis The meiotic division in zygote. It results in formation of haploid individuals.

zymase An enzyme extract of Brewer's yeast that can perform fermentation. It contains a mixture of glycotic enzymes.

zymogen An inactive enzyme precursor that after secretion is chemically altered to active form, e.g. protein digesting enzyme trypsin is secreted by pancreas as trypsinogen (zymogen).

zymogenic

1. Producing or causing fermentation.
2. Relating to a zymogen.

zymology The scientific study of fermentation.

APPENDICES

Appendix-1
Important Branches of Biology

Branch	Description
Acarology	Study of mites and ticks
Agriculture	Cultivation of useful plants
Agrobiology	Study of quantitative science of plant nutrition
Agrology	Study of soil science related to crop production
Agronomy	Management of farms and cultivation of field crops
Agrostology	Study of grasses
Algology (Phycology)	Study of algae
Anatomy	Study of internal organs as revealed by dissection and observed by unaided human eyes
Andrology	Deals with disorders of male reproductive system. It is opposite of Gynaecology
Angiology	Study of blood vessels or circulatory system in animals
Anthology	Study of flowers
Aphidology	Study of aphids (plant lice)
Apiculture	Study of bee-keeping for extracting honey and wax
Araneology	Study of spiders
Arboriculture	Study of shrubs and tree cultivation
Arthrology	Study of joints
Bacteriology	Study of bacteria
Biochemical Taxonomy	Study of taxonomy of related plants based on chemical reactions
Biochemistry	Study of chemicals constituting the body and chemical reactions
Biomedical Engineering	Designing and production of artificial body parts (like artificial limbs, valves, vessels, iron lungs, heart-lung machines, pacemaker, artificial kidneys, etc) for man to overcome defects
Biophysics	Physical aspects of living systems
Biotechnology	Utilisation of living organisms or of substances obtained from them in industrial processes
Bonsai	Study of art and hobby of growing dwarf forms of trees and shrubs in pots

Branch	Description
Bryology	Study of bryophytes
Carcinomatology	Study of cancers
Cardiology	Study of heart functioning and diseases
Cell Biology	Structure, function, reproduction and life history of cells
Cetology	Study of whales
Chondrology	Study of cartilage
Craniology	Study of skull
Ctetology	Study of origin and development of acquired characters
Cytology	Detailed structure of cells
Cytogenetics	Study of structure and function of the cell, especially chromosome
Dactyology	Communication system as for deaf using signs made by fingers
Dactylography	Science of fingerprints for identification
Dairy Science	Maintenance of systematically bred milk producing animals, collection and preservation of milk and production of milk products
Dendrochronology	Counting and analysing annual growth rings of trees
Dendrology (Xylology)	Study of trees
Dentistry	Study of teeth and its problems
Dermatology	Study of skin and its diseases
Desmology	Study of ligaments
Dietetics/Sitology	Study of nutrition/diet
Ecology	Scientific analysis and study of interactions among organisms and their environment
Economic Botany	Study of commercially useful plants
Economic Zoology	Study of useful and harmful activities of animals
Edaphology (Pedology)	Study of soil
Ethology	Study of behaviour of animals in response to the environmental stimuli
Eugenics	Study of science dealing with genetic factor which improve successive generations of human race
Embryology	Study of development of embryo from fertilisation of ovum to foetus stage
Entomology	Scientific study of insects
Enzymology	Study of enzymes, their kinetics structure, function as well as relation to each other

Branch	Description
Ethnobotany	Scientific study of the relationships that exist between people and plants
Ethology	Scientific and objective study of animal behaviour
Euphenics	Science of making phenotypic improvements to humans after birth
Euthenics	Study of improvement of human functioning and well-being by improvement of living conditions
Fishery	Study of occupation of collecting and catching fish and other animals of sea, lakes or rivers
Floriculture	Cultivation of flower yielding plants
Forensic Science	Application of scientific knowledge to questions of civil and criminal law. Fingerprints, blood typing, etc., are studied to detect crime
Forestry (Silviculture)	Study of development and conservation of forests and their products
Genetics	Study of genes, heredity and genetic variations
Genetic Engineering	Manipulation of genes at will
Gene Therapy	Removal of defective genes and their actual replacement
Gerontology	Study of ageing
Gynaecology	Study of female reproductive organs, diseases and hygiene of women
Haematology	Study of blood and blood disorders
Helminthology	Study of parasitic worms (helminths)
Hepatology	Study of liver
Herpetology	Study of amphibians and reptiles
Horticulture	Development and management of orchards and gardens
Hydroponics	Soilless culture of plants
Hypnology (Somnology)	Scientific study of sleep
Ichthyology	Study of fishes
Immunology	Study of natural and acquired resistance of body against diseases
Karyology	Study of structure and function of cell nuclei
Kinesiology	Study of body movements
Laryngology	Study of larynx
Leprology	Study of leprosy
Lichenology	Study of lichens

Branch	Description
Limnology	Study of freshwater ecology
Malacology	Study of molluscs
Malariology	Study of malaria
Mammalogy	Study of mammals
Mastology	Study of breasts (mammary glands)
Medicine	Science of treating diseases with drugs or other curative substances
Melanology	Study of skin pigments
Microbiology	Study of microscopic organisms
Mycology	Study of fungi
Myremecology	Study of ants
Nematology	Study of nematodes
Neonatology	Medical care of newborn infants
Nephrology	Study of kidney
Neurology	Study of brain/nervous system
Nosology	Classification of diseases
Nursing	Care of ill and injured person
Obstetrics	Deals with care of women during pregnancy, childbirth and the recuperative period
Occupational Therapy	The teaching and training of handicapped and disabled (both physically and mentally) persons to lead a normal life
Odonatology	Study of dragonflies
Odontology	Study of teeth and gums
Olericulture	Study of cultivation of vegetables
Oncology	Study of cancer/tumours
Ontogeny	Study of the entirety of an organism's lifespan
Oology	Study of bird eggs, nests and breeding behaviour
Ophilogy (Serpentology)	Study of snakes
Ophthalmology	Anatomy, physiology and diseases of the eyes
Ornithology	Study of birds
Orthopaedics	Diagnosis and repair of disorders of bones, joints and muscles
Osphresiology (Oneirology)	Study of dreams
Otolaryngology	Study of ear and larynx
Otology	Study of ears

Branch	Description
Otorhinolaryngology	Study and treatment of diseases of ENT (Ear, Nose and Throat)
Paediatrics	Branch of medicine dealing with diseases of children
Palaeobotany	Study of plant life of past, based on fossil studies
Palaeozoology	Study of fossils of past animals which have been extincted and their distribution in time
Palynology	Study of pollen grains
Parasitology	Study of parasites
Parazoology	Study of sponges
Pathology	Study of diseases, their symptoms, causes, spread and control
Pharmacognosy	Use of plant and their parts in drug industry
Pharmacology	Study of preparation of drugs and their effect on organisms
Pharmacy	Study of compounding and dispensing of drugs
Phenology	Recording and study of periodic biotic events like flowering, leaf fall and migration
Phrenology	Study of feelings
Physiology	Study of functions and processes of life
Physiotherapy	Treatment of diseases, bodily weaknesses or defects by physical remedies such as massage, exercises, etc.
Phytogeography	Study of distribution of plants on Earth
Phytography	Branch of botany dealing with descriptive features of plant parts
Phytosociology	Science of plant communities, their structure and composition
Piggery	Study of keeping pigs for meat
Pisciculture	Cultivation (breeding and rearing) of fishes
Pomology (Carpology)	Science dealing with fruits and fruit yielding plants
Prosthesis	An artificial replacement or substitute for a loss in part or full of a limb due to acquired or congenital cause
Protistology	Study of protists (unicellular plants and animals)
Protozoology	Study of protozoans (acellular eukaryotes)
Psychiatry	Science dealing with diagnosis and treatment of mental disorders
Psychology	Study of nature of behaviour and working of mind
Pteridology (Pterology)	Study of pteridophytes/ ferns

Branch	Description
Radiobiology (Actinobiology)	Effect of radiations on body
Radiology	Deals with X-rays and imaging techniques for medical diagnosis
Radiotherapy	Treatment of diseases with X-rays and radioactive substances
Rhinology	Study of nose/olfactory organs
Saurology	Study of lizards
Sericulture	Study of rearing silkworms for silk
Serology	Study of serum/antigen-antibody reactions in the blood/immunological reactions
Sonography	Ultrasound imaging
Sonology	Study of hearing
Spermology (Seed Biology)	Study of seed
Sphygmology	Study of arterial pressure and pulse
Splanchnology	Study of visceral organs
Stomatology	Study of foregut (buccal cavity and stomach)
Surgery	Treating injuries, deformities and diseases by operations
Syndesnology	Study of joints and ligaments
Taxidermatology	Study of skin for stuffing and preservation of animals for museums
Tectology	Study of structural organisation/functional anatomy of the body
Teratology	Study of malformation and abnormal growth in foetus due to teratogens during embryogenesis, e.g. phocomelia
Therapeutics	Science of treatment of the diseases
Toxicology	Study of toxic effects of drugs on body
Torentology	Study of diseases of embryo
Traumatology	Study of wounds
Trichology	Study of hairs
Trophology	Study of food and nutrition
Urology	Study of urinary disorders
Venereology	Study of VDs (Venereal Diseases) and STDs (Sexually Transmitted Diseases)
Virology	Study of viruses
Zoogeny	Origin and development of animals

Branch	Description
Zoogeography	Deals with distribution of animals in different parts of the world on Earth
Zoonosology	Study of handicapped animals
Zoopathology	Study of diseases of animals and their control
Zoophytology	Study of drifting organisms (zooplanktons) like diatoms
Zootechny	Science of breeding and domestication of animals
Zootomy	Dissection and anatomy of animals
Zymology	Study of zymoses (fermentation)

Appendix-2
Fathers of Various Biological Branches

Branch	Father
Anatomy	Andreas Vesalius
Antibiotics	Alexander Fleming
Antiseptic surgery	J Lister
Ayurveda	Charaka
Bacteriology	Robert Koch
Biology	Aristotle
Birdman of India	Salim Ali
Blood circulation	William Harvey
Blood Groups	K Landsteiner
Botany	Theophrastus
Botany (modern)	K Bauhin
Bryology	Johann Hedwig
Chemotherapy	Paul Ehlrich
Cloning	Ian Wilmut
Comparative Anatomy	G. Cuvier
Concept of Evolution	Empedocles
Conditional Reflex	Ivan Pavlov
Cytogenetics	H J Muller
Cytology	Robert Hooke
Cytology (modern)	Carl P Swanson
DNA Fingerprinting	Alec Jeffrys
ECG	Einthoven
Embryology	Kaspar Friedrich Wolf

Branch	Father
Embryology (modern)	Karl Ernst von Baer
Endocrinology	Thomas Addison
Epidemiology	John Snow
Ethology	Konrad Lorentz
Eugenics	Francis Galton
Experimental Genetics	TH Morgan
Experimental Physiology	Galen
Fermentation	Louis Pasteur
Genetic Engineering	Paulberg
Genetics	GJ Mendel
Genetics (modern)	William Bateson
Gerontology (Ageing)	Korenchevsky
Gene therapy	Anderson
Green Revolution	N Borlaug
Haploid Genetics	Dodge (Neurospora genetics)
Histology	Francois Xavier Bichat
Homoeopathy	Samuel Hahnemann
Human genetics (biochemical genetics)	Archibald Garrod
Immunology	Edward Jenner
Indian Algology (phycology)	MOP Iyengar
Indian Botany and Indian Herbaria	William Roxburgh
Indian Bryology	SR Kashsyap
Indian Ecology	R Mishra
Indian Embryology	P Maheshwari
Indian Green revolution	MS Swaminathan
Indian mycology and plant pathology	EJ Butler
Indian Palaeobotany	Birbal Sahni
Indian Palynology	PKK Nair
Indian Physiology	RN Dastur
Indian Systematic botany (Taxonomy)	H Santapu
Indian Zoology	Col Stevenson
Medicine	Hippocrates
Microbiology	Louis Pasteur
Microscopic Anatomy	Marcello Malpighi
Microscopy and Microbiology	Av Leeuwenhoek
Mosaic theory	Wilhelm Roux

Branch	Father
Mycology	PA Micheli
Palaeontology	Leonardo-da-vinci
Palaeontology (modern)	Cuvier
Palynology	Erdtman
Pathology (modern)	R Virchow
Plant Anatomy	N Grew
Plant Ecology	Warming
Plant Pathology	De Bary
Plant Physiology	Stephan Hales
Polio Vaccine	J Salk
Protozoology	Av Leeuwenhoek
Quantitative or Polygenic Inheritance	Kolreuter
Radiation Biology	Muller
Regulative Theory of Embryonic Development	Hans Driesch
Special Creation Theory	Father Suarez
Stress Physiology	Hans Selye
Surgery	Susruta
Taxonomy	Carolus Linnaeus
Tissue Culture Virology	Gottlieb Haberlandt Martinus Beijerinck
White Revolution	V Kurien
Zoology	Aristotle

Appendix-3
Human Diseases & Disorders

Genetic Disorders

I.

Gene mutation	Mode of inheritance	Main symptom
Albinism	Autosomal recessive	Little or no pigment in the hair, eyes or skin; decreased vision or blindness, skin cancer.
Alkaptonuria	Autosomal recessive	Darkening of urine, joint pain and dyspnoea.
Colour blindness	X-linked recessive	Difficulty in distinguishing between colours.

Gene mutation	Mode of inheritance	Main symptom
Cystic fibrosis	Autosomal recessive	Unusually thick mucus clogs lungs, liver and pancreas.
Duchenne Muscular Dystrophy (DMD)	X-linked recessive	Fatigue, learning difficulties, muscle weakness, progressive difficulty, walking.
Haemophilia	X-linked recessive	Blood does not clot.
Huntingdon's Chorea	Autosomal dominant	Disorganised muscular movements.
Marfan's syndrome	Autosomal dominant	Disproportionately long arms, legs, fingers and toes, heart murmurs.
Phenylketonuria (PKU)	Autosomal recessive	Intellectual disabilities or mental retardation, seizures, stunted growth.
Sickle-cell anaemia	Autosomal recessive	Anaemia and interference circulation.
Tay-Sachs disease	Autosomal recessive	Deafness, blindness, paralysis or loss of muscle action.
Thalassemia	Autosomal recessive	Anaemia, weakness, fatigue, shortness of breath.

II.

Chromosomal mutations	Chromosome affected	Symptoms
Alzheimer's disease	19 and 21	Dementia leading to loss of memory, confusion, hallucinations.
Cridu chat syndrome	5	Widely spaced eyes, small head, moon-like face, cry-like kitten, congenital heart disease.
Down's syndrome (21-trisomy)	21	Decreased muscle tone, flat face, slanting eye, reduced intelligence.
Edward's syndrome (18-trisomy)	18	Tightly clenched fingers, small jaws, deformed ears, small mouth, nose and fingers.
Klinefelter's syndrome	XXY (trisomy)	Feminised male with female like sparse body hair, knock knees, enlarged breasts (gynaecomastia).

Chromosomal mutations	Chromosome affected	Symptoms
Patau's syndrome (13-trisomy)	13	Hare-lip, cleft palate, sloping forehead, small head and brain and defective eye development
Superfemales	One or more extra X-chromosome	Abnormal sexual development and mental retardation.
Supermales	Extra Y-chromosome	Abnormal height, mental retardation, criminal bent of mind and more aggressive males.
Turner's syndrome	XO (monosomy)	Sterile females with short stature, webbed neck, abnormal intelligence.

Infectious Diseases

I.

Bacterial disease	Causative organism	Symptoms
Anthrax	*Bacillus anthracis*	Swelling in sore and lymph nodes (cutaneous), nausea, vomiting, abdominal pain (gastrointestinal), shortness of breath, flu-like symptoms (pulmonary)
Cholera	*Vibrio cholerae*	Reduction in body's water level, slow down of blood circulation, vomiting, acute diarrhoea, muscular cramps, etc.
Diphtheria	*Corynebacterium diphtheriae*	Mild fever, sore throat, patches on false membrane in the throat.
Gonorrhoea (sexually transmitted disease)	*Neisseria gonorrhoeae*	Joint pain redness, swelling, pus discharge through urethra, frequent and burning urination.
Leprosy (Hansen's disease)	*Mycobacterium leprae*	Chronic infection of skin and tissues, formation of ulcers, deformities of fingers, toes.
Pertussis (whooping cough)	*Bordetella pertussis*	Constant cough, vomit, frothy discharge from mouth and nose.
Plague (black death)	*Pasteurella yersinia pestis*	Acute body pain, reddish eyes, increase in body temperature, inflammation of glands in neck, armpit and thigh.
Pneumonia	*Streptococcus pneumoniae*	Shaking, chill, fever, chest pain, production of sputum, breathing problems.

Bacterial disease	Causative organism	Symptoms
Syphilis (sexually transmitted disease)	*Treponema pallidum*	Red spots on penis and vagina after sexual intercourse, serious tissue disruption on any part of the body, painless sore on the genitalia, skin eruptions, etc.
Tetanus (lock jaw)	*Clostridium tetani*	Painful muscular spasms (especially neck and jaw) difficulty in chewing and swallowing, fever and headache.
Tuberculosis	*Mycobacterium tuberculosis*	Constant cough in severe cases, sputum with blood, pain in chest, loss of body weight.
Typhoid (enteric fever)	*Salmonella typhi*	Inflammation of ileum, abdominal pain, constant fever, weakness.

II.

Viral disease	Causative organism	Symptoms
Chicken pox	*Varicella zoster virus*	Fever (appearing about 14 days after the contact of disease), headache, appearance of reddish spots turning into the pearly blisters on the chest, back and face, abdomen, scabs from 4-5th day of the fever.
Chikungunya	*Chikungunya virus*	Sudden onset of fever, crippling joint pain.
Dengue or breakbone fever	Dengue virus (*Flavi ribo virus*)	High fever, rashes on face and head, severe pain in eyes, weakness, muscle and joint pain.
Ebola	Ebola viruses	Fever, sore throat, muscle pain, headache followed by vomiting, diarrhoea and decreased function of liver and kidney. Internal and external bleeding.
Hepatitis	Hepatitis virus	Jaundice, fever, nausea, pains in muscles and joints.
Herpes	*Herpes simplex virus*	Typical skin sores.
Influenza or flu	Influenza viruses (A and B)	Fever, chill, muscular aches, uneasiness, etc.
Measles	*Rubeola virus*	Fever, inflammation of respiratory mucous membranes, sensitive to light, loss of appetite, vomiting, rashes on the skin, which are eliminated in nearly 10 days.

Viral disease	Causative organism	Symptoms
Mumps	*Paramyxo virus*	Fever, swelling in glands particularly salivary gland, parotid gland or organs, severe pain when mouth is opened, incubation period is 15-25 days.
Rabies (hydrophobia)	*Rabies virus*	Central nervous system effected, severe headache, high fever with alternating stages of excitement and depression, difficulty in swallowing, paralysis, convulsions, etc.
SARS	SARS *corona virus*	Fever and cough, lethargy symptoms, sore throat, breathing problem.
Small pox	*Variola virus*	High fever, headache, backache, skin rashes on the third day passing through bright red spots, pustules, papules and vesicles scabs which fall on leaving pitted pock marks.
Swine flu	H1N1 virus	Cough, fever, sore throat, bodyache followed by severe vomiting, abdominal pain, pneumonia and respiratory failure.
Yellow fever	Yellow fever virus (live as parasite in animals) *Flavi virus*	Fever begins after an incubation period of 3-6 days, fever, headache, chills, back pain, loss of appetite, nausea, vomiting, liver damage resulting in severe jaundice yellowing of the skin, human death in severe cases.
Poliomyelitis	Polio virus	Effect on central nervous system, cells of the backbone and intestine become completely damaged. Paralytic type of indication follows and reaches maximum within 2-3 years. Difficulty in forward bending movement of head.

III.

Protozoan disease	Causative agent	Symptoms
Amoebic dysentery or amoebiasis	*Entamoeba histolytica*	Mild diarrhoea, in severe cases, mucosa and blood appears in the stools.
Giardiasis	*Giardia intestinalis*	Epigastric pain, abdominal discomfort, headache, mild diarrhoea with foul smelling and greasy stool.

Protozoan disease	Causative agent	Symptoms
Kala-azar	*Leishamania donovani*	Enlargement of organs like spleen, liver, fever, jaundice and skin pigmentation, fever, headache.
Malaria	*Plasmodium* sp.	Fever at intervals, acute chillness, shivering, nausea, headache.
Pyorrhoea	*Entamoeba gingivalis*	Pus in teeth emerges through the gums of the teeth.Sometime wounds also occur in the roots of the teeth.
Sleeping sickness	*Trypanosoma brucei* (protozoan parasite)	Enlargement of glands at the back of the neck, rashes on the back and chest, pain in the joints, swelling in ankles and hands, mental disturbance, lethargy, etc.

IV.

Fungal diseases	Causative agent	Symptom
Athlete's foot	*Trichophyton*	Skin problems.
Dhobie itch or crotch itch or Tinea cruris	*Candida albicans*	Red rashes with itching usually in groin.
Ringworm	*Micosporum trichophyton*	Dry scaly lesions, intense itching.
Scabies	*Acarus scabies*	Some irritation occur on skin making it dry with white rashes.

V.

Helminthic diseases	Causative agent	Symptoms
Ascariasis	*Ascaris lumibricoides*	Internal bleeding, muscular pain, fever, anaemia, blockage of intestinal passage.
Filariasis/ Elephantiasis	*Wuchereria bancrofti*	Enlargement of certain part like legs in elephantiasis, lymph channel affected.
Hookworm infection	Roundworm that affects small intestine	Blood loss, anaemia, heart burn and constipation, etc.
Schistosomiasis	Blood flukes *Schistosoma mansoni, S. japonicum, S. haematobium*	Asthmatic attacks, hepatitis, weight loss, lack of appetite.
Tapeworm	*Taenia saginata, Taenia solium*	Gastric problems anaemia.

Appendix-4

Common Diseases of Plants

Bacterial Diseases

I.

Name of Diseases	Causal Organism
Bacterial blight of rice	*Xanthomonas oryzae*
Black rust of crucifers	*Xanthomonas campestris*
Blight of beans	*Xanthomonas phaseoli*
Canker of tomato	*Agrobacterium michiganese*
Citrus canker	*Xanthomonas citri*
Crown gall of rose	*Agrobacterium tumefaciens*
Crown gall of tomato	*Agrobacterium tumefaciens*
Crown gall of *Rubus sp.*	*Agrobacterium tumefaciens*
Fire blight of pear and ***apple***	*Erwinia amylovora*
Ring rot of potato	*Corynebacterium sepidonicum*
Soft rot of vegetable	*Erwinia carotovora*
Wilt of cucurbits	*Erwinia tracheiphila*
Wilt of pepper	*Pseudomonas solanacearum*
Wilt of potato	*Pseudomonas solanacearum*
Wilt of tobacco	*Pseudomonas solanacearum*
Wilt of tomato	*Pseudomonas solanacearum*

Fungal Diseases

II.

Name of Diseases	Causal Organism
Black rust of wheat	*Puccinia triticina*
Brown rust of wheat	*Puccinia recondita*
Brown leaf spot of rice	*Helminthosporium oryzae*
Club root of cabbage	*Plasmodiophora brassicae*
Covered smut of barely	*Ustilago hordei*
Covered smut of oat	*Ustilago kolleri*
Downy mildew of grapes	*Plasmophora viticola*
Downy mildew of pea	*Peronospora pisi*
Early blight of potato	*Alternaria solani*
Powdry mildew of grapes	*Unicinula necator*
Powdry mildew of pea	*Erysiphe polygoni*

Name of Diseases	Causal Organism
Powdery mildew of wheat	*Erysiphe graminis*
Red rot of sugarcane	*Glomerella tucumanensis*
Smut of bajra	*Tolyposporium penicillariae*
Stem gall of coriander	*Protomyces macrosporus*
Tikka disease of groundnut	*Cercospora personata*
Wart disease of potato	*Synchytrium endobioticum*
White rust of crucifers	*Albugo candida*
Late blight of potato	*Phytophthora infestans*

Viral Diseases

III.

Disease	Causal Virus
Abutilon mosaic	*Abutilon* mosaic virus
Bunchy top of banana	Banana bunchy top virus
Cucumber mosaic	Cucumber mosaic virus
Little leaf of brinjal	Brinjal little leaf virus
Little leaf of cotton	Cotton little leaf virus
Tomato leaf curl	Tobacco leaf curl virus
Papaya mosaic	Papaya mosaic virus
Potato leaf roll	Potato leaf curl virus
Potato mild mosaic	Potato virus X
Potato rugose mosaic	Potato virus X, potato virus Y
Ratoon stunt of sugarcane	Ratoon stunt virus
Rosette of groundnut	Groundnut mosaic virus
Sugarcane mosaic	Sugarcane mosaic virus
Tobacco mosaic	Tobacco mosaic virus
Tristeza of citrus	*Citrus tristeza virus*

Appendix-5

Common Diseases of Animals

Viral Diseases

Diseases	Affected Animal	Causal Organism
Blue tongue	Cow and sheep	Transmitted by mosquito
Foot and Mouth disease	All ruminants (cows, buffaloes, sheep, goat and pigs)	Virus–O, A, C, Asia-1, SAT-1, SAT-2 and SAT-3

Diseases	Affected Animal	Causal Organism
Mucosal	Cow	Pesty virus (Togo viridi group)
Pox	Cows, buffaloes, goat, sheep and camel	Vaccinia (Orthopox virus)
Rinderpest	All ruminants (cow, buffaloes, sheep, goat and pigs)	Measles and distemper viruses

Bacterial Diseases

Diseases	Affected Animal	Causal Organism
Anthrax	Cattle, buffalo, goat, sheep, horses, pigs, camel, dogs and elephant	*Bacillus anthracis*
Black quarter	Low-aged cow, buffalo and goats	*Clostridium chauvoei*
Bovine mastitis	Cow, buffalo, goat and sheep (female)	*Streptococci agalactiae, S. Uberis, S. dysgalactial and Straphylococcus pyogenes*
Haemorrhagic septicaemia	Cow, buffalo, goat and pig	*Pasteurella multocida*
Tuberculosis	Animals and birds (man too)	*Mycobacterium tuberculosis*

Appendix-6

Major Mass Extinctions

Extinction event (s)	Date (millions of year ago)	Organisms most affected	Cause (s)
Ordovician silurian	443	Graptolites, enhinoderms, brachiopods, trilobites, ostracods, nautiloids.	Glaciation climate change and fall in sea levels.
Late Devonian (series)	359	Cephalopods, corals, brachiopods, bryozoans, echinoderms, trilobites, ammonites, agnathans, armoures fishes, three quarter of species dies out.	Global cooling, reduced oxygen levels in deeper water, asteroid impacts.

Extinction event (s)	Date (millions of year ago)	Organisms most affected	Cause (s)
Permian (The Great Dying)	248	Corals, crinoids, ammonites, brachiopods, bryozoans, echinoderms, trilobites, land plants, insects, terrestrial vertebrates, 96% species died out.	Asteroid impact, flood basalt eruptions, catastrophic methane release, drop in oxygen levels or some combination of these.
Jurassic Triassic	200	Bivalve and cephalopod molluscs, marine reptiles, conodonts, large amphibians, insects.	Climatic changes due to continental drift, flood basalt eruption and asteroid impacts.
Cretaceous-Tertiary (K-T)	65	Dinosaurs, ammonites, fish, brachiopods, ptero- saurs, planktonic organisms many flowering plants.	meteorite collision
Sixth mass extinction	Ongoing	The Hawai'l chaff flower, the golden Coqui Puerto Rican tree frog, the Martinique Parrot, the Yuman box turtle, the Madagascan Pygmy hippo and the Japanese sea lion are amongst the list that include the 784 species of plants and animals that have recently vanished from earth because of human activities.	Human activities like habitat destruction, invasive species, pollution, human overpopulation and over harvesting (hunting, fishing and gathering).

Appendix-7

Nobel Laureates in Medicine (Biology)

Laureate	Contribution
John O'Keefe, May-Britt Moser and Edvard I. Moser (2014)	For their discoveries of cells that constitute a positioning system in the brain.
James E. Rothman, Randy W. Schekman and Thomas C. Südhof (2013)	For their discoveries of machinery regulating vesicle traffic, a major transport system in our cells

Laureate	Contribution
Sir John B. Gurdon and Shinya Yamanaka (2012)	For the discovery that mature cells can be reprogrammed to become pluripotent
Bruce A. Beutler and Jules A. Hoffmann (2011)	For their discoveries concerning the activation of innate immunity
Ralph M. Steinman (2011)	For his discovery of the dendritic cell and its role in adaptive immunity
Robert G. Edwards (2010)	For the development of in vitro fertilization
Elizabeth H. Blackburn, Carol W. Greider and Jack W. Szostak (2009)	Discovered how chromosomes are protected by telomeres and the enzyme telomerase
Herald Zur Hausen (2008)	Postulated that oncogenic Human Papilloma Virus (HPV) caused cervical cancer
Francoise Barre-Sinoussi and Luc Montagnier (2008)	Discovered Human Immuno Deficiency Virus (HIV)
Mario Capechi, Sir Martin Evans and Oliver Smith (2007)	Discovered the introduction of specific gene modifications in mice by use of embryonic stem cells
Andrew Z. Fire (USA) and Craig C. Mello (2006)	Discovered RNA interference gene silencing by double-stranded RNA
Barry J. Marshall (Australia) and J. Robbin Warren (Australia) (2005)	Discovered the bacterium **Helicobacter pylori** and its role in gastritis and peptic ulcer disease
Emil Adolf von Behring (First Nobel Laureate) (1901)	For the development of serum therapies against diptheria and tetanus